Computer Supported Cooperative Work

Also in this series

Rich Ling and Per E. Pedersen (Eds)

Mobile Communications

Re-negotiation of the Social Sphere

With 19 Figures

 Springer

Rich Ling, PhD
Telenor R&D, Fornebu, Norway

Per E. Pedersen, PhD
Department of Information and Communication Technology, Agder University College,
Grimstad, Norway

Series Editors
Dan Diaper, PhD, MBCS
Senior Research Fellow, School of Computing Science, Middlesex University, UK

Colston Sanger
Shottersley Research Limited, Little Shottersley, Farnham Lane
Haslemere, Surrey GU27 1HA, UK

British Library Cataloguing in Publication Data
A catalogue record for this book is available from the British Library

Library of Congress Control Number:
2005925180

CSCW ISSN 1431-1496
ISBN-10: 1-85233-931-4
ISBN-13: 978-1-85233-931-9
Springer Science+Business Media
Springeronline.com

Typesetting: Gray Publishing, Tunbridge Wells, Kent, UK
Printed and bound in the United States of America
34/3830-543210 Printed on acid-free paper SPIN 10991800

Preface

Mobile telephony has arrived on the scene. According to statistics of the International Telecommunications Union, in the mid-1990s, less than one person in 20 had a mobile telephone; as of 2003, this had risen to on person in five. In the mid-1990s, the GSM system was just being commercialized, there were serious coverage and interoperability issues that were not yet sorted out and handsets were only beginning to be something that did not require a car to transport them. In the mid-1990s, if a teen owned a mobile telephone it was likely an indicator of an over-pampered rich kid rather than today's sense that it is a more or less essential part of a teen's everyday identity kit.

Hence, in less than a decade, this device has established itself technically, commercially, socially and in the imagination of the people. It has changed the way we think about communication, coordination and safety and it has changed the way we behave in the public sphere. The mobile telephone has become an element in our sense of public and private space and in the development of our social and psychological personas. It has become an arena wherein the language is being played with, morphed and extended. Finally, it is reaching out into ever-new areas of commerce and interaction.

All of this is, of course, interesting to social scientists. As brought out by Woolgar later, this is, in some ways, a type of experiment writ large that has engendered serious insight into the functioning of the social group and the individual in society.

During the past 5 years, there has been a growing focus on scientific research having to do with mobile communications. Scholars from Europe, North America, Asia and countries such as Israel have investigated the ways in which mobile communication interacts with social dynamics. Collections include *Perpetual Contact* edited by Katz and Aakhus (2002), the *Wireless World* book edited by Brown et al. (2002), the work by Nyíri (2003), and that of Fortunati et al. (2003). In addition, there are starting to be some book-length analyses of mobile telephony. These include the work of Kopomaa (2000) and the analysis of the social consequences of mobile telephony by Ling (2004).

The existing corpus of material examines mobile telephony from a range of perspectives. However, several dimensions of the work on mobile telephony have not been as completely developed. In addition to the social issues of mobile telephony, there is a need to examine the psychological, linguistic, work-oriented and commercial aspects of mobile telephony. In addition, there is a need to focus analysis on the non-voice uses of mobile communication such as text messages and data transfer. These are areas that receive focus, often for the first time, in this volume. This book contributes to this discussion using the work of authors from around the globe, including a particularly strong contribution from East Asian scholars. The chapters represent the work of authors that come from Europe, Asia, North America and the Middle East.

The book is organized into five parts: (1) an introductory part that focuses on the placement of mobile communication in the broader context of information and communication research, (2) the interplay between the public and the private (often drawing on the Goffmanian front stage/backstage metaphor) and (3) the psychological (4) the linguistic dimensions of mobile telephony and finally (5) the commercialization of mobile communication applications.

In the first part of the book there are four chapters that stretch out the canvas of research on mobile communication. These chapters look into the way in which mobile communication fits into the broader issues of communication and digitally mediated communication (Haddon and Woolgar), an examination of how one might approach the interactions between traditional mobile telephony and Wi-Fi (Sawhney), and finally a piece by Katz and Sugiyama that moves away from the instrumental examination of mobile communication to examine its expressive dimensions.

Part 2 is entitled Public and Private Spaces. In many respects this part forms the center point of the book. It contains a series of chapters that specifically look into how the mobile telephone is being adopted, used (and sometimes resented) in various public spaces. The articles range from analyses of how the mobile telephone is being used in open office landscapes in Norway (Julsrud), to transport systems in the Philippines (Paragas), to teens' use of mobile telephony in Japan (Ito). Beyond the geographic diversity of the situations described in the chapters, the contributors examine the broader interaction between the social sphere, space and mobile communication. Many of the contributions in this part of the book underscore the tensions and turbulence associated with introducing the device into pre-existing social situations. In effect, the mobile telephone brings up for reconsideration previously taken for granted assumptions as to when and where mediated communication should take place.

Beyond the social dimensions of mobile telephony, the book takes up several other less visited areas. These include the psychological, linguistic and the commercial dimensions of mobile telephony. Much of the analysis of mobile telephony has taken place in the context of communication departments in universities and in the research departments of telecom operators. This has perhaps resulted in a focus on the social dimension, since it is often sociologists and their ilk that make up the personnel in these institutions. Thus, there is a sense in which the analysis of mobile telephony is only starting to be explored. There is clearly much work that has been done in the areas of, for example, user gratifications (Leung and Wei, 2000), computer-mediated communication (Grinter and Eldrige, 2001), information systems research (Manning, 1996) and various aspects of design (see, for example, Chuang et al., 2001). Hence the map is not completely blank. However, there are significant areas that we are in the process of filling in. It is in this spirit that we address this issue by including three topic areas, those of the psychological, linguistic and commercial/applications of mobile communications.

It is clear that a psychological understanding of mobile telephony will enhance our understanding of the phenomena: the degree to which mobile telephony is a disturbing influence, the ways in which users can develop psychological dependence, the links between loneliness and ritualistic behavior and the ways in which social attribution are used in order to make sense of mobile telephone use. In a similar way, there are clear contributions to be made when examining the linguistic dimensions of mobile communication. In many ways, the spontaneous nature of SMS challenges our sense of the voice/written dichotomy. Thus, studying texting (and also language in instant messaging) provides insight into how people are using and developing the language in real-life situations when confronted by the need to form communications using new types of mediation. Beyond the broader issues of language formulation, there is also insight into which subgroups are at the forefront here. Finally, there is the meta-language surrounding the ownership and marketing of mobile communication devices that is open for consideration.

The marketing issues and the psychological dimensions are also drawn on in the final part of the book, where adoption and commercial issues are considered. If one considers mobile communication and service adoption through the glasses of gratifications and domestication research, one finds that gratifications obtained from mobile services go far beyond the instrumentality of flexibility and availability. This finding pushes the realm of analysis beyond the traditional issues associated with adoption research that usually focuses instrumentality in the form of usefulness and ease of use. It is possible that even if services have been designed to meet the functional needs of the subscriber, derived motivations of expressiveness are just as important as perceived usefulness to the individual end-user's adoption of the services. These findings are also seen in the adoption and use of various applications in a broad variety of settings.

Hence we are starting to trace the broader impacts of mobile communication in a variety of areas. Beyond the simple adoption of the mobile telephone, the devices impacts are starting to be seen in a broad variety of settings (as evidenced by the broad geographic distribution of the authors represented here). In addition, the adoption and use of mobile telephony is being studied by a broad selection of disciplines.

References

Brown, B., Green, N. and Harper, R. (2002) *Wireless World: Social and Interactional Aspects of the Mobile Age*. Springer, London.

Chuang, M.C., Chang, C.C. and Hsu, S.H. (2001) Perceptual factors underlying user preferences toward product form of mobile phones. *International Journal of Industrial Ergonomics*, 27, 247–258.

Fortunati, L., Katz, J.E. and Riccini, R. (eds) (2003) *Mediating the Human Body: Technology, Communication and Fashion*. Lawrence Erlbaum, Mahwah, NJ.

Grinter, R.E. and Eldridge, M. (2001) Y do tngrs luv 2 txt msg? In Prinz, M., Jarke, Y., Schmidt, K. and Wilf, V. (eds), *Proceedings of the Seventh European Conference on Computer-Supported Cooperative Work ECSCW'01*, Kluwer, Dordrecht, pp. 219–238.

Katz, J.E. and Aakhus, M. (2002) *Perpetual Contact: Mobile Communication, Private Talk, Public Performance*. Cambridge University Press, Cambridge.

Kopomaa, T. (2000) *The City in Your Pocket: the Birth of the Mobile Information Society*. Gaudeamus, Helsinki.

Leung, L. and Wei, R. (2000) More than just talk on the move: Uses and gratifications of the cellular phone. *J&MC Quarterly*, 77, 308–320.

Ling, R. (2004) *The Mobile Connection: the Cell Phone's Impact on Society*. Morgan Kaufmann, San Francisco.

Manning, P.K. (1996) Information technology in the police context: the 'sailor' phone. *Information Systems Research*, 7, 52–62.

Nyíri, K. (2003) Mobile communication: essays on cognition and community. *In Communications in the 21st Century*. Passagen Verlag, Vienna.

Acknowledgments

We would like to thank various people and institutions for their support in the preparation of this book. First we wish to thank all the authors who have agreed to go through the process of preparing manuscripts for this volume. Their time, dedication and willingness to accept the endless e-mails from the editors is, of course, what made this book possible. We also wish to thank Telenor Research and Development, Agder University College and the Norwegian Research Council (in the form of their KIM project) for supporting the work carried out here.

Rich Ling
Per Pedersen

Contents

List of Contributors

Naomi S. Baron

Naomi S. Baron is Professor of Linguistics at American University in Washington, DC. Her research interests include computer-mediated communication, the influence of technology on spoken and written language, child language acquisition and the history and structure of English. Currently, she is investigating the relationship between language used in instant messaging, face-to-face speech and offline writing.

Jeffrey Boase

Jeffrey Boase is a doctoral student in the department of sociology at the University of Toronto and currently a visiting doctoral fellow at Harvard University's National Center for Digital Government. His dissertation research examines the role of the Internet in maintaining relationships, and the implications this has for social capital and social tolerance. To examine this topic, he designed a telephone questionnaire in collaboration with Barry Wellman and the Pew Internet and American Life Project. This questionnaire was given to 2200 randomly selected respondents throughout the continental USA, during the spring of 2004. He is also heavily involved in the design of the Connected Lives study in collaboration with Barry Wellman and other members of NetLab.

Akiba A. Cohen

Akiba A. Cohen (PhD, Michigan State University, 1973) is Professor of Communication at Tel Aviv University, Israel, and the founding chair of the department. He is a former president of the International Communication Association and a Fellow of the Association. His recent research interests include the study of news, comparative research on media and the measurement of mobile phone use. He has published several books, including *The Holocaust and the Press: Nazi War-Crimes Trials in Germany and Israel* (with T. Zemach-Marom, J. Wilke, and B. Schenk), Hampton Press, 2002; and *Global Newsrooms, Local Audiences: a Study of the Eurovision News Exchange* (with M.R. Levy,

M. Gurevitch and I. Roeh), John Libbey, 1996. He has also published several articles and presented a number of convention papers on mobile phone use.

Kathleen M. Cumiskey

Dr Kathleen M. Cumiskey is an Assistant Professor in the Psychology Department at the College of Staten Island, which is part of the City University of New York. She was granted her PhD from the Social-Personality Psychology Program at the City University of New York's Graduate Center in 2003. Her research engages classic theories of psychology in the study of mobile communication technology. Her focus is on the social meaning of mobile communication technology and how it impacts face-to-face interaction. She lives on Staten Island and utilizes the many public spaces of New York City to observe how Americans respond to the public use of mobile phones.

Leopoldina Fortunati

Leopoldina Fortunati teaches Sociology of Communication and Sociology of Cultural Processes at the Faculty of Education of the University of Udine. She represents Italy in the COST Technical Committee for Social Sciences and Humanities and is part of the action COST A20 "The Impact of Internet on the Mass Media in Europe". She has conducted many research projects in the field of gender studies, cultural processes and communication technologies. She is the author of many books, among which are *The Arcane of Reproduction* (Autonomedia, 1995) and *I Mostri nell'Immaginario* (Angeli, 1995), and is the editor of *Gli Italiani al Telefono* (Angeli, 1995) and *Telecomunicando in Europa* (1998), and with J. Katz and R. Riccini *Mediating the Human Body. Technology, Communication and Fashion* (2003). She has published many articles in journals such as *Information, Communication, Society, Réseaux, Trends in Communication, Revista de Estudios de Juventud, Widerspruche and Personal and Ubiquitous Computing*. Her works have been published in seven languages.

Leslie Haddon

Dr Leslie Haddon is a Visiting Research Associate at the University of Essex and teaches part-time at the London School of Economics. With a doctorate on the development of home computing, he has for nearly two decades worked chiefly on the social shaping and consumption of information and communication technologies. These have included a range of studies on various aspects of telecommunications, including the Internet and mobile telephony, and involved work in a number of pan-European studies. He has been part of a European network of researchers (COST269) and recently was co-organizer of the conference *The Good, the Bad and the Irrelevant: the User and the Future of Information and Communication Technologies*. He has written a wide range of articles and more recently (2004) published *Information and Communication Technologies in Everyday Life: a Concise Introduction and Research Guide* (Berg, Oxford). See http://members.aol.com/leshaddon/Index.html for further details and links to publications.

Ylva Hård af Segerstad

Ylva Hård af Segerstad is a researcher at the department of linguistics at Göteborg University in Sweden, where she also earned her PhD in 2002. Her dissertation analysed language use in four modes of computer-mediated communication: e-mail, web chat, instant messaging and SMS. She is now involved in a research project called "Learning to Write in the Information Society" at the same department, which investigates children's use of writing in the school environment and during leisure time.

Mizuko Ito

Mizuko Ito is a cultural anthropologist of technology use, focusing on children and youth's changing relationships to media and communications. Her current research is on Japanese technoculture, and she is completing a study on how children in Japan and the USA engage with post-Pokemon media mixes. She is co-editor of *Personal, Portable, Pedestrian: Mobile Phones in Japanese Life*. She is a Research Scientist at the Annenberg Center for Communication and a Teaching Fellow at the Anthropology Department at the University of Southern California. Past workplaces include the Institute for Research on Learning, Xerox PARC, Tokyo University, the National Institute for Educational Research in Japan and Apple Computer. Her web page is at http://www.itofisher.com/mito.

Marianne Jensen

Marianne Jensen holds a degree in Media and Communication from the University of Oslo, Norway, and has worked in the field of new media in everyday life at Telenor R&D since 1994. Her studies include themes such as ICTs and families, media use and gender, the use of multimedia in learning and future home oriented issues. She has been actively involved in several projects related to the building of the "House of the Future" at Telenor, as the youth room of the future and technology to support communication in distributed families.

Tom Erik Julsrud

Tom Erik Julsrud is currently a PhD student at the University of Trondheim (NTNU) working with organizational identity in mobile and dispersed workgroups. He has studied sociology, psychology and politics at the University of Oslo, and holds a Norwegian master's degree in communication studies from the University of Oslo. Working as a research scientist at Telenor R&D since 1994, he has specialized in organizational communication and organizational transformations related to introduction of telecommunications and general ICT. He has most recently been managing a project studying the introduction of multimedia messaging among groups of mobile professionals, and a Nordic project on workplace design (www.dekar.org). He is the co-author of a book of Marshall McLuhan (*Den elektroniske Nomade*, Spartacus, 1998) together with John Willy Bakke.

James E. Katz

Dr. James E. Katz is a Professor in the Department of Communication at Rutgers University. He is also the director of the Center for Mobile Communication Studies there. Dr Katz has published widely in the field of mediated interpersonal communication. Among his books that discuss mobile communication are *Connections* (Transaction, 1999), *Perpetual Contact* (Cambridge, 2002, co-edited with Mark Aakhus) and *Social Consequences of Internet Use* (MIT Press, 2003, co-authored with Ron Rice). His work on telephony has been translated into French, German, Dutch, Spanish, Chinese and Japanese. Major media outlets interview him frequently about his research. He has also been active in public policy debates about cell phone usage and has been a member of the New Jersey State Legislature's Council of Academic Policy Advisors.

Dafna Lemish

Dafna Lemish (PhD, The Ohio State University, 1982) is Associate Professor and Chair of the Department of Communication, Tel Aviv University, Israel. Her research and teaching interests include gender-related issues of media representations and consumption as well as children, media and leisure. She has published extensively in academic journals and books on these and other topics. Her forthcoming book (with Götz, Aidman and Moon, Lawrence Erlbaum Pubishers) is entitled *Media and the Make-Believe Worlds of Children: When Harry Potter meets Pokéman in Disneyland*. Her interest in mobile phones focuses on issues of technology and construction of identities, gender and cultural, as well as on the integration of the mobile into family and social life.

Rich Ling

Rich Ling is a sociologist at Telenor R&D located near Oslo, Norway. He is an Adjunct Professor at the University of Udine and a Visiting Professor at the University of Michigan. He received his PhD in Sociology from the University of Colorado, Boulder, in his native USA. Upon completion of his doctorate, he taught at the University of Wyoming in Laramie before coming to Norway on a Marshall Foundation grant. Since that time he has worked at the Gruppen for Ressursstudier (The resource study group) and he has been a partner in a consulting firm, Ressurskonsult, which focused on studies of energy, technology and society. For the past 10 years, he has worked at Telenor R&D and has been active in researching issues associated with new information communication technology and society with a particular focus on mobile telephony. He has led projects in Norway and participated in projects at the European level. He has published numerous articles, lectured at Universities in Europe and the USA and has participated in academic conferences in Europe, Asia and the USA. He has been responsible for organizing scholarly fora and editing both academic journals and proceedings from academic conferences. He is a founding member and the first chair of the Society for the Social Study of Mobile Communication and the author of the book *The Mobile Connection: the Cell Phone's Impact on Society*.

Steve Love

Steve Love is a Lecturer in the School of Information Systems, Computing and Mathematics at Brunel University, West London, UK. His research interests include looking at the social impact of mobile phone behavior in public places and effects of individual differences in relation to user interface design for mobile phone applications.

Louise Mifsud

Louise Mifsud is a PhD student at Agder University College, Kristiansand, Norway. Her research interests include digital literacy, mobile and hand-held technology in education. Publications include: "Alternative learning

Arenas – Pedagogical Challenges to Mobile Learning Technology in Education", in *The Proceedings of the IEEE International Workshop on Wireless and Mobile Technologies in Education* (2002); "Are mobile technologies an integral part of learning culture?" in *Learning Technology Newsletter*, 5(2), 2003; "Digitalising Literacy?", paper presented at the NOKOBIT 2003, Oslo; "Learning 2go: making reality of the scenarios?", in Attewell, J., and Savill-Smith, C. (2004), *Learning with Mobile Devices. A Book of Papers*, LSDA, London.

Kakuko Miyata

Kakuko Miyata is Professor of Social Psychology at Department of Sociology, Meiji Gakuin University in Tokyo. She received her PhD from Department of Social Psychology, the University of Tokyo. She was a visiting scholar at the Centre for Urban and Community Studies, University of Toronto in 2002. She has written several books in the field of the use and social impact of computer networking, including *The Society of Electronic Media* (in Japanese), which won the Social Science Award from the Japan Telecommunications Advancement Foundation and which was translated into Korean. Currently her works focus on a comparative study of the impact of Internet activities on social capital in Japan and North America, including "Social Capital and Internet Use in Japan" in *Social Capital: Advances in Research*.

Siri Johanne Nilsen

Siri Johanne Nilsen has a Cand. Polit. from the Department of Media and Communication, University of Oslo. She is currently a research scientist at Telenor R&D. She specializes in analysing new trends in residential interactive services and the Internet.

Herbjørn Nysveen

Herbjørn Nysveen (herbjorn.nysveen@nhh.no), PhD, is an Associate Professor at the Norwegian School of Economics and Business Administration. His research activities focus on relationship marketing, e-marketing and mobile commerce. His research has been published in journals such as *Decision Support Systems, Journal of Interactive Marketing, Journal of Information Technology* and *International Journal of Research in Marketing*. His research on mobile commerce is focused on demand-side issues, such as end-user adoption and consumer–brand relationships. Nysveen is affiliated with the Institute for Research in Economics and Business Administration in Bergen, Norway.

Fernando Paragas

Fernando Paragas is an Assistant Professor at the College of Mass Communication of the University of the Philippines Diliman. He is currently a Fulbright fellow, pursuing his PhD degree in Communication at Ohio University. His papers on communication technologies have been presented in Thailand, Korea, Spain, Norway, the USA, Italy, and Hungary. He was the recipient of a research grant from the Asian Federation of Advertising Associations and Newsweek Magazine.

Woong Ki Park

Dr Woong Ki Park received a BA in Communications from the University of Michigan, Ann Arbor (USA) in 1991, MA in Advertising and PR from Sogang University (Korea) in 1994 and PhD from Temple University (USA) in Mass Media and Communications in 1999. His experience includes having worked for the Merit/Burson-Marstellear and LG/BBDO agencies. He taught at University of Southern Maine in Portland, Maine, before going to Soongsil University. His main research interest is in media psychology. Recently, he has received funding from Korea Telecommunication Cultural Foundation and published a book titled *Mobile Communication and Culture*.

Per E. Pedersen

Per E. Pedersen is Professor of Information Management at the Agder University College. He holds a Dr Oecon degree (PhD equivalent) from the Norwegian School of Economics and Business Administration, where he also holds an adjunct professorship. His current research areas are consumer behaviour in electronic markets, user adoption of mobile and cross-media services, social and commercial effects of mobile services and the organization of telecommunication value chains. He is currently also a member of the E-markets research group at the Foundation for Research in Economics and Business Administration and has been responsible for several research projects in this research group funded by industry partners, the Research Council of Norway and the European Commission. Pedersen has several publications in international scientific journals and in well-known conference proceedings.

Carole Rivière

Carole Rivière studied history and politics at the Sorbonne in Paris before studying sociology. She attained her PhD in 1999, writing on "Phone sociability: a contribution for the analysis of personal networks and social change". She is now working at the research center of France Telecom in the sociological lab., "Sociology of Uses and Statistical Information Process", that specializes in the analysis of the new behaviors of communication. She has done surveys of SMS use and mobile phone use among the Japanese. In addition, she has examined the anonymous relationships on the web.

Harmeet Sawhney

Harmeet Sawhney is Associate Professor in the Department of Telecommunications at Indiana University, Bloomington. He brings a diverse background that includes engineering, business, and communication to the study of telecommunications policy-related issues. His research focuses on the processes that shape the development of telecommunications infrastructure and other large-scale networks such as canals, railroads and highways. His research articles have appeared in *Telecommunications Policy, Journal of Broadcasting and Electronic Media, Media, Culture, and Society, Info, Entrepreneurship and Regional Development and The Information Society* and in book chapters in edited volumes. He is currently serving as the Editor-in-Chief of *The Information Society*.

Lauren Squires

Lauren Squires graduated with University Honors from American University in 2003, concentrating in philosophy and communication. She is now a graduate student in linguistics at the University of Virginia. Her research interests lie at the intersection of sociolinguistics, media studies, communication theory and computer-mediated communication.

Satomi Sugiyama

Satomi Sugiyama is a doctoral candidate in the Department of Communication at Rutgers, The State University of New Jersey, and a MacArthur Fellow in the Department of Philosophy at Colgate University. She is particularly interested in the influence of mobile communication technologies on social behaviors in a global context, and on various forms of aesthetic expression via everyday communication. She has collaborated with Dr James E. Katz on several studies and conference papers. Her most recent conference presentation is Katz, J.E. and Sugiyama, S. (2003) "Mobiles and fashion: a survey of US and Japanese youth", National Communication Association Annual Meeting, Miami, November 20. Her publications include: Sugiyama, S. and Katz, J.E. (2003) Social conduct, social capital and the mobile phone in the US and

Japan: a preliminary exploration via student surveys, in N. Kristof (ed.), *Mobile Democracy. Essays on Society, Self and Politics.* Passagen Verlag, Vienna, pp. 375–386.

Alex S. Taylor

Alex Taylor is a researcher at Microsoft Research in Cambridge, UK. He draws heavily on the ethnographic tradition to investigate the real-world use of technological and non-technological artefacts. With this orientation, particular emphasis is given to embodied practices and the ways in which the properties of material things are bound up in the organization of social life. His work is predominantly directed towards recovering the detail of everyday (inter)action and using the relevant findings to inform the design of interactive and collaborative systems. In addition to examining the use of mobile phones, Alex has participated in various research programs studying, for example, home life, legal practice and healthcare.

Sara Tench

Sara Tench graduated with University Honors from American University's School of International Service in 2002. She is currently pursuing a graduate degree at Johns Hopkins University. She is interested in the impact of communications technology on culture, especially with respect to international communication.

Marshall Thompson

Marshall Thompson graduated with University Honors from American University in 2003 with a BA in Anthropology. After working briefly in London, he returned to Washington, DC, to pursue a career in the travel industry. Besides traveling, his interests include music, films, politics and vintage motor scooters.

Helge Thorbjørnsen

Helge Thorbjørnsen, PhD, is an Associate Professor at the Norwegian School of Economics and Business Administration in Bergen, Norway. His research activities are in the fields of relationship marketing, brand management, e-marketing and mobile commerce. He is the author of several papers published in, e.g., *Journal of Service Research, Journal of Interactive Marketing* and *Journal of Brand Management*. The findings reported in his chapter in this book are based on a research project in mobile commerce at the Institute for Research in Economics and Business Administration in Bergen, Norway.

Kristin Thrane

Kristin Thrane is an ethnologist (Nordic folk life research) at Telenor R&D. She obtained her Cand. Philol. degree from the Institute of Cultural Studies, University of Oslo, were she focused on material culture and shaping of the domestic sphere. At Telenor R&D her research has focus on the adoption and use of new broadband services and technology in the private sphere.

Barry Wellman

Sociology professor Barry Wellman heads the NetLab at the University of Toronto's Centre for Urban and Community Studies. He is the chair of the Communications and Information Technology section of the American Sociological Association, Chair Emeritus of the ASA's Community section and the founder of the International Network for Social Network Analysis. His most recent edited books are *The Internet in Everyday Life* (with Caroline Haythornthwaite) and *Networks in the Global Village*.

Steve Woolgar

Steve Woolgar is a sociologist who is Director of Research and holds the Chair of Marketing at the Saïd Business School, University of Oxford. He was formerly Professor of Sociology, Head of the Department of Human Sciences and Director of CRICT at Brunel University. He took his BA (First Class Honours), MA and PhD from Emmanuel College, Cambridge University. From 1997 to 2002 he was Director of the UK national ESRC Programme *Virtual Society? – The Social Science of Electronic Technologies*. He has published widely in social studies of science and technology, social problems and social theory. *Virtual Society? – Technology, Cyberbole, Reality* (Oxford University Press) was published at the end of 2002. His work has been translated into Dutch, French, Greek, Japanese, Portuguese, Spanish and Turkish. He is currently working on governance and accountability relations in mundane technologies, on the social shaping of eScience and eSocial Science, and on the dynamics of provocation.

Vicki Yung

Vicki Yung is a consultant for international branding and mediated communication in Washington, DC. She holds a PhD in Linguistics from

Georgetown University. Her dissertation "Mobile Phones in Interaction: Discourse Analysis and the Material World" is based on a study of mobile phone use in Hong Kong and Beijing. Prior to her pursuit of a doctorate degree, she was a lecturer at City University of Hong Kong, where she taught courses in intercultural and professional communication. She has conducted research in entertainment and news media, mediated communication in an interactive television (ITV) learning environment and the sociocultural aspects of mobile communication.

Part 1
The Dimensions of Mobile Communication

Introduction

Rich Ling and Per E. Pedersen

In the first part of the book, we draw out some of the broader issues associated with mobile communication, and in particular its placement in the broader context of research on information and communications research. The part contains, in order of presentation, chapters by Leslie Haddon, Steve Woolgar, Harmeet Sawhney, and James Katz and Satomi Sugiyama.

The first two chapters look into the mobile telephone as one of several types of communication media. The authors work to place the mobile telephone into an increasingly complex communications landscape. Haddon starts by posing the question of how people manage their mediated communication at a basic daily level? Specifically, how are decisions made when confronted with the possibility of communicating via, for example, landline telephony, SMS, e-mail voice mobile and soon Wi-Fi? Each of these can be used, but research shows that some are use in certain situations for certain purposes. Hence there is an active evaluation of how to best and most appropriately communicate. Clearly, previous generations have had the opportunity to choose communications media (letter vs telegram vs telephone call). The difference here is that the media landscape is increasingly complex. Haddon then looks into the research questions that these developments suggest along with a discussion of their methodological implications.

Steve Woolgar takes up the discussion at this point and sets it into the context of how theorizing associated with mobile telephony will be framed. The popular use of mobile telephony has arisen in the shadow of the Internet. Thus, it can be seen as the theoretical "little brother" of the Internet. This means that the first impulse is often to alter the work done on the Internet and see if – using a type of hand-me-down approach – it fits mobile telephony. Woolgar looks into this issue to examine the potentials and the tensions of using Internet theorizing in the case of mobile communication. Although there are some advantages to this approach, there are also dangers. Woolgar concludes by suggesting increased attention be paid to what he calls "interpretive flexibility of the theory–technology relation" in a way that will allow new purchase in our ability to understand the use of new technology.

Harmeet Sawhney rises to Woolgar's challenge of theorizing and also Haddon's call to avoid seeing communication systems in isolation when the presents what he calls the Infrastructure Development Model (IDM). Drawing on historical parallels with the interaction between canal and railroad development, he looks into the potential interaction between Wi-Fi and existing telephony systems including mobile telephony. The sense here is that Wi-Fi, and perhaps more generally Internet Protocol (IP)-based applications such as IP telephony, will slowly develop and eventually replace the larger and more entrenched telephony system. Increasing accessibility to Wi-Fi has allowed enthusiasts to create their hot spots and to develop what Sawhney calls "an entirely new kind of a bottom-up network". He goes on to describe these developments as a type of libertarian impulse that was also evident in the development of the original Internet. This approach is also often seen in contrast to the centrally planned and commercial telephony system that is today's locus of the Internet, telephony and mobile telephony. He suggests, "We are seeing an insurrection by libertarian bands of sorts who take pleasure in subverting corporate frameworks." Starting as isolated areas, Sawhney suggest that Wi-Fi will slowly link and grow in order allow one what approaches universal access. Clearly, this type of development will have an impact on issues such as service quality, the engineering of mobile radio systems, infrastructure maintenance, governmental regulation and the like. The perspective of the telecom industry would be that while certain portions of traffic will migrate to these ad hoc systems, there are still advantages to be enjoyed – and indeed engineering exigencies – that are a part and parcel of the traditional centrally planned networks. Clearly, time will tell. Sawhney's contribution helps to focus our attention on these types of issues.

The final chapter in this part looks in a somewhat different direction and indeed a direction that challenges the assumption that mobile telephony is only about bits and bytes or only about communication. James Katz and Satomi Sugiyama point to the way in which the mobile phone is intentionally a type of fashion or art. They develop an analysis that shows how in the eyes of some users the mobile telephone plays into the modern sense of a streamlined fashion statement and how this is further formed into the reflection of the individuals sense of taste, social position and aesthetics. It is increasingly being integrated into the way in which we present ourselves to society. The facts that the device is worn and that we are concerned directly with the aesthetics of the actual physical device help us to understand that although there are the techno/functional/regulatory issues as outlined by the previous authors, there are also issues of personal style and identity. Indeed, these same issues come up in the final section of the book in Pederson's analysis of commercialization of certain mobile services.

Hence the papers in the first part of the book set the scene by examining how mobile telephony fits into the repertoire of communication possibilities for the individual, how these potentials might be looked at in the development of theoretical models, the way in which they may develop vis-à-vis other technologies and finally what mobile telephony may mean in terms of our sense of self and identity.

Research Questions for the Evolving Communications Landscape

Leslie Haddon

2.1 Introduction

The reader should note that the mobile phone is not mentioned in the title of this chapter. This is intentional. This is the point. To complement the increasing amount of research focusing on mobile phones, this chapter aims to explore the merits of looking at mediated communication in general, seeing the mobile as just one element of that context.

Our communications options are becoming increasingly complex with relatively more major and minor options becoming available. At one time, to study mediated interpersonal communication meant studying the fixed-line phone. Mobile telephony and communication via the Internet may have been the more general, outstanding recent additions, but we might think also of the various ways we have of sending and receiving voice messages, text messages and images, manipulating them and controlling communication.

Increasingly we can ask not only why we *use* a particular channel or function but why we *choose* it from amongst the possibilities. And when we use one medium, what implications has that for the use of another? To answer such questions requires thinking holistically, conceptualizing communication practices as an ensemble, a repertoire. Indeed, we already see some researchers addressing such relations between the different elements of those repertoires when they refer to processes of substitution or complementarily and to specialized use of certain channels – to be discussed below.

This chapter attempts to use existing studies, mainly of the telephone, the mobile phone and e-mail, to address systematically issues about that repertoire of practices. What new questions, or reformulations of questions, do we ask when this is the focus rather than the individual medium? In part, this exploration also has an 'applied' social science agenda behind it, in terms of suggesting what those working to develop ICTs might consider when making sense of new emerging practices and thinking about possible future ones.

1. The chapter starts by asking a general question about the limits of the object of study: what might be the boundaries of what we consider to be a communication practice within this repertoire?
2. Next, it asks the historical question of where do practices come from? Specifically, what are the continuities between the different elements of the repertoire? How do some of the ways in which we communicate draw upon practices developed using other media or indeed from other routines in our daily lives?
3. Turning to how we mange this repertoire, how do we make choices among the elements of that repertoire? What types of factors shape choices, not just in terms of technical features, but also in terms of social relationships? These are not always easy decisions and, indeed, we might ask how much the term 'choice' adequately captures the process.
4. Finally, the chapter asks about the dynamics of the whole media repertoire over the longer term. What factors contribute to changes in communications choices? How does the repertoire evolve?

2.2 The Boundaries of the Object of Study: Communication-related Practices

The first question concerns the scope of any enquiry into communication repertoires: how broad a vision should we have of what elements count as a communication? Specifically, this section argues why we should always try to imagine what could count as communication-related practices that go beyond but help to make sense, of the more detailed patterns of communication.

Although we often think of communication as the exchange of words or information between people, there are also a broader set of actions we take that can modify the communication act. Thinking of the mobile phone more generally, we might include the way in which we control mobile use, such a controlling who the mobile number is given out to, switching it off and switching it to voice mail (Haddon, 2004). These can all either shape 'use' or may be considered to be a part of an expanded definition of 'use'. We might also include how people talk about communications, such as the way they exchange information about how best to exploit mobile tariff structures. Then there are practices such as changing SIM cards or else people borrowing someone else's mobile phones if the mobile phone network of the person being called means this is cheaper or effectively 'free'. And perhaps some of the best recent examples of communications-related practices are those related to texting, especially by youth, which have been documented in a number of studies. These include saving messages and talking with others when composing them (Johnsen, 2002; Kasesniemi and Rautianen, 2002).

Equivalent, in effect boundary defining, questions can, and have, been asked of other means of communications such as the phone and Internet, and could be asked or emerging ones. In what ways can these practices take on a wider importance? To take an older example, research on early British home computing in the mid-1980s, including an analysis of game-playing, argued that it was important to look beyond the moment of sitting in front of the screen to consider all the acts related to 'computing' (Haddon, 1992). These acts included reading the magazines about the new hobby of computing and about the latest games. Knowledge about these games and tips about how to play them were especially important for those, mainly male youth, who followed games as they developed into a cultural industry. Meanwhile, computer-related activities included talk about the computer, comparing experiences, discussing possibilities, both inside and especially outside the home, primarily in the informal culture of school. And they included sharing and copying software.

One reason why it was important for us as researchers to take such practices into account was that they had a bearing upon our very evaluation of what constituted the activities of 'home computing' and 'game playing'. For example, at the time there were widespread social concerns – not unlike ones currently related to the Internet (Nie, 2001) – about people, especially males, becoming isolated in front of screen. This involved fears about the computer making them anti-social (Shotton, 1989). However, by looking at these other computer- and game-related activities, it became clear that in certain senses both home computing and games playing could actually be very sociable activities.

In fact, the very interest in these early home computers derived partly from these wider social practices, as did decisions that people took about the time they allocated to the computer and to games. Moreover, those interactions and activities away from the screen influenced what people did when they then played games or explored the potential of these early computers. Finally, appreciating the role played by computer-related practices was important for the analysis of patterns of actual use. For example, this contributed to the explanation of gender differences. On the whole, girls did not partake in the range of activities noted above, which meant that although they may have used a computer or played games, their experience was not the same as that of the boys who were involved in computing and games playing in these other ways.

To sum up this section, when encountering new elements to the communications repertoire, we need to ask what may count as communication-related practices. This is important because it may make a difference to how we conceptualize that communication and to wider claims made about what it means for society. However, these practices may also influence the experience of communicating, decisions when choosing among the communications repertoire and hence the patterns of communication that emerge.

2.3 Continuities Between Media

Jouet, reviewing French research, notes that the adoption of new ICTs takes place against a backdrop of pre-existing techniques. Further new uses are often an extension of what has gone before. The use of the new tools is grafted on to the practices associated with older ones (Jouet, 2000). A similar point is made in an empirical study of Internet use by teenagers, where Quebec researchers argue that we can better understand how people use and interact around media and technologies by locating this behavior in the context of their wider and pre-existing cultural practices. In this sense we can think of a "continuum of uses" as new practices are "inscribed within" (or built upon) old ones (Millerand et al., 1999).

In fact, similar observations occur elsewhere. For example, if we look at the social shaping of media, Winston shows how much the form of early TV was based on the broadcasting principles that had evolved in relation to radio (Winston, 1989). This formed part of an argument to show how the emergence of an even apparently dramatically "new" medium is to some extent an evolutionary process. Arguably this is one of the contributions of both an historical and social science memory, always challenging the discourses of technological revolutions.

To take another example, computer games had numerous lineages back to older technologies, to the video arcade machines, to TV games consoles, even to pinball (Haddon, 1992, 1999a). Once again, it was important to appreciate such continuities in order to understand the historical construction both of interest in these technologies and of the particular practices relating to them. For example, in the 1970s, games arcades were mainly male domains, and so it is easier to appreciate the greater interest of male youth in early home games machines. Looking for continuities can, in some circumstances, help explain patterns of use.

If the discussion above was intended to establish why it is important to consider continuities in general, we can now go further using the examples of existing studies to map at least some different types of continuities in relation to our communication repertoires.

First, there are continuities in communications related to specific events. For example, this might include using e-mail to announce the birth of a child, whereas previously one might have made phone calls or sent cards (Manceron et al., 2001). It could include male partners acquiring a mobile to be reachable at childbirth, whereas in the past they might have given a fixed phone number where could be found (Manceron et al., 2001). And it could cover sending e-mails or electronic Christmas cards at Christmas instead of sending Christmas cards.

Second, there are continuities from more routine practices. One example might be people using the mobile phone at work for private purposes (Mercier, 2001) – whereas in the past one might have used the work fixed line (De Gournay, 1997). Another would be children chatting on-line after school, or using a mobile phone, whereas in the past they might have used a fixed line. We have the 'gift' of mobile phone calls or text messages, whereas in the past some fixed line calls could have equally well been conceptualized as gifts (Johnsen, 2002; Nafus and Tracey, 2002). Also, the practice of teenagers texting each other in the classroom is really an updated, and less visible, version of passing paper notes around without the teacher seeing (Ling and Yttri, 2002).

Finally, we have continuities from previous practices not necessarily considered in the past as communication. One illustration of this is the case of teenagers "hanging out" on-line via instant messaging to hear of something interesting, to hear if something was "happening", compared with hanging out in physical public spaces such as shopping malls (Rainee, 2001).

On the one hand, looking at such continuities also raises the question: how "new" are new practices? But if we always look just for continuities, we may be asking a conservative question. Clearly, we also need to appreciate how different new practices are, what they lead to which is a new departure and where and in what ways they make a difference. For example, evidence has been cited to support the argument that using mobiles at work has led to extra social communications by males that might otherwise not have taken place, and that they may have increasingly become "ambassadors for family" (Mercier, 2001). Meanwhile, e-mail sent to those at the periphery of social networks instead of the occasional letter or Christmas card may have led to more communication and, indeed, maintaining contact with older social networks that might otherwise have been lost. Finally, Ling (and others) draw attention to the opposite scenario, asking whether people are actually communicating more with fewer, but closer, social network members via the mobile phone at the cost of weaker social ties (Ling, 2004).

In sum, when studying both current and future innovations, we can ask whether we can see any types of continuity from previous practices. If so, what types of continuities exist and what types of past precedent are we talking about? On the other hand, to what extent are new practices different from older ones and how does this make a difference to how we experience them? Finally, there are the analytical challenges that such questions might pose: how much can one only appreciate continuities

Another question concerns the status of the items on a list such as the one above. How much technical fluidity is there? Which technical parameters are more core to a particular medium, relatively unchanging, and where is there more scope for innovation that may change the possibilities and constraints associated with that option over time (as in the way mobile phones have become more personalizable, to take one of the examples above).

The second set of considerations are the social factors favoring certain communication choices. These could include the purpose and content of communication, such as communications that are gifts or communications that provide a sense of security. They could cover the urgency of communication, varying from communications in the event of emergencies and new contingencies and communications to finalize arrangements to meet through regular contact with close family, to occasional contact with extended family and distant friends to keep in touch.

We would have to consider the social relationship to the other communicator: for example, contact with immediate family, communications with extended family, contact with close friends, "mates", old friends, acquaintances, colleagues, etc. Then we have the physical proximity of communicators, meaning the degree of distance within a country in addition to communications abroad. One further factor would be the "social location" of the persons communicating, for example, whether they are at home, in the work place or in various public spaces. And finally there are the communication norms of social networks, an example being the fact that texting, e-mail and/or chat are the most commonly used media of some networks of young people.

On the one hand, various studies suggest that these are indeed all considerations in making choices. However, we should bear in mind that all such checklists are simplifications. For example, consider the communication norms of social networks. Within social groups or networks there can be rules and expectations about the appropriateness of different media for different circumstances. One study of texting by youth showed that there are understandings about when it is inappropriate to use texting as opposed to using other means of communication, including face-to-face communication. For instance, the study showed how it is not considered right to end a relationship, to dump someone, through sending a text message (Taylor and Harper, 2001). However, even here there is not always consensus, as some network members make choices that others think are inappropriate.

The third set of social considerations consists of the social constraints affecting the choice of medium. These include the regulation of one's communications by others, as when parents attempt to regulate outgoing calls on the fixed line, mainly but not exclusively because of cost considerations (Haddon, 1994, 1998) – which may privilege the use of the mobile in certain cases. Nor should we forget the formal regulation of certain media in other social spaces (e.g. e-mail for social purposes at work; Haddon, 2002).

Social constraints are also revealed in what is felt to be inappropriate communication in certain social spaces and reactions to this, such as informal pressures to restrain mobile use in certain public spaces. We would have to consider social commitments to other people at certain times, for example, not using the Internet at certain times in the evening in order to keep the family phone line free for incoming calls (Lelong and Beaudouin, 2001). And we would want to include the strategies that people use for controlling their own communication, such as young people steering calls to their mobiles, or the mobile's voice mail, rather than to the domestic fixed line (or household answering machine) or the reluctance to give out mobile numbers to work colleagues, as has been noted in several studies.

One key point to derive from this discussion of constraints is that it actually provides a useful antidote to some of the misleading connotations of emphasizing choice amongst the communication repertoire. It reminds us that although we do indeed ultimately make choices, these are made within social constraints, sometimes external, as in the case of the social pressures from other people, and sometimes partly of our own making, as in our obligations to others and our own communication agendas.

Finally, to complete the picture, we need to consider how people link different forms of communication when making choices. After considering various examples (and echoing themes from previous French studies), de Gournay and Smoreda conclude that despite claims about technological convergence some communication tools are used for only certain types of communication, i.e. there is a degree of specialization (de Gournay and Smoreda, 2001). This may be true generally, but people do also sometimes shift between different parts of the communication repertoire in the short term. This means that for a more complete picture of the complexity of repertoire management, we also need to consider the dynamic dimension.

Hence we have another level of relationship between the elements of the communication repertoire: the short-term movement from one to another. This can include reconfirming a message made in one medium through another, such as e-mailing someone to confirm a phone decision. Then we have using one medium to set up another, for example making a phone call to ask for an attached file to be sent by e-mail.

Shifts between media may result when the first choice fails or is not available, as in the case of sending a text message when a phone call does not get through or phoning someone on the mobile when a fixed line is blocked (Klamer et al., 2000; Haddon and Vincent, 2004a, b). Or there is the example of using one mode when someone failed to reply by another – such as the example of a teenager phoning on the mobile to ask why someone had not replied to their text message, i.e. the latter had not met their expectations about how to use one channel of communication properly in terms of giving a timely reply (Taylor and Harper, 2001). Also, shifts may be necessary to sort out problems through one medium that were created through using another, for example making a phone call to clarify an e-mail (Mante-Meijer and Haddon, 2002).

To summarize, when we start to outline the considerations affecting choice, which are more nuanced than can be contained in the discussions above, then we can start to appreciate the ways in which social actors as communication managers make complex decisions. In fact, they are having to make more complex decisions than previous generations, who had fewer options. A second observation can be derived from the examples related to shifts between communications, especially the latter showing communications problems. These draw attention to the fact the managing the communications repertoire does not always run smoothly. A typology of factors is a start, but that does not in itself tell us who does what, in what circumstances, i.e. the patterning on communication choices. For example, might we expect to find differences by generation, with older generations feeling more at home using a more limited repertoire, and younger ones experimenting more? Or might we expect to find the influence of other standard socio-demographic variables, such as gender, life stage or class? There is clearly scope for further research here.

2.5 Longer Term Dynamics of the Communication Repertoire

This last section completes the review by considering longer term changes that people make in relation to their communications repertoire. Of course, this can occur in terms of changes in how people communicate with specific individuals, as their relationships develop (Ling, 2000). However, here we consider the more general question of people developing new routines and changing their practices for handling certain types of communication.

The dynamics of how we develop our use of individual ICTs had been tackled in a number of different research traditions. There have been studies of apprenticeships with new ICTs (Lelong and Thomas, 2001). The process of integrating ICTs into daily life and their subsequent careers has been considered with the domestication framework (Haddon and Silverstone, 1994; Bakardjieva, 2001). Also, there has been research into the effects on communication of major changes in life stages (Manceron et al., 2001).

One change in emphasis that arises when we look at the communications repertoire as a whole rather than individual means of communications is that as new communications possibilities come on the scene, people already have an existing set of options that they are using. This means that they may simply decide that they can manage with those existing options, or that there are particular reasons for sticking to their existing practices. For example, the quotation below shows how someone tried out e-mail for organizing meetings, but then abandoned it to go back to the older practice of using the phone.

> If it's something like someone sends you a message about where to meet that night it's quicker just to pick up the phone. E-mails can be terribly delayed. It's a real problem actually, it really screws things up, I can think of loads of arrangements that have been totally screwed up by e-mails not getting through in time. I've missed meeting people and I've not known about things that I've been invited to because they didn't come through. (Haddon, 2000)

However, even if at one point in time one option seems to be have disadvantages compared with another (or have other implications), this does not prevent it from being adopted at a later stage. Or, as the example below shows, the experience of problems associated with a new element of the repertoire may lead to it being modified.

> I feel like I can get a bit overwhelmed with all the e-mail I receive (…) I'm writing much shorter mails because of that, there's just too many, and I don't put as much effort in as I used to. It's got to the point where I just send something, however short, just to maintain a presence. I just say hello rather than give much detail about what I'm doing. (Haddon, 2000)

The last point to make by way of general observation is that within the repertoire different practices within the communications repertoire will also have different degrees of inertia. Certain ones may be relatively short-lived, such as some people's experience of pagers in the few years before the mobile phone and texting options became popular. By comparison, for certain purposes many people feel comfortable using the fixed phone line, despite the arrival of more possible alternatives. If we look beyond just electronically mediated communication, for some letter writing has been more drastically affected by the advent of e-mails, whereas for others it has retained a place in life. For some youth, chatting on the phone after school remains important, whereas for others on-line chat has taken on some of that role.

Having established some general sense of the changes that can occur over the medium to longer term, we can start to chart some of the factors that can lead to such change.

One such factor is changing individual or household circumstances (e.g. life stage, work, commitments), as shown in a study where people first used e-mail and mobiles with the birth of the first child (Manceron et al., 2001). Another factor is the changes in the form of communications within social networks or within a cohort of people of the same generation, such as the rise of texting amongst youth in various countries.

Then we have to consider wider societal changes, or changes within particular institutions, as when in the UK, the BBC promoted e-mail addresses and texting, or when social debates were covered in the media about the need to regulate mobile phone usage (e.g. because of health, because of disturbances of public space). The changing regulation of mobile telephony use in certain public spaces is another example. Finally, there are changes in communications options and promotions (e.g. new products, new pricing, new technical options, new marketing), such as pre-paid cards for mobile phones and the marketing of mobiles as fashion items.

For research purposes, it is worth noting one methodological issue here. In many respects the macro level, the wider societal changes, leaves more historically visible traces, e.g. when events are captured in the media, the appearance of new adverts, in the documentation of the launch of innovations and in institutional memories of when a decision was made. How practices emerged, were tried out, perhaps the role of chance events at the other levels, the individual and group levels, can be more easily forgotten. People can usually remember the process of acquiring technologies and services, because that is a major decision. They might remember usage associated with significant events, such as the decision to use e-mail to tell people of the birth of a child. However, there are many, many smaller practices, now routine, when it is more difficult to remember the details of how they emerged, why choices were made, why some things were and were not considered.

The problem is that because such small changes in practice escape the research eye, it is difficult to say how much the pattern that did emerge could easily have been otherwise – be that in terms of an individual's repertoires or the usage of some means of communication within wider social groups. For example, at one stage the use of the fixed line for social purposes was not anticipated within the telecommunications industry. To give a more recent example, with hindsight one can give reasons why texting might appeal, especially to youth. Indeed, contemporary youth can articulate these reasons now. On the other hand, texting still, arguably, represented a major shift in practices over a relatively short period of time. It is perhaps a surprising move to have the type of socializing we see on SMS handled by text instead of orally (Fortunati, 2000). It is not to clear that one would have thought this a likely development in advance, which has implications for the limits of analysts' ability to predict the take-up of new innovations.

This problem of invisibility raises some questions for researchers and product developers trying to understand patterns of adoption. It raises the question of the contingency of the patterns that exist now. Although it is commonplace for analysts to be able give reasons for why a practice was not taken up by individuals or groups, should we question whether rejection was automatic or whether that failure to adopt was inevitable? In other words, where might non-use have been contingent and it could have been otherwise if other things had occurred, including chance events?

2.6 Conclusion

In a world where the communication options open to us are becoming ever more varied, this chapter has explored what can be gained through seeing the totality of communications practices as a repertoire and analyzing the relationships between its elements. It has put this into some historical context by showing parallels with previous questions raised about the relationships between other ICTs. Also, it has posed some questions we might ask, or a framework for thinking about, not only current practices but also ones related to emerging and future innovations.

In considering the scope of the repertoire, we looked first at examples of communication-related practices, broadening our viewpoint to ask what the object of analysis is when studying communication, and what this could include. The chapter also reflected upon why these practices might be important in terms of how we characterize and evaluate communications practices and understand patterns of use and choice.

Next we considered the relationship between repertoire elements in terms of the evolution of new practices from existing ones. This looked at the continuity between practices, as identified by some previous research, a focus on which helps avoid emphasizing too much their uniqueness and novelty and stresses evolution over revolution. But of course, this does not mean ignoring their difference and the implications that may follow from this.

We moved on to the factors affecting choice between the elements of the repertoire, exploring ways in which that choice was complex. Thinking in terms of repertoires helps move the emphasis from user to communications manager. Within such choices technical qualities play a role, but the emphasis was on charting key important social processes and indicating the limits of choice – how choices are made within constraints.

Finally, we looked at the dynamics of the elements in the repertoire, showing the types of influence that can lead to changes in the balance of practices. What the chapter has not done is to discuss the social consequences of that changing balance. That is starting to be addressed, as noted earlier when some researchers ask about the implications of certain Internet use or mobile phone use. But what the chapter has attempted to do is to provide some general guidelines for understanding the take-up of new communications options, for understanding why they are used in the way that they are and our degree of commitment to them.

There is one further question to end on, and that is one of how settled the choices from that repertoire are for different people at any one time. It takes some time to learn to use a technology. It takes some time to learn when it can be useful. However, it also takes some time to learn to manage the combination of options available to us, one that has become increasingly complex. This means time to experiment, to find what works best in what circumstances, to discover the implications of certain practices and circumstances when they prove to be problematic. But this process itself takes place within a dynamic environment. The mobile phone and Internet may have started as mass markets in the mid-1990s, but there have been further new developments throughout (the rise of texting, the complexities of mobile pricing, the growth of spam e-mail) in addition to the uneven and staggered take-up of new communications options among the wider population. Without even thinking in terms of the longer-term dynamics described above, in empirical studies we might always ask how much the patterns of use that we uncover are relatively settled, even in the shorter term.

2.7 References

Bakardjieva, M. (2001) Becoming a domestic internet user. Paper for the conference e-Usages, Paris, 12–14 June.

De Gournay, C. (1997) C'est personnel ... La communication priveé hors de ses murs. *Reseaux*, 82/83, 21–40.

De Gournay, C. and Smoreda, Z. (2001) La sociabilité téléphonique et son ancrage spatio-temporel. Paper for the conference e-Usages, Paris, 12–14 June.

Fortunati, L. (2000) The mobile phone between orality and writing. Paper for the conference e-Usages, Paris, 12–14 June.

Haddon, L. (1992) Explaining ICT consumption: the case of the come computer. In Silverstone, R. and Hirsch, E. (eds), *Consuming Technologies: Media and Information in Domestic Spaces*. Routledge, London, pp. 82–96.

Haddon, L. (1994) The Phone in the Home: Ambiguity, Conflict and Change. Paper presented at the COST 248 Workshop The European Telecom User, April 13–14, Lund, Sweden.

Haddon, L. (1998) Il controllo della comunicazione. Imposizione di limiti all'uso del telefono. In Fortunati, L. (ed.), *Telecomunicando in Europa*. Franco Angeli, Milan, pp. 195–247.

Haddon, L. (1999a) The development of interactive games. In Mackay, H. and O'Sullivan, T. (eds), *The Media Reader: Continuity and Transformation*. Sage, London, pp. 305–327.

Haddon, L. (1999b) European perceptions and use of the internet. Paper for the conference Usages and Services in Telecommunications, Arcachon, 7–9 June.

Haddon, L. (2000) Old and new forms of communication: e-mail and mobile telephony, a report for BT.

Haddon, L. (2004) *Information and Communication Technologies in Everyday Life: a Concise Introduction and Research Guide*. Berg, Oxford.

Haddon, L. and Vincent, J. (2004a) Making the most of the communications repertoire – mobile and fixed. Paper for the conference The Global and the Local in Mobile Communication: Places, Images, People, Connections, Budapest, 10–11 June.

Haddon, L. and Vincent, J. (2004b) Managing a communications repertoire: mobile vs landline. Paper for the 5th Wireless World Conference Managing Wireless Communications, Surrey University, Guildford, 15–16 July.

Haddon, L. and Silverstone, R. (1994) The careers of information and communication technologies in the home. In Bjerg K. and Borreby, K. (eds), *Proceedings of the International Working Conference on Home Oriented Informatics, Telematics and Automation*, Copenhagen,

27 June–1 July, pp. 275–284.

Johnsen T (2002) The social context of the mobile phone use of Norwegian teens. In Katz, J. and Aakhus, R. (eds), *Perpetual Contact: Mobile Communication, Private Talk, Public Performance*, Cambridge University Press, Cambridge, pp. 161–170.

Johnsson-Smaragdi, U. (2001) Media use styles among the young. In Livingstone, S. and Bovill, M. (eds), *Children and Their Changing Media Environment. A European Comparative Study*, Lawrence Erlbaum, Mahwah, NJ, pp. 113–140.

Jouet, J. (2000) Retour critique sur la sociologie des usage. *Réseaux*, **100**, 486–521.

Kasesniemi, E. and Rautianen, P. (2002) Mobile culture of children and teenagers in Finland. In Katz, J. and Aakhus, R. (eds), *Perpetual Contact: Mobile Communication, Private Talk, Public Performance*. Cambridge University Press, Cambridge, pp. 170–192.

Klamer, L., Haddon, L. and Ling, R. (2000) The qualitative analysis of ICTs and mobility, time stress and social networking. Report of EURESCOM P-903, Heidelberg.

Klamer, L., Mante-Meijer, E., Heres, J., Pierson, J., Petterson, C., Thrane, K. and Turk, T. (2003) Capabilities in action: what people do. Report of the Cost 269 Capability Group. http://www.cost269.org.

Lelong, B. and Beaudouin, V. (2001) Usages d'internet, nouveaux terminaux et hauts debits: premier bilan après quatre années d'expérimentations. Paper for the conference e-Usages, Paris, 12–14 June.

Lelong, B. and Thomas F (2001) L'apprentissage de l'internaute: socialisation et autonomisation. Paper for the conference e-Usages, Paris, 12–14 June.

Ling, R. (2000) Direct and mediated interaction in the maintenance of social relationships. In Sloane, A. and van Rijn, F. (cds), *Home Informatics and Telematics: Information, Technology and Society*. Kluwer, Norwell, MA, pp. 61–86.

Ling, R. (2004) *The Mobile Connection. The Cell Phone's Impact on Society*. Morgan Kaufmann, San Francisco.

Ling, R. and Helmersen, P. (2000) "It must be necessary, it has to cover a need": the adoption of mobile telephony among pre-adolescents and adolescents. Paper presented at the seminar Sosiale Konsekvenser av Mobiltelefoni, Oslo, 16 June.

Ling, R. and Yttri, B. (2002) Hyper-coordination via mobile phones in Norway. In Katz, J. and Aakhus, R. (eds), *Perpetual Contact: Mobile Communication, Private Talk, Public Performance*. Cambridge University Press, Cambridge, pp. 139–169.

Livingstone, S. (2002) *Young People and New Media: Childhood and the Changing Media Environment*. Sage, London.

Livingstone, S. and Bovill, M. (eds) (2001) *Children and Their Changing Media Environment. A European Comparative Study*. Lawrence Erlbaum, Mahwah, NJ.

Manceron, V., Leclerc, C., Houdart, S., Lelong, B. and Smoreda, Z. (2001) Processus de hiérarchisation au sein des relations sociales et diversification des modes de communication au moment de la naissance d'un premier enfant. Paper for the conference e-Usages, Paris, 12–14 June.

Mante-Meijer, E., Haddon, L., Concejero, P., Klamer, L., Heres, J., Ling, R., Thomas, F., Smoreda, Z. and Vrieling, I. (2001) Checking it out with the people – ICT markets and users in Europe. Report for EURESCOM, Heidelberg.

Mante-Meijer, E. and Haddon, L. (2002) Working in international research groups. Report for COST269. http://www.cost269.org.

Mercier, P.-A. (2001) Nouveux moyens de communication interpersonelle et partage des rôles en matière de sociabilité au sein des couples. Paper for the conference e-Usages, Paris, 12–14 June.

Millerand, F., Giroux, L., Piette, J. and Pons, C. (1999) Les usages d'internet chez les adolescents québécois. Paper for the conference Usages and Services in Telecommunications, Arcachon, 7–9 June.

Nafus, D. and Tracey, K. (2002) Mobile phone consumption and concepts of person. In Katz, J. and Aakhus, R. (eds), *Perpetual Contact: Mobile Communication, Private Talk, Public Performance*, Cambridge University Press, Cambridge, pp. 206–222.

Nie, N. (2001) Sociability, interpersonal relations and the internet. Reconciling conflicting findings. *American Behavioral Scientist*, **45**: 420–435.

Rainee, L. (2001) Technology and the social world of American teens. Presentation at the Workshop 'Domesticating the internet, commercializing the family: a comparative look at families, the internet and issues of privacy', Haifa, 4–6 June.

Segalen, M. (1999) La téléphone des familles. *Réseaux*, **76**, 15–44.

Shotton, M. (1989) *Computer Addiction: a Study of Computer Dependency.* Taylor and Francis, London.

Taylor, A. and Harper, R. (2001) The gift of the gab? A design oriented sociology of young people's use of 'MobilZe!' Working Paper, Digital World Research Centre, University of Surrey, available at http://www.surrey.ac.uk/dwrc/papers.html.

Winston, B. (1989) The illusion of revolution. In Forester, T. (ed), *Computers in Human Context: Information Technology, Productivity and People.* Blackwell, Oxford, pp. 71–81.

3

Mobile Back to Front: Uncertainty and Danger in the Theory–Technology Relation

Steve Woolgar

3.1 Introduction

Such is the speed of recent technological change that it is already hard to recall the extent of consternation and surprise associated with the first appearance of mobile phones. Uncertainties about how, when, where and why to use one's mobile have all but disappeared. For a fleeting moment in their early history, uncertainty about what they were for was tantamount to a kind of Garfinkelian breaching experiment (Garfinkel, 1967). Our familiarity with appropriate modes of communication and social interaction was momentarily stood on its head.

By now we have also accumulated a large and still rapidly growing range of research reports about the uptake and use of new electronic communications, Internet and mobile communications included. Arguably, however, we as yet have little clarity about the range of theoretical frameworks for making sense of all this, or of the relation between the different perspectives on offer. Although we have many diverse suggestions – from such as Castells (1996, 2001, 2004), Hine (2000), Jones (1998), Kitchin (1998), Slevin (2000), Wellman and Haythornthwaite (2002) and Woolgar (2002) – we have rather little sense in depth of how these perspectives overlap and where they differ. We are at a stage in the development of social studies of technologies when we now need to reflect on the different theories available and on their relative utility.

An obvious way to approach this would be to undertake a critical comparative review of the main arguments available. However, an important complementary task is to consider what is involved when we apply (social science) theoretical frameworks and perspectives to new technologies. The contention of this chapter is that we need better to understand the theory–technology relation. We need to consider what happens to our analyses and what shapes our theoretical and other concerns, our approaches to research, as new technologies appear and old ones disappear.

The emergence of the new mobile technologies provides an important opportunity to consider the connections between the nature of the new technologies we seek to understand and the range of options at our disposal for understanding them. We know little about how responsive existing theories are to new technologies as they come on stream. For example, to what extent is the main thrust of recent social science arguments about mobile telephony just a rewarmed version of all that has been said previously about the Internet? Are mobiles just a new excuse for repeating well worn arguments about the genesis and effects of new technologies? Or, alternatively, can a new technology offer radical challenges to existing theory and, if so, how? This chapter examines these aspects of the relation between technology and theory with particular reference to the emergence of the mobile phone. Although it has excited a lot of attention, it could be argued that the mobile is already becoming a familiar new technology. So the mobile provides a timely focus for examining how our preconceptions about a new technology – the attitudes and assumptions of the analytic stances we bring to bear – impinge on the range of questions we ask and how they shape the kinds of answers we are prepared to accept.

The argument is organized as follows. The first section outlines some general considerations of the relation between theory and technology and raises concerns about the extent to which we neglect the interpretive flexibility of the technologies in question. It is suggested that much greater emphasis be given to understanding "new technology" as a shifting, interpretively flexible resource for theory change. The second section focuses in particular on the implications of this flexibility when we are dealing with "new" technologies as they emerge and disappear. The third section then examines one particular aspect of social interaction – as epitomised by Goffman's notion of "front stage, back stage" – in the light of the contention that mobiles are significantly changing the bases of social action. I argue that one particular sense of this metaphor should remind us of key aspects of our approaches to technology, viz. the uncertainty, ambiguity, essential openness and indefiniteness of what these things are and what they do. Then, in the final section, I consider how this affects the portability (we can say the mobility) of theoretical findings. This is worked through by examining the relation between theoretical claims derived from studies of the Internet – the Five Rules of Virtuality (Woolgar, 2002) – and the possibly analogous Five Rules of Mobility.

3.2 The Theory–Technology Relation

In order to highlight some key features of the relation between theory and technology, let us first consider some recent treatments of Internet technologies. A notable characteristic of many recent Internet studies publications is that they are organized primarily around familiar themes and issues – concerns which occupied social science researchers before the advent of the new technology. This is reflected in book titles such as *The Governance of Cyberspace* (Loader, 1997); *Digital Democracy* (Hague and Loader, 1999); *The Politics of Cyberspace* (Toulouse and Luke, 1998); *Cyberpower* (Jordan, 1999); *Cybercrime* (Thomas and Loader, 2000); *Communities in Cyberspace* (Smith and Kollock, 1999); and *Digital Capitalism* (Schiller, 2000). We thus see that theoretical precepts and categories such as governance, democracy, politics, power, crime, communities and capitalism are worked around "digital" and the realm of the "cyber". Even allowing for the fact that publishers often overplay these terms for marketing purposes, this trend raises an important question about the extent to which research on the Internet is merely following and reaffirming preconceived theoretical preferences. Could the same be happening with our efforts to apply existing precepts to mobile telephony?

What accounts for this phenomenon? Drawing on insights from the history and sociology of science into ways in which researchers (natural scientists) treat new and unexpected observations, we learn that the research process is characterized by an inherent conservatism or what psychologists have called a "confirmation bias". The situation can be caricatured as follows. Especially under conditions of what Kuhn (1971) calls "normal science" (or "hack science"), natural scientific activity is a routine process of fitting square plugs (observations) into round holes (theories). When this does not work, the standard conventional assumption is that there must be something wrong with the plug (observation), not with the hole (theory). In other words, in the event of potentially disconfirming data, the conservative tendency of most research is to find fault with the (methods of) observation, *not* with the theory. An important aspect of this, to which we return below, is that the thing observed can not be what it appears (or is presented) to be. The working injunction is to hang on to the theory at all costs! The "actual character" of the newly observed phenomenon becomes more or less incidental to the renewed articulation of pre-existing analytic frameworks and perspectives.[1]

What encourages this tendency in those of us confronted with new(ly observed) technologies? A common rationale for maintaining pre-existing theoretical frameworks in the face of new technical phenomena is

---

[1] Of course, one acknowledges here the tendency for every writer (theorist, analyst) to bring to bear their own theoretical predilection to the discussion, and this chapter is no exception. The question is whether and to what extent is this tendency inevitable?

that the new technologies are massively overhyped. According to this point of view, it is important that we revert to steady application of theories as a counterbalance to the hype. Thus, as far as the Internet is concerned, it is said that we social scientists should contribute good solid empirical work in the face of the wild (unsupported) imaginings about its supposedly transformative qualities. For example, in his preface to the recent collection by Wellman and Haythorthwaite (2002), Castells writes:

> This book is precious. It provides us with reliable, scholarly research on the hows and whats of the Internet as it relates to people's lives. The critical importance of the Internet as a new medium of communication is only surpassed by the amount of fantasy and gossip that surround its development. (Castells, 2002, xxix)

This is characteristic of the views of authors who offer a somewhat revisionist history of the stages of social science attitude to the Internet. According to this account, our understanding and treatment of the Internet went through a first stage of unbridled hype (or "cyberbole"[2]), then moderated balance and only now, latterly, good old academic empirical investigation. What a relief! The valiant social scientists have corrected the populist and naïve media accounts. Normal (social scientific) service has been resumed! What remains safe and unchallenged in this account is the theoretical attitude. This perspective maintains a clear distinction such that the application of existing theoretical apparatus – the reliable scholarly research – remains uncontaminated by the empirical observations and results.[3]

Similarly, a rationale for research on mobile phones are that their production and use are a huge phenomenon; it is affecting peoples' lives in all kinds of ways; and we do not know enough about this (e.g. Katz and Aakhus, 2002). This is entirely reasonable. The main imperative for research is the discovery of more information about uptake and use, about cultural and national differences and, more generally, about variations in relation to the whole standard apparatus of social science categories: gender, work, education, policy, youth and so on. Thus, for example, Katz and Aakhus (2002, p. 317) envisage a relation between theory and the mobile phenomena whereby theorists subsequently (or at least, antecedently) provide a conceptual justification for observed consistencies in peoples' interactions with technologies.

---

[2] "Cyberbole" denotes the exaggerated (hyperbole) depiction of characteristics pertaining to cyber phenomena; see Woolgar (2002).
[3] It is a well-documented characteristic of natural science that researchers tend to preserve this distinction wherever possible, and where anomalies arise, it is the empirical results which are first subject to critical scrutiny. Lakatos even formalises this in his notion of the "protective belt". Modifying the theory is (Lakatos suggests, should be) a last-ditch measure.

However, the downside of this kind of rationale is that it tends to discourage theoretical change and development. The theory remains intact and unchallenged as it is applied to the new technological phenomenon. This somewhat conservative (epistemologically speaking) research strategy is caricatured in the caustic remark ascribed to Harold Garfinkel (1967, 2002) that sociology is, in the main, a "no news, no lose" enterprise. In other words, Garfinkel is suggesting, the kinds of finding we social scientists unearth are ultimately unsurprising, and we deal with them in such a way that our theoretical frameworks remain undisturbed. The alternative situation at which Garfinkel is hinting is one where our engagement with the new technology has a profound, destabilizing, challenging and perhaps transforming effect on our theoretical assumptions.[4] We return to this alternative possibility below.

A central difficulty here is the schematic assumption that we are applying a given theoretical perspective to a fixed, singular and knowable entity: the technology. Against this, there is considerable work, especially as arising from science and technology studies (STS), which emphasizes the *interpretive flexibility* of technology. In brief, this body of literature emphasizes that the particular form of a technology, its technical capacity and effects are historically and socially contingent. The form and capacity are not given and, in particular, they cannot be straightforwardly extrapolated from preceding technologies. They are, instead, the upshot of processes of social construction. In short, the technology *could be otherwise*.

A rather limited sense of interpretive flexibility was initially applied mainly to the socio-historical circumstances of the production of new technology. Thus, contributions to the social construction of technology (SCOT) emphasized, for example, the contingent social relations which eventually gave rise to the well-known contemporary standard form of the two wheeled bicycle (Pinch and Bijker, 1984), or the various circumstances which came to determine what counts as a sufficient measure of nuclear missile accuracy (MacKenzie, 1990). However, it subsequently became clear that interpretive flexibility is also usefully understood in relation to the ways in which technology is interpreted and used. In other words, interpretive flexibility not only characterizes the different developmental paths that a technology can follow over time, it also refers to variations in understanding and use of any "given" technology. It is this latter sense of interpretive flexibility that finds expression in a large number of ethnographic studies of the apprehension, reception, use,

---

[4] It is ironic that although in one reading of his work Garfinkel seems to advocate the courting of risk and danger in theoretical development, much of his radical program of ethnomethodology has since been transformed into a routinized machinery for the analysis of conversation. For a discussion of the problems of sustaining radicalism in social science theory, see Woolgar (2005).

deployment, depiction and representation of technologies. This later mode of ethnographic (or ethnographically inclined) research on technology is more appropriately designated "technography", since this term signals the need to maintain a sceptical ethnographic attitude towards the technical object at the very heart of the study, that is, towards claims about and representations of technical capability and effect.

So, for our purposes, a first sense of interpretive flexibility arises in relation to the historical emergence of new technology and it reminds us that in researching new technologies, we have a moving target. The technology we study is evolving and changing. In the beginning, few are absolutely certain what it is for and what it can do. It has often just recently come on the scene, its effects and implications are the subject of widespread speculation.[5] Thus, in the case of the mobile, we already know that initial views about its nature and effects are changing; that early ideas about its primary mode of usage, as a device for voice communication, turned out to be mistaken because few predicted its massive role in text messaging; that current estimates of its all pervasiveness may come to be seen as overblown. We have to understand that any one technology will very likely pass through a cycle of cyberbole and subsequent accommodation, just in the way that Sawhney (2005) has shown for Wi-Fi technologies. So a coda to the need for detailed empirical information about uptake and use is that there are most appropriate points in the cycle when this can best be done.

We should also note the part we play as researchers in the cycle of emergence and disappearance of new technologies. Thus, the academic study of a new technology – even, ironically, those efforts devoted to its deconstruction – can play a part in prolonging its shelf life. For example, part of the persistence of claims about the possible impacts of new technology can be attributed to the extent of academic investment in its investigation. Conferences about "the mobile"; research centers devoted to the study of "the Internet"; research programmes entitled "Virtual Society? – even with the question mark – can ironically prolong the life of the very concept of the virtual. The remark-on-able properties of the technology in question are given extended life as a result of our efforts to research them.

More generally, the application of existing theory to a new technology can ignore, or give little significance to, the fact that technologies both emerge and disappear. Routine usage of a technology over a period of time may even lead to a situation where we are no longer conscious that the artefact in question is "a technology". For example, we might imagine

---

[5] As is suggested below, it is important not merely to set out simply to "counter" such widespread speculation. One reason is that such speculation is often the rationale for providing resources for social science investigation.

that pencils may once have caused consternation and concern, both in their use and in the meta commentary that surrounded them. But worried about, for example, how they might deskill experts in the oral tradition have clearly since diminished.[6]

At the same time, the application of theory to "the technology" may give undue emphasis to one particular interpretation of what the technology is. The difficulty here is that this may unduly fetishize the technology. It solidifies and perpetuates a singular contingent version of what the technology is and what it can do. And by supporting this version, we support the particular discursive rendering of the technology advanced by a particular social constituency. It further implies the given centrality of the technology for everything that happens around it. Haddon (2005) is right to insist that we should situate the mobile within a broader conception of communication. Our problem may be much more to do with communication than it is to do with mobiles per se. In short, we should try to avoid fetishizing the mobile (or indeed any technology).[7]

A second sense of interpretive flexibility further counters the view that technology can (or should) be treated like any definite singular object, by stressing that interpretation, reading and making sense of technology are a constant feature of social life. Technologies are not given. They are instead discursive moves in a never ending cacophony of efforts at social ordering. So the straightforward application of analytic formulae to the new technology (Internet, mobile) misses this important feature of "technographic" STS.[8] This, I think, is the value of STS slogans such as "technology is congealed social relations", or that "technology comprises new social arrangements". Such slogans require some unpacking. However, it is fairly easy to see how the definition of technology as social arrangements allows a general understanding of the relatively robustness of systems as diverse as waste collection, the government, technologies of representation, standards of evidence and so on. From this point of view, the main difference with things that we more commonly think of as technologies is that the latter often take a material (or electronic) form, and the fact of their apparently greater robustness is what causes all the excitement and concern. These are social arrangements rendered material. The social arrangements are thereby blackboxed, the inner workings of which are known only to technical experts, and which in virtue of their

---

[6] Any "consternation and concern" is now limited to the process of, e.g., a novice/child learning how to use an established technology.

[7] An alternative extreme view is provided by one (especially "high-church") form of ethnomethodological perspective. For example, Button (1992) argues that the imposition of theoretical perspectives (in this case social constructivism) tends to lose sight of what is specific about the technology itself.

[8] It is curious that, with few exceptions, work in science and technology studies, itself forming a powerful tradition in the social analyses of technology, has not yet been widely recognized as a resource for social studies of the Internet and of mobile technologies.

material construction attain an apparent robustness, resist deconstruction and endure.

The distinction made by Hine (2000), among others, is helpful here: we need to remember that the technology exists both as a culture and as cultural artefact. The latter refers to the thing or the system under a narrow (perhaps technical?) definition of technology. Importantly, it is the upshot of a process of social and cultural construction. The former includes knowing how to engage competently with discourses about the technology even if you do not use it, what terms of description, ranges of opinion, parameters of debate and knowledge make sense in speaking of the new technologies. An interesting example is the mode of recitation used by some presenters of BBC Radio 4's "Today" programme when they read out a website address: "w w w dot b b c dot u k forward slash today" is delivered with a slight sneer. The clear implication is that this information is for technies and nerds, not for people such as themselves! So the mere intonation of a representation of the technology can do boundary work! Yung (2005) similarly provides an insightful articulation of the "ideotechnic values" present in the Hong Kong media's discursive representation of mobile technologies.

In this section we have identified a major problem with the standard conception of the technology theory relation. An inherent conservatism mitigates against challenging existing theory in the face of possibly "anomalous" new technological phenomena. And the preservation of an undisturbed theoretical schema is aided by inattention to the varieties and extent of interpretive flexibility.

3.3 Uncertainty and the Reflexive Ironic Attitude

We need to be alert to the problematic connection between object (technology) and theory and, in particular, we need to take seriously the proposal that neither is, nor need be, fixed. Of course we can use existing theoretical frameworks about technological impact to "shed light on" new technologies: what can the social construction of technology (SCOT) tell us about mobiles, what can the Five Rules of Virtuality (Woolgar, 2002) tell us about mobiles (see below)? Certainly we can use the new technologies to interrogate and sharpen our thinking about our theories. However, we should resist the temptation to smother the uncertainty of new technologies by imposing "theoretical" frameworks which "make sense of" the phenomenon. Can we go further and entertain the possibility that the technology might shape the kinds of theoretical scheme we deploy? Instead of simply applying pre-existing frameworks to the new technology, can the technology change our theory? Can the new technologies get us thinking at a different analytic (and thereby perhaps at a different theoretical) level?

This section considers how this might be done by exploiting some of the uncertainties, ironies and ambiguities associated with our experiences of new technology.

Technologies should be understood as sets of more or less congealed social arrangements. However, the key point is that newly envisaged social arrangements associated with a new technology are only ever a proposed reading, a discursive move, which may or may not have the desired effect (or indeed have any effect). I think this is what gives sense to the relationship between technology and new ways of theorizing which is hinted at by Cooper et al. (2002, p. 300; my emphasis; cf. Hine, 2000): "… as is the case with other recent technologies, *the mobile device invites questions* about what constitutes the empirical, and what is the most appropriate unit of analysis."

The suggestion here is that new technology can give rise to new social science, not through a determinist relation, but in response to "invited questions". The "invitation" arises through the apprehension of the mobile as a possible concatenation of alternative social arrangements. In particular, I suggest, the invitation arises at the point of uncertainty in reading what the envisaged new social relations might be. Hence the point of uncertainty in our reading is most definitely a moment we must hang on to. We should at all costs avoid the temptation to smother uncertainty by beating it over the head with imported theories.

How might we sustain greater sensitivity to this issue? We need an approach which neither adopts the hype about new technologies nor straightforwardly rejects it. Of course, that's easier said than done! In my own work I have suggested that it requires a reflexive ironic attitude to the technology. That is, we have to find a way of embracing all that is said about the new technologies *at the same time as* questioning it. Hence, we need both to embrace statements such as:

> The advent of constant Internet connectivity and mobile communication has transformed the ways in which many businesses, organizations and industries function. (Lindgren et al., 2002)

and to question and interrogate them. In other words, we need also to make clear that this is a *claim*, so that we might begin to highlight the circumstances of its genesis and to ask what generated and sustains the claim:

> *The advent of constant Internet connectivity and mobile communication has transformed the ways in which many businesses, organizations and industries function (Lindgren et al., 2002). (Woolgar, 2005b)*

The point is that neither adoption nor rejection of the claim will suit our purposes. Pronouncements about the character and effects of new technology are always situated, in that they are made with (certain) readers in mind. In this sense, depictions of new technology constitute a text which "performs community". The technology text inscribes and prescribes certain identities, roles and groups and the relations between them. The technology text makes available a moral universe which depicts the rights, responsibilities and expectations associated with the entities which populate it. It follows that the adoption (or rejection) of a claim about a particular technology is equivalent to the adoption (or rejection) of a performed community. We risk losing audiences. Instead, we need to take the more agnostic path and, in line with the ethnographic (technographic) sensibility, to focus on the processes whereby such claims and depictions gain currency. We should, so to speak, try to run with both straight and italicised versions at the same time.[9]

A wide range of types of technographic encounter with mobiles might be used to interrogate the nature and extent of interpretive flexibility. Examples might include the urban legends about mobile phone use which have circulated in recent years. These can be understood as stories about likely effects on social order in the face of rank uncertainty about the capabilities and impacts of new technology. Urban legends are frequently moral tales about the dire consequences of transgressing established (and valued) social organizational boundaries, and are played out in relation to new technologies as potentially threatening and destabilizing accepted divisions.

3.4 Front Stage, Back Stage: Stipulative and Technographic Models of Social Action

In the interests of modifying the theory–technology relation, I have thus far argued for a perspective which emphasizes the interpretive flexibility and uncertainty associated with the construction, interpretation and use of new technology. This section examines the implications of this perspective for arguments about the ways in which social interaction is affected by new technology. In particular, we focus on Goffman's notion of "front stage, back stage" in the light of the contention that mobiles are significantly changing the bases of social action. In terms of the rubric which informs the contributions to this volume: "... we wish to frame the discussion around mobile communication's impact on our front stage facades, our back stage interactions – as well as the blurring of these two" (Ling, 2003). Given what we have already argued about interpretive flexibility and about the theory–technology relation, to what extent does the

---

[9] A similar attempt to maintain agnosticism with respect to claims about the social impacts of Internet technologies centered on the inclusion of a question mark in the title of a major UK national research program: "Virtual Society? the social science of electronic technologies"; see Woolgar (2002).

Goffman metaphor usefully illuminate what we can say about mobile communications?

Goffman draws upon and develops a dramaturgical model of social interaction. The "I" and the "me" of symbolic interactionism are the stimulus for Goffman's thinking on social interaction more generally. Especially in the presentation of self in everyday life (Goffman, 1971), social beings manage the tension between the "I" and the "me" by performing. They act out the difference between them; they are social actors who act out a role. The tension is arguably multiple when we consider all the kinds of situations and interactions through which people pass in the course of a day. Goffman's solution is a picture of individuals switching roles. As adept participants in society, they swap different personae and scripts as a routine part of their interactions with others. Importantly, each of these roles, either by implication or sometimes quite explicitly in Goffman's account, stands in contrast[10] with the non-acting, real self, to which the actor returns at the end of the day, perhaps when the demands for performance are finally relaxed. The real self is the self of the back stage.

By contrast, in his work on total institutions, Goffman (1968) focuses on the processes of degradation of self, wherein individuals have tokens of identity removed as part of the process of induction. Goffman stresses that one's concept of self is dependent on interactions with others so that, deprived of these interactions, and deprived also of one's name, appearance and possessions, one's sense of self is removed. Here there seems a slightly different emphasis, in that such interactions and token of interactions are not mere role play, beneath which lies another (real) self. In this account, the interactions are constituent of self. In total institutions, then, the self which is being degraded is the last refuge, there is nothing beneath this singular presented self.[11] In other situations, by contrast, the social actor switches between many different roles.

Of interest here is the extent to which Goffman intended a contrast between the superficial and the real, between the appearance and the underlying reality. There are references to letting the mask slip, and discussion of ideas such as role distance and face work which suggest that actors are busy engaging in *deceptions*, i.e. misrepresentations of the actual hidden self.

We thus see the essentially binary structure of Goffman's depiction of social interaction. Front stage is separate from the back and the two stand in a relation similar to what Garfinkel (1967), after Mannheim, calls the

---

[10] At some points of his argument, Goffman suggests a less marked contrast, for example when he describes how individuals style their activities, the processes whereby individuals present themselves in particular ways so as to portray a desired self image.

[11] Although Goffman (1968) does describe modes of resistance, for example, small rituals of behavior, whereby members of total institutions try to preserve their identity (cf. Cohen and Taylor, 1972).

documentary method of interpretation. The surface appearance is an act that to a greater or lesser extent conceals the underlying reality. The surface act can in principle be related to the underlying person in an inspectable fashion. The best actors, it is sometimes said, reveal nothing of themselves but there is nonetheless a discoverable connection between act and self, a path between the two which becomes evident when concealment falters or when the mask slips. And the promise, at least, is that what becomes evident is the underlying reality. The Goffman acting analogy depends on a binary distinction between front and back stage.

This analogy then informs discussions about alleged changes in interaction concomitant with the advent of new technologies. Thus goes the argument. Goffman stresses the importance of face-to-face interaction, which sustains front-stage work. New electronic technologies, notably the Internet and the mobile, deprive us of this face-to-face contact. Since, it is supposed, anyone can thereby assume any identity on the Internet (or at the end of a mobile, or as an SMS sender/receiver), the immediacy of face-to-face contact is removed. Thus, with a mobile, you can be communicating anytime, anywhere. You can be multitasking, having a face-to-face conversation with one person while texting with another; or having a face-to-face conversation with one person while talking on the phone to another.

As a result, it is argued, audiences are less certain about what/who is being performed. This also means that it is easier to guess who is the real you. It is more difficult to maintain face work. So front stage is no longer distinct from back stage. So the boundaries between them blur. The overall picture is one of a relatively stable pattern of social interactions that becomes disrupted by the arrival on the scene of a new technology. In particular, the stability lies in routine, understood and accountable ways of making out that something (even if we do not know what) lies beneath the surface.

The central assumption here is that negative effects result from withholding or diminishing face-to-face interaction. However, there is interesting evidence that reduced bandwidth can actually *reverse* the presumed impoverishment of social interaction. Thus, for example, Watt et al. (2002) show how the removal of visual communication channels led to an increased identification with group goals and a heightened sense of group membership, among a sample of people communicating by computers. The implication is that bandwidth does not correlate positively with ease of social interaction. Or, more to the point, that attempts to mimic face-to-face interaction through electronic media may be misplaced. In computer-mediated communication, at least, there may be forms of sociality which depend on suppressing or reducing just those channels of communication traditionally associated with face-to-face interaction.[12]

---

[12] Cf. the fourth rule of virtuality, discussed below: "the more virtual, the more real".

So there is empirical evidence to doubt the application of the Goffman model to the use of mobile technologies. It could be that the model is simply outdated (Fortunati, 2005). However, there is the additional point that this reading of Goffman underplays the experience of learning to see things in a different light. As Harvey Sacks notes, "... when you read a book like *The Presentation of Self in Everyday Life*, on pretty much any page you come across something that's news to you, which you hadn't noticed, which you could notice, which you can thereafter more or less see going on." (Sacks, 1992, vol. I, p. 619). In other words, Sacks stresses the experience of surprise and revelation when reading Goffman, the suggestion that things are other than they seem. Sacks has little time for the static sociological purposes to which these observations are then put, preferring instead to emphasize their power of surprise and revelation.

Let me suggest instead a different way of understanding the import of Goffman's dramaturgical vision, a different view of social players on the stage. Take, for example, the very contrasting views of a playwright such as Luigi Pirandello.[13] For Pirandello, the settled, fathomable outcome of narrative and plot is anathema. For example, *Absolutely {perhaps}*,[14] currently enjoying a popular revival in London's West End, is precisely concerned with the absence of a solution to puzzles about the relation between the act and its underlying reality, between imagination and reality.

> Mine is a serious theatre. It requires the total participation of that moral entity – man. It is not comfortable theatre. It is a difficult theatre, even a dangerous one. (Pirandello, 1935)

By "the total participation of that moral entity – man", I think Pirandello means to explore the situation where there is no escape, no boundaries which can keep safe, divided off, stable, contained, the phenomenon being discussed. Indeed, towards the end of the play, several presumed boundaries are shown to be not what they appear, the distinction between actors and audience (when several "members of the audience" join in the discussion with those on stage) and the distinction between actors and playwright, when it eventually becomes obvious that the professor character is actually (playing) the playwright. Right at the end, it is made clear that consternation and irresolution stem from the fact that the three principals, the focus of accusation and counteraccusation of madness throughout, form an impenetrable circle, from which all others are excluded.

> I think that life is a very sad piece of buffoonery because we need to deceive ourselves constantly by creating a reality (one for each and never the same for

---

[13] The following section follows Woolgar (2005a).
[14] The recent English translation of *Cosi e {se vi pare!}*.

all) which from time to time is discovered to be in vain and illusory. (Pirandello, 1935)

In Pirandello, then, the instability and unresolvability of the reality/illusion couple are paramount.[15] In many senses, Pirandello injects a fluidity and uncertainty into the idea of "front stage back stage" that far surpasses Goffman. A sign of our post-modern times? Well, no, since Pirandello wrote *Absolutely {perhaps}* in 1913! So was he "ahead of his time"? A postmodernist stuck in the 1900s? Perhaps there is something in this since he is now hailed as a forerunner of existentialism, the theatre of the absurd, and Ionesco, and direct lines are drawn from him through surrealist comedy to Monty Python.

The potential importance, then, of the "front stage, back stage" metaphor is not that it provides another convenient way for analysts to divide up the world, but that in suggesting that the world is other than it seems, it signals a moment of uncertainty, and instability, of suggesting that what seems to be the case is not the case. On the whole, our typical response to these moments of uncertainty is to try to get through them as quickly as possible, to get to the other side, to return to normal order. Indeed, the extent of our dependence on order and normality is indexed by the extent of consternation (alarm, upset, or humor) associated with disruptive moments.[16] However, the point of the Pirandellan, as opposed to the Goffman, model of social interaction is that we glimpse the possibility of consternation and disruption. The relatively stable binary opposition between self and action is replaced by instability and uncertainty about what counts as front and what as back.

So what we have seen in this section is the difference between relatively static (or stipulative) and relatively dynamic (or technographic) models of social action. The latter encourages us to emphasize the uncertainties and danger associated with interpretation, and leaves behind the former's model of social relations premised on a binary divide between appearance and reality.

3.5 From Five Rules of Virtuality to Five Rules of Mobility?

In this final section, we use the main themes developed thus far in the chapter to consider the application of an existing theoretical schema to a new technology. The existing schema comprises the "Five Rules of Virtuality" (Woolgar, 2002, 2004), and mobiles are the new technology.

---

[15] This is, of course, just perfect for a Professor of Marketing, where reality/illusion issues are paramount (cf. Woolgar and Simakova, 2003).

[16] The resonance with Garfinkel's (1967) "breaching experiments", mentioned earlier, is clear. Their importance is not so much in showing that people dislike bizarre situations as in demonstrating the difficulty they experience if outcomes (resolutions) are withdrawn, deferred or made unavailable.

At the time of their first announcement, many of the research results of the Virtual Society? programme were counter-intuitive. The research demonstrated that the new technologies were not being taken up at the rate we had been led to believe, or by the people and groups we had anticipated, nor were they being used in the ways or for the purposes envisaged.[17] In response to the changing environment of expectations and views about new technologies, the overall research findings were presented so as to draw attention to the enduring tension between claims about the supposedly transforming character of the technologies ("virtuality") and about the alleged actuality of their use ("reality"). The "rules" are intended as thematic research highlights deriving from the combined results of several research projects. More importantly, this way of organizing the findings was intended to make it possible to extrapolate the results to other technologies. We should note, however, that these are not determinative rules, but rather rules of thumb. In particular, they are supposed to encourage the possibility that things might be much less certain than they seem, that, in particular, they may be counter-intuitive.[18]

How far it is possible and/or useful to apply this theoretical schema, derived from empirical research on Internet technologies, to issues around mobile technologies? To what extent do the "rules" travel across technologies, from Internet to mobile.

3.5.1 Rule 1: The Uptake and Use of the New Technologies Depends Crucially on Local Social Context

Different aspects of 'context' bear upon the reception and deployment of electronic technologies. The importance of these 'non-technical' circumstances is that they explain, for example, why the current rate of straightforward rapid expansion may not continue, e.g. Wyatt et al. (2002) on the discovery of large cohorts of teenage Internet drop-outs, or Swann and Watts (2002) on the lack of take-up of Virtual Reality technologies. Close social psychological study of the comparison between computer-mediated communication and face-to-face communication shows that reduced bandwidth and increased anonymity can actually accentuate feelings of group belonging and identification, e.g. Watt et al. (2002). By taking specific senses of context into account, we glimpse the basis for an accentuated, perhaps even novel, form of sociality arising from the use of internet communications. Studies of the different kinds of 'e-gateway' which promise access to and participation in the virtual world (cybercafes,

---

[17] In a relatively short time, many of these outcomes became received wisdom, partly under the influence of the Virtual Society? program.

[18] The following depiction of the five rules follows Woolgar (2004).

telecottages) suggest that 'third place' characteristics of local social context – a social setting separate from both domestic and economic spheres – provide a key to the successful integration of the real and the virtual, e.g. Liff et al. (2002).

3.5.2 Rule 2: The Fears and Risks Associated with New Technologies are Unevenly Socially Distributed

The research demonstrates that views about new technology, the anticipations, concerns, enthusiasm and so on, are unevenly socially distributed. The research shows, for example, the transformative power of expectations about, and performances of, technological artefacts in social action, e.g. Knights et al. (2002) on financial services; how views about technology are constantly "at stake", e.g. McGrail (2002) on the use and reception of CCTV and related surveillance technologies in high-rise housing; and that a variety of counter-intuitive usages of technology at work are not easily classifiable as either conformity or resistance to surveillance capable technologies. Thus, for example, Mason et al. (2002) find that, against expectation, their respondents accorded a markedly low priority to the question of privacy in relation to the impact of surveillance capable technologies on social relations at work. Mason et al. use this finding as the basis for challenging some common assumptions about the privacy impacts of new technologies at work, especially in those literatures influenced by the labor process tradition. Instead of starting with the assumption that relations between management and employees are intrinsically oppositional, Mason et al. stress the importance of examining the actual usage and experience of new technologies in complex social situations

3.5.3 Rule 3: Virtual Technologies Supplement Rather than Substitute for Real Activities

Our research showed that the virtual tends to sit alongside the real that, in much popular imagination, it is usually supposed to supplant. Thus, against the prospects for "virtual learning" (one part of the vision of "virtual universities") our research found that the mere ability of students to access ICT failed to re-mediate the communal dimensions of learning, e.g. Crook and Light (2002); that virtual social life provides a further dimension to a person's real social life, not a substitution for it, e.g. Nettleton et al. (2002); that sources of virtual support via the Internet were used together with other resources and became enmeshed into peoples' social lives, in some cases thereby transcending the boundaries of real and virtual life.

3.5.4 Rule 4: The More Virtual the More Real!

This rule is an extension of the previous one. Not only do new virtual activities sit alongside existing 'real' activities, but the introduction and use of new 'virtual' technologies can actually stimulate more of the corresponding 'real' activity. The results of "Virtual Society"? Research demonstrates the interplay of real and virtual connectivity, e.g. Wittel et al. (2002) on the new media sector in London's 'silicon alley', and the ways in which e-mail generates more real meetings, and even meetings to resolve disputes generated by e-mail communication, e.g. Brown and Lightfoot (2002). This rule has important implications for business practice, specifically for the claim that networking computing can fundamentally change the nature and management of organizational memory.

3.5.5 Rule 5: The More Global the More Local!

Virtual technologies are famously implicated in the much discussed phenomenon of globalization. In one (of many possible) interpretations of the term, globalization means the rapid movement and spread of symbolic and financial capital. Electronic technologies facilitate the rapid traffic in communication, the instantiation of activities and institutions at widespread locales and the insinuation of standardized identities and imagery (especially brands) in multiple locations. Globalization is about the death of distance and new technologies are claimed to be "space defying, boundary crossing and ubiquitously linking". Against this, however, much research has found that the realization of the ideal of a "virtual organization" was actually set aside in favor of more trusted business solutions to the organizational problems of coordination. Workers' efforts are directed primarily at "making the new technology at home" within existing work practices. In so doing, it is local relevance that is crucial rather than global dimensions, e.g. Hughes et al. (2002). Relatedly, it has been shown that the use of ICTs can reiterate spatial divisions and distinctiveness rather than helping to ameliorate them, e.g. Agar et al. (2002). To a large extent, this rule is a consequence of the preceding four rules.

So do the rules travel? How far it is possible and/or useful to apply this theoretical schema, derived from empirical research on Internet technologies, to questions about the genesis and use of mobile technologies? It could be reasonably argued that in each case the real/virtual distinction could be replaced by a fixed/mobile contrast. Thus:

1. The uptake and use of the mobile technologies depend crucially on local social context.
2. The fears and risks associated with mobile technologies are unevenly socially distributed.

3. Mobile technologies supplement rather than substitute for fixed activities.
4. The more mobile the more static/fixed!
5. The more global the more local!

It has been commented that the five rules are pitched at such a general level that they could be taken for a description of the general dynamics of social change (S. Livingstone, Personal communication, 2002). This might account, in part, for their applicability to mobile technologies. It is consistent with the view that social change occurs without respect to the particular technologies in question. The rules travel across technologies, from Internet to mobile, because they are rules about technologies in general, construed as constellations of more or less stable social arrangements.

3.6 Conclusion

"Front stage, back stage" provides a useful metaphor for initially organizing and assessing our efforts to make sense of the uptake and use of mobiles. However, I have argued in this chapter that it also provides the occasion to take a hard look at the relation between new technologies and available theoretical models of social interaction. In particular, I have suggested that the process we implicitly take for granted, of "applying" existing theoretical frameworks and perspectives to new technologies, requires critical reassessment. This process seriously underplays the interpretive flexibility which characterizes technology, it fails to take advantage of the insight that technologies can be construed as temporarily stable sets of social arrangements and it misses the opportunity of capitalizing on the uncertainties involved in technology use.

We are beginning to see a definite shift in analytic mood, from the constant efforts to map and measure use, penetration and "impact", to new efforts to understand how new technologies emerge into the spotlight and recede into the background, how they move from front stage to back stage, and sometimes back again, and how the boundaries between front and back are reconstituted in the process. I suggest that in our efforts at new understanding, we need to eschew the straightforward application of existing, off-the-shelf, theories. Our recognition of the importance of interpretive flexibility and uncertainty, and the deployment of an appropriately reflexive ironic attitude, can instead foster a radical reappraisal of our theoretical assumptions.

If our own (social scientific) theater comprises unidimensional, static, predictable and safe applications of theory, we will doubtless continue to pull in the crowds. Audiences will continue to be attracted, perhaps by the promise of becoming better informed about the social dimensions of new technology. However, if we also want to make them think, to disturb existing preconceptions and assumptions, our theater will have to become much more dynamic, unpredictable, unstable, provocative and, above all, dangerous.

> The audience was outraged at the play's conclusion. They were shocked by it. There had never been an ending like that in the history of drama. It broke all the rules. At the curtain call when the author appeared some of the audience cheered. But some of them yelled obscenities. One irate customer tore his theatre seat from its moorings and hurled it onto the stage. It narrowly missed Pirandello's head. (Zeffirelli, 2003)

If social analyses of mobile communications are to provide the basis for fashioning a more theoretically dangerous attitude to technologies in general, we shall need to encourage, recover and preserve just these kinds of Garfinkelian moment.

3.7 References

Agar, J., Green, S. and Harvey, P. (2002) Cotton to computers: from industrial to information revolutions. In Woolgar, S. (ed), *Virtual Society? – Technology, Cyberbole, Reality*. Oxford University Press, Oxford, pp. 264–285.

Brown, S.D. and Lightfoot, G. (2002) Presence, absence and accountability: e-mail and the mediation of organisational memory. In Woolgar S. (ed), *Virtual Society? – Technology, Cyberbole, Reality*. Oxford University Press, Oxford, pp. 209–229.

Button, G. (1992) The curious case of the disappearing technology. In Button, G. (ed), *Technology in Working Order: Studies of Work, Interaction and Technology*. Routledge, London.

Castells, M. (1996) *The Rise of the Network Society*. Vol. 1 of The Information Age: Economy, Society and Culture. Blackwell, Oxford.

Castells, M. (2001) *The Internet Galaxy: Reflections and Society*. Oxford University Press, Oxford.

Castells, M. (2002) The Internet and the network society. In Wellman, B. and Haythornwhaite, C. (eds), *The Internet in Everyday Life*. Blackwell, Oxford, pp. xxix–xxxi.

Castells, M. (ed.) (2004) *The Network Society: a Global Perspective*. Edward Elgar, London.

Cohen, S. and Taylor, L. (1972) *Psychological Survival: the Experience of Long Term Imprisonment*. Penguin, Harmondsworth.

Cooper, G., Green, N., Murtagh, J. and Harper, R. (2002) Mobile society? In Woolgar, S. (ed.), *Virtual Society? – Technology, Cyberbole, Reality*. Oxford University Press, Oxford, pp. 286–301.

Crook, C. and Light, P. (2002) Virtual society and the cultural practice of study. In Woolgar, S. (ed.), *Virtual Society? – Technology, Cyberbole, Reality*. Oxford University Press, Oxford, pp. 153–175.

Fortunati, L. (2005) In Ling, R. (ed), *Mobile Communications: Re-negotiation of the Social Sphere*. Springer-Verlag, London.

Garfinkel, H. (1967) *Studies in Ethnomethodology*. Prentice Hall, Englewood Cliffs, NJ.

Garfinkel, H. (2002) *Ethnomethodology's Program: Working Out Durkheim's Aphorism*. Rowman and Littlefield, London.

Goffman, E. (1971) *The Presentation of Self in Everyday Life*. Penguin, London.

Goffman, E. (1968) *Asylums: Essays on the Social Situation of Mental Patients and Other Inmates*. Penguin, Harmondswirth.

Haddon, L. (2005) Research questions for the evolving communications landscape. In Ling, R. (ed.), *Mobile Communications: Re-negotiation of the Social Sphere*. Springer-Verlag, London.

Hague, B.N. and Loader, B. (1999) *Digital Democracy: Discourse and Decision Making in the Information Age*. Routledge, London.

Hine, C. (2000) *Virtual Ethnography*. Sage, London.

Hughes, J.A., Rouncefield, M. and Tolmie, P. (2002) The day-to-day work of standardisation: a sceptical note on the reliance on IT in a retail bank. In Woolgar, S. (ed.), *Virtual Society? – Technology, Cyberbole, Reality*. Oxford University Press, Oxford, pp. 247–263.

Jones, S.G. (ed.) (1998) Cybersociety 2.0: Revisiting Computer-mediated Communication and Community. Sage, London.

Jordan, T. (1999) *Cyberpower*. Routledge, London.

Katz, J. and Aakhus, M. (eds) (2002) *Perpetual Contact: Mobile Communication, Private Talk, Public Performance*. Cambridge University Press, Cambridge.

Kitchin, R. (1998) *Cyberspace: the World in the Wires*. Wiley, Chichester.

Knights, D., Noble, F., Vurdubakis, T. and Willmott, H. (2002) Allegories of creative destruction: technology and organisation in narratives of the e-economy. In Woolgar, S. (ed.), *Virtual Society? – Technology, Cyberbole, Reality*. Oxford University Press, Oxford, pp. 99–114.

Kuhn, T.S. (1971) *The Structure of Scientific Revolutions*. Chicago University Press, Chicago.

Liff, S., Steward, F., and Watts, P. (2002) New public places for Internet access: networks for practice-based learning and social inclusion. In Woolgar, S. (ed.), *Virtual Society? – Technology, Cyberbole, Reality*. Oxford University Press, Oxford, pp. 78–98.

Lindgren, M., Jedbratt, J. and Svensson, E. (2002) *Beyond Mobile: People, Communications and Marketing in a Mobilized World*. Palgrave, Basingstoke.

Ling, R. (2003) Conference announcement: "Front Stage – Back Stage", Grimstad, Norway, 22–24 June.

Loader, B. (1997) *The Governance of Cyberspace*. Routledge, London.

MacKenzie, D. (1990) *Inventing Accuracy: a Historical Sociology of Nuclear Missile Guidance*. MIT Press, Cambridge, MA.

Mason, D., Button, G., Lankshear, G. and Coates, S. (2002) Getting real about surveillance and privacy at work. In Woolgar, S. (ed.), *Virtual Society? – Technology, Cyberbole, Reality*. Oxford University Press, Oxford, pp. 137–152.

McGrail, B. (2002) Confronting electronic surveillance: desiring and resisting new technologies. In Woolgar, S. (ed.), *Virtual Society? – Technology, Cyberbole, Reality*. Oxford University Press, Oxford, pp. 115–136.

Nettleton, S., Pleace, N., Burrows, R., Muncer, S. and Loader, B. (2002) The reality of virtual social support. In Woolgar, S. (ed.), *Virtual Society? – Technology, Cyberbole, Reality*. Oxford University Press, Oxford, pp. 176–188

Pinch, T.J. and Bijker, W.E. (1984) The social construction of facts and artefacts: or, how the sociology of science and the sociology of technology might benefit each other. *Social Studies of Science*, **14**, 399–441.

Pirandello, L. (1935) Interview. In Zeffirelli, F. (2003) "The division between reality and imagination: Martin Sherman talks to Franco Zeffirelli about Pirandello" in Programme Notes for *Absolutely! {perhaps}*. Wyndhams Theatre, London.

Sacks, H. (1992) *Lectures in Conversation*.

Sawhney, H. (2005) Wi-Fi networks and the rerun of the cycle. In Ling, R. (ed.), *Mobile Communications: Re-negotiation of the Social Sphere*. Springer-Verlag, London.

Schiller, D. (2000) *Digital Capitalism: Networking the Global Market System*. MIT Press, Cambridge, MA.

Slevin, J. (2000) *The Internet and Society*. Polity, Cambridge.

Smith, M.A. and Kollock, P. (eds) (1999) *Communities in Cyberspace*. Routledge, London.

Swann, G.M.P. and Watts, T. (2002) Visualisation needs vision: the pre-paradigmatic character of virtual reality. In Woolgar, S. (ed.), *Virtual Society? – Technology, Cyberbole, Reality*. Oxford University Press, Oxford, pp. 41–60.

Thomas, D. and Loader, B. (2000) *Cybercrime: Law Enforcement, Security and Surveillance in the Information Age*. Routledge, London.

Toulouse, C. and Luke, T.W. (eds) (1998) *The Politics of Cyberspace: a New Political Science Reader*. Routledge, London.

Watt, S., Lea, M. and Spears, R. (2002) How social is internet communication? A reappraisal of bandwidth and anonymity effects. In Woolgar, S. (ed.), *Virtual Society? – Technology, Cyberbole, Reality*. Oxford University Press, Oxford, pp. 61–77.

Wellman, B. and Haythornthwaite, C. (eds) (2002) *The Internet in Everyday Life*. Blackwell, Oxford.

Wittel, A., Lury, C. and Lash, S. (2002) Real and virtual connectivity: new media in London. In Woolgar, S. (ed.), *Virtual Society? – Technology, Cyberbole, Reality*. Oxford University Press, Oxford, pp. 189–208.

Woolgar, S. (2002) Five rules of virtuality. In Woolgar, S. (ed.), *Virtual Society? – Technology, Cyberbole, Reality*. Oxford University Press, Oxford, pp. 1–22.

Woolgar, S. (2004) Reflexive Internet? – the British experience of new electronic technologies. In Castells, M. (ed.), *The Network Society: a Global Perspective*. Edward Elgar, London, pp. 125–142.

Woolgar, S. (2005a) Ontological disobedience – definitely! {maybe}. In Sica, A., and Turner, S.P. (eds), *A Disobedient Generation*. Chicago University Press, Chicago.

Woolgar, S. (2005b) Mobile back to front: uncertainty and danger in the theory-technology relation. In *Mobile Communications*. R. Ling and P. Pedersen (eds.) London: Springer. Pp. 23–43.

Woolgar, S. and Simakova, E. (2003) Marketing marketing. Paper presented at Skebo conference, Stockholm School of Economics, Sweden, 14–16 June 2003, and at 4S/EASST conference, Paris, 25–28 August 2004.

Wyatt, S., Thomas, G. and Terranova, T. (2002) They came, they surfed, they went back to the beach: conceptualising use and non-use of the internet. In Woolgar, S. (ed.), *Virtual Society? – Technology, Cyberbole, Reality*. Oxford University Press, Oxford, pp. 23–40.

Yung, V. (2005) Ideotechnic values of the mobile phone in Hong Kong. In Ling, R. (ed.), *Mobile Communications: Re-negotiation of the Social Sphere*. Springer-Verlag, London.

Zeffirelli, F. (2003) The division between reality and imagination: Martin Sherman talks to Franco Zeffirelli about Pirandello in *Programme Notes for Absolutely! {perhaps}*. Wyndhams Theatre, London.

4

Wi-Fi Networks and the Reorganization of Wireline–Wireless Relationship

Harmeet Sawhney

4.1 Introduction

New technologies often help subvert the established order.[1] The mysterious process by which they bring about unexpected change is at best partially understood. Zuboff (1988) likens it to the turning of a kaleidoscope. "Imagine a hand nudging the kaleidoscope's rim until hundreds of angles collapse, merge, and separate to form a new design" (p. 387). She suggests that technological change has the same transformational power as the hand that turns the kaleidoscope. Christensen (1997) makes a distinction between "sustaining technologies" and "disruptive technologies". The former improve the performance of established products and systems and are thereby readily accepted by the mainstream customers. The latter, often with poor initial performance characteristics, are at first embraced by fringe groups for their peculiar reasons and later, after the technology has improved our time, an entrepreneur introduces it into the mainstream market and proceeds to disrupt it. The potential for subversion is especially high in the case of communications technologies because "new media embody the possibility that accustomed orders are in jeopardy, since communication is a particular kind of interaction that actively seeks variety. No matter how firmly custom or instrumentality may appear to organize or contain it, it carries the seeds of its own subversion" (Marvin, 1988; quoted in Giese, 1996, p. 139).

---

[1] The paper that was presented at the Front Stage/Back Stage: Mobile Communication and the Renegotiation of the Social Sphere conference, Grimstad, Norway, June 2003 was published in *Info* (Sawhney, 2003). This chapter builds on the *Info* paper. The introduction and the last third of the chapter are new. The concluding section offers far more specific predictions than the earlier paper.

Today, in the telecommunications arena, one of the big questions is whether Wi-Fi, an unexpected grassroots uprising of sorts, will disrupt the prevailing network configuration. This chapter will draw on the past experience with infrastructure networks – canals, railroads, highways, telegraph, telephone and other large-scale networks – to assess the subversive potential of Wi-Fi. Correspondingly, whereas new technology-enabled subversion occurs across the entire spectrum of communication technologies, the ensuing analysis will be limited to infrastructure networks – a particular type of communication technology.

The physical aspect of the infrastructure networks is usually described through the concepts of "nodes" and "links". They take very different physical forms depending on the nature of the infrastructure network. In the case of the telephone network, the switches are the nodes and the transmission lines the links. But then, in the case of the railroad network, the junction points, which are also incidentally called switches, are the nodes and the railroad tracks the links. Even though the telecommunications gear is tangibly very different from railroad equipment, these networks show similarities in their development patterns. These similarities stem from the fact that the way they are organized or the "system of relationships" is a bigger determinant of their character than the discrete technological pieces – switching, transmission, railroad tracks, etc. – that constitute them. These "system of relationships" are not determined merely by technocratic considerations but to a very large degree by socioeconomic forces and reflect the society that creates them.

The infrastructure networks are different from communications technologies such as radio, TV, film and recorded music that tend to attract much attention. Although infrastructure networks provide the technological platforms that allow communication at a distance, they are not tied to the content that flows on them. This point is important for conceptual clarity because it explains how infrastructure networks are different from other communications systems and thereby delineates the scope of the analysis presented in this chapter. Interdependencies created between communications systems because of shared content allow for symbiotic relationships. In the 1950s, when television networks started to attract audiences, Hollywood refused to allow the broadcasting of its films because it saw the new technology as a mortal threat. Later, Hollywood realized its folly and opened up a new stream of revenue by forging a mutually profitable relationship with the television networks. The pattern was repeated in the 1970s when Hollywood saw VCRs as a major threat and then it went on to use the new technology to distribute its products. So now we have film, television and VCRs co-existing in a symbiotic relationship with contents of one being "poured" into the other. On the other hand, in the case of infrastructure networks, the key issue is interconnection of networks. A new infrastructure network can only be commercially viable if it can interconnect with entrenched infrastructure networks and thereby move its traffic towards the intended destination.

In this day and age, a new technology does not strike roots and grow on a virgin ground. Instead, it encounters a terrain marked by old technologies. The situation is peculiarly complex in the case of infrastructure networks. The terrain is usually straddled by an entrenched "backbone network" – an extensive system of nodes and links that provides connectivity between *distant* points. A new infrastructure technology has to fit into this framework. It typically starts as a "feeder" network – a relatively smaller system of nodes and links that collects traffic from a *local* area and takes it to the backbone network for onward transmission. Infrastructure Development Model (IDM) models how this initial relationship between the old and new infrastructure networks changes over time with the latter eventually displacing the former as the backbone network (Sawhney, 1992).[2] The following analysis employs IDM to assess the likely impact of Wi-Fi on the prevailing network configuration.

The first section provides an overview of IDM and thereby sets the framework for the rest of the paper. The next section examines the evolution of wireless technologies before the advent of Wi-Fi. It discusses why for a while it seemed likely that wireless technologies would break out of the mold and not follow the development pattern indicated by IDM. The third section examines the development of Wi-Fi and finds that it seems poised to follow the IDM pattern. The concluding section ponders on the future development trajectory of Wi-Fi technology.

4.2 Infrastructure Development Model

The infrastructure networks in the USA tend to develop in a decentralized, uncoordinated, and bottom-up manner. Their development is not guided by a blueprint, a grand plan or a vision of any sort. They sort of emerge. This peculiarly American process is a reification of the country's political ethos. As Anderson (1985) points out, "the American political and problem-solving style is incremental, not synoptic – this country is wary of large-scale blueprints" (p. 280). While the polycentricity of the American socio-political environment energizes the infrastructure development process by encouraging entrepreneurial activity, it creates major problems when the time comes to forge the disparate networks into a

---

[2] IDM captures the motif that recurs whenever America creates its infrastructure networks. Whereas research on other countries is limited, studies on infrastructure networks in France and Canada indicate that patterned growth of infrastructure networks also occurs in other socio-cultural contexts. However, each country has its own infrastructure development pattern (Sawhney, 1993, 1999). Since this paper uses the analytical framework provided by IDM which is derived from the American experience, the analysis of Wi-Fi networks in this paper will be limited to the USA.

unified national system. Yet, like in other arenas of American life, operating underneath the chaos created by the push and pull of multiple centres of initiative there are unifying forces that eventually create an integrated network. This underlying pattern which gives a distinctive character to the development of infrastructure networks in the USA is captured by IDM (Figure 4.1).

The reader will find a detailed explanation of IDM in Sawhney (1992). In this section, only the pattern underlying the development of railroads will be discussed in its entirety. In the case of other networks, only the major points of their development will be mentioned to highlight its defining characteristics.

The first major infrastructure network in the USA was the inland waterways. The man-made canals were used to link rivers, lakes and oceans to create an extensive network of inland waterways. Considering the fact that the waterways, with relatively minor intervention in the form of canals, were essentially nature's creation, they do not fall within the ambit of IDM. At the same time, the waterways provided the spine around which the railroads first sprouted. Therefore, they form an important part of the analysis.

Today's gigantic railroad systems started as small experimental networks far removed from each other (Stage 1). The first major commercial application of railroads was as feeders for the canal system. Typically, they were used to transport coal from the mines to a nearby canal. In essence, they filled in gaps in areas where the terrain did not allow an extension of the canal system (Stage 2). The canal companies invested heavily in the development of railroads because they saw them as complementary networks which extended their reach and thereby fed traffic from the outlying areas into the canal systems (Stage 3). At this time, long-distance railroads were beyond imagination as it seemed fanciful to suppose that the railroads would ever displace rivers and canals as the chief means of transporting bulky cargo over long distances. Eventually, the development of long-distance capabilities tied the once isolated railroad islands into an integrated system (Stage 4). With the emergence of an integrated system, the railroads became competitors to the canals, and the earlier complementary relationship was decoupled (Stage 5). The railroads became the dominant system, and the canals saw a rapid decline (Stage 6). Now, only those elements of the erstwhile canal system survive which can play a useful role in the new order (Stage 7). After the collapse of the canals, the railroads almost totally dominated other modes of transportation for many decades. However, the cycle did not stop here as a new technology appeared on the horizon – automobiles (Stage 8).

The railroads developed an expansive national network but they could not get around the feeder problem similar to that of the canals. The railroad network simply could not be extended to each farm or factory and the farmers and factory owners had to haul their goods to the nearest

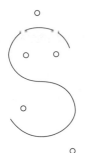

Stage 1: Sprouting of infrastructure islands

The infrastructure technologies first appear as technological islands. There is no interconnection among them and the islands are isolated from one another. These islands are hotbeds of entrepreneurial energy. They are basically demonstration projects which test out revolutionary ideas. Their actual commercial potential is still uncertain.

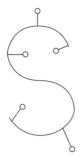

Stage 2: Development as a feeder

The new technology is found to be viable and its basic potential is seen in its role as a complement to the old system. The new technology can reach into areas which were inaccessible to the old technology due to its different technological base. At this stage it is still a short-haul technology.

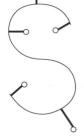

Stage 3: Encouragement by the old system

The new technology in its role as a feeder generates additional traffic for the old system. In effect, the new technology increases the old system's catchment area by extending its reach. The old system encourages and aids the development of the new technology.

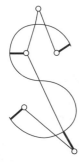

Stage 4: Long-distance capabilities and system formation

The long-distance capabilities are developed. The isolated bits of new technology become directly interconnected and start bypassing the old system. The interconnection process creates problems of coordination and standardization. Eventually the long-distance capability results in an integrated system.

Figure 4.1 Infrastructure development model (IDM). The different S shapes are mnemonics for the concept "system". Reprinted from *Telecommunications Policy*, Vol. 16. H. Sawhney, Public telephone network: stages in infrastructure development, pp. 538–552, copyright (1992), with permission from Elsevier Science.

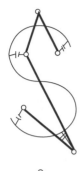

Stage 5: Competition with the old system

The old system finds itself threatened. At first it goes on an offensive but soon adopts a defensive posture. Charges are made about "unfair" subsidy. The need to protect the franchises of the old system is stressed. Emerging competition is depicted as something wasteful. Finally an attempt is made to accommodate the new technology within the existing order.

Stage 6: Subordination of the old system

The old system's rearguard action is unable to withstand the onslaught of new technological developments. Eventually the old system caves in or is pushed into a subservient role.

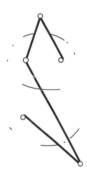

Stage 7: Reversed feeder relationship

The old system disintegrates into fragments, and only those fragments which can serve a unique niche survive. They either fill in a gap where for economic or technological reasons it is not attractive to extend the new system, or they supplement it along routes where there is a specialized kind of traffic. In effect these fragments now serve as feeders for the new system.

Stage 8: Rerun of the above cycle

The new system dominates until another technology appears. The newer technology then grows along the same cycle. Eventually it replaces the new system. The cycle goes on. Just when it seems the ultimate technological plateau has been reached, another technology appears.

Figure 4.1 (*cont'd*)   Infrastructure development model (IDM). The different S shapes are mnemonics for the concept "system". Reprinted from *Telecommunications Policy*, Vol. 16. H. Sawhney, Public telephone network: stages in infrastructure development, pp. 538–552, copyright (1992), with permission from Elsevier Science.

railroad track. Within this context, the automobile generated much enthusiasm because of its potential to extend the railroad's catchment area by moving goods and passengers from the surrounding area to the railroad tracks. The railroads made a concerted effort to encourage the development of roads to expand the use of automobiles. So much so, they started "good roads" trains that took road building equipment into the countryside and helped communities build roads. In an eerie parallel to their own history, the railroads encouraged the development of roads without ever imagining that the automobile would one day develop long-distance capability and thereby become competitors. As they say, the rest is history.

In the case of telecommunications networks, we see similar patterns. The telegraph network could not be extended to every office and home because of the nature of the technology, especially the need for trained operators to send and receive messages. Consequently, the problem of getting the message from the central office to the designated person was the weakest link in the telegraph system. The telephone was seen as the feeder technology that would fill in this gap. At that time, the conventional wisdom was that telephone would never develop long-distance capabilities because of the technical challenges involved. Furthermore, even if long-distance telephony did become possible, there would be very little demand for it as the main users of long-distance communications were businessmen who need written records of their communications. The possibility of conducting business transactions verbally over the telephone was beyond imagination. Eventually, the development of long-distance telephony displaced telegraph as the dominant communications technology.

In all the examples discussed above, we see a consistent pattern. A new technology strikes roots as a feeder to the established system and thereby is seen as extending its reach. The relationship between the new and old technology seems symbiotic and thereby stable and enduring. In effect, the new technology appears to have strengthened the entrenched paradigm. It is, however, eventually shattered with the unanticipated development of an independent system based on the new technology. IDM, which captures this pattern as an abstraction, provides a conceptual foil for analyzing the relationship between wireline and wireless networks in the next section.

4.3 Wireline–Mobile Relationship

Mobile communication first became possible in the form of walkie-talkies. They were in effect little islands of communication (Stage 1). With the advent of cellular, mobile communication became a networked phenomenon. Cellular plugged the gap between the wireline network and

51

automobiles. In effect, it extended the reach of the wireline network into the mobile environment and thereby became a feeder technology (Stage 2). The wireline companies saw cellular as a complementary technology and invested in its development (Stage 3). At the time of the publication of IDM in 1992, it was apparent that mobile communication had followed the pattern outlined in IDM in the initial stages of its development. The big question then was whether or not the development process would move on to next stage (Stage 4) where the new system starts to decouple from the old one. At that time, it seemed unlikely, for the following reasons.

One, the inter-modal compatibility – the capacity of two or more technological systems to work together as a larger virtual system that appears almost seamless to the users – of modern communications technologies was likely to change the dynamics of network development. In the case of transportation systems, inter-modal compatibility was poor, as goods had to be physically transferred from one system to another when they were complementary to one another. For example, coal that was shipped by the railroads from the mines to the canals had to be unloaded from railroad wagons and carried to the barges. Within this context, the benefits of disrupting the complementary relationship between the old and new technology and creating a new integrated system around the new one were great. On this score, the situation is very different in the case of modern communication technologies. The information can relatively easily be transferred from one mode to another, mobile to wireline to mobile. This was not always the case with earlier communication technologies. For example, telegraph messages could not be seamlessly carried over to telephone and vice versa. Today, digitization and the accompanying development of translation technologies have greatly mitigated inter-modal transfer problems and thereby removed one important impetus for creation of an integrated system.

Two, the new technology in this particular case, mobile, required a scarce resource – spectrum. On the other hand, the old technology, wireline network, conserved this very same resource. In such a case, a complementary relationship is advantageous because it optimizes the use of spectrum. The cordless phone, a relatively simple commonplace technology, illustrates the point. The low-powered transmitter on the base station allows for mobile communication within a 75-meter range. On the other hand, the base station itself is connected to the wireline network. In this configuration, the wireline network provides connectivity with the rest of the world without consuming the spectrum, whereas the low-powered transmitter provides mobility where it is needed and yet limits the use of the spectrum to a small area. If the system were to go entirely wireless, it would use far more spectrum than a hybrid arrangement. The same archetypal pattern can be found in the architectures of PCS networks, wireless LANs and other wireless networks. Hence it would seem that wireless technologies, because of the mobility they offer and lower implementation costs, would become the primary interface between the subscribers and the telephone network. The public telephone network itself would remain by and large a wireline network that provides high-bandwidth transports and universal connectivity. While the wireline network will be pushed to the background as mobile interfaces grow, it will remain the vital backbone of the entire network.

Considering these factors, it seemed unlikely that mobile would develop along the path suggested by the IDM. But then, IDM also suggests that one should expect a surprise somewhere down the line. As we shall see in the following section, this surprise shows up in the form of Wi-Fi technology.

4.4 Wi-Fi Networks

While the corporate world has either stumbled with its mobile Internet projects or created pseudo-Internets such as I-Mode, a grassroots phenomenon in the form of Wi-Fi network seems poised to overturn the corporate apple cart. Interestingly, the cordless phone once again provides a good model for understanding this new technology. Initially, the Wi-Fi technology was directed at creating mobility within a building. Just as a cordless user can walk around the house or office while talking over the telephone, the early Wi-Fi networks sought to create similar mobility for Internet access via laptop computers. Instead of directly connecting a high-speed line (telephone or cable) to a computer, it was connected to a low-powered transmitter that communicated with laptops within its range. The technology took a new turn when users started leaving these Wi-Fi transmitters unsecured. This allowed their neighbors and others within the unsecured transmitter's range to use the Internet connection for free. The people who left their transmitters unsecured did not care because they were paying a flat monthly fee for Internet access. Unauthorized use by others did not cost them anything. In fact, many take pleasure at the subversive nature of their actions. As Negroponte (2002) explains:

> Depending on the intervening materials, a vanilla Wi-Fi can radiate more than 1000 feet. Since I live in a high-density area, my system reaches perhaps 100 neighbors. I do not know how many use it (totally free) – frankly, I do not care. I pay a fixed fee and am happy to share.
>
> Because further down the street, beyond the reach of my system, another neighbor has put in Wi-Fi. And another, and another. Think of a pond with one water lily, then two, then four, then many overlapping, with their stems reaching into the Internet (online).

The water lily imagery nicely captures the complementary relationship between Wi-Fi and wireline Internet.[3] The flower symbolizes the hot spot, the circular area generated by a Wi-Fi transmitter within which Internet can be accessed, and the stem represents the telephone or cable line that connects the transmitter to the Internet. The overlapping water lilies allude to a high concentration of hot spots in parts of Silicon Valley, Boston and other urban areas that overlap to create fairly large Wi-Fi patches.

So far, Wi-Fi's development pattern has stayed within the mold of earlier mobile technologies, i.e. it has not gone past Stage 3 of IDM. However, it seems poised to go on to Stage 4 in a totally unexpected way.

> In the future, each Wi-Fi system will also act like a small router, relaying to its nearest neighbors. Messages can hop peer-to-peer, leaping from lily to lily like frogs – the stems are not required. (Negroponte, 2002, online)

What Negroponte is suggesting is not mere fantasy but actually possible. As explained later, "lily to lily" communication has already started to occur on a small scale. If the trend continues, we could start to see a bypass of the old technology that is characteristic of the Stage 4 of IDM. Interestingly, the industry's reaction to the emergence of Wi-Fi networks is already that one usually sees in the Stage 5 of IDM. Clearly, the industry perceives a threat in Wi-Fi technology.

> Until now Wifi (sic) has been viewed by many technology analysts as an upstart from-the-bottom technology that has the potential of upsetting other capital-intensive technology deployments, like the expensive next-generation data-oriented cellular networks known as 2.5G and 3G that are being established by companies such as AT&T Wireless, Cingular, Nextel, T-Mobile, Sprint and Verizon. (Markoff, 2002b, pp. C1 and C4).

The industry's response to this threat is even more telling. Like the canals, railroads and telegraph, it seeks to accommodate the new technology within the existing order by strengthening the complementary relationship between the old and the new technology.

> It (VoiceStream Wireless) announced in mid-March that it will integrate Wi-Fi technology … with its existing network to provide "seamless service" – an Internet connection that switches automatically from Wi-Fi to 3G and back – starting early next year. Sprint PCS is working on something similar, although it hasn't unveiled an offering. "There is no question that Wi-Fi will be complementary to 3G wireless," says Sprint spokesman Dan Wilinsky. (Stone, 2002, online)

---

[3] Negroponte credits Alessandro Ovi, technology adviser to European Commission President Romano Prodi, for the lily analogy.

When viewed from within the existing framework, a complementary relationship between cellular and Wi-Fi makes a great deal of sense because the strengths and weaknesses of the two technologies complement each other. Wi-Fi offers much greater bandwidth and therefore higher transmission speeds mainly because it employs low powered transmitters for communication over short distances. It is limited, however, to high population density areas. Cellular, on the other hand, has an extensive network already in place. Its downside is significantly lower transmission rates. Therefore, a seamless service that automatically switches back and forth between the two networks would be a win–win solution for everybody. While in the downtown area, a subscriber could use higher bandwidth Wi-Fi networks and then switch over to cellular service when driving into the outlying areas.

In sum, the logic of a complementary relationship is indeed compelling. IDM, however, prompts us to remember that what seems very rational within the existing framework may not remain so in a new network paradigm.

4.5 The Question of the Future

As mentioned earlier, at the time of the publication of IDM in 1992, it seemed unlikely that wireless would move beyond Stage 3 of the model, for two reasons: (1) inter-modal compatibility and (2) spectrum scarcity. Subsequently, with the emergence of Wi-Fi, wireless is today on the verge of entering Stage 4. The big question now is whether it will complete Stage 4 and go to Stage 5.

When the Wi-Fi transmitters start directly routing messages to their neighbors or, as Negroponte (2002) poetically says, "lily to lily", wireless will enter Stage 4. However, that does not mean that it will necessarily complete Stage 4. Direct communication between neighboring Wi-Fi transmitters would indeed result in the bypass of the entrenched wireline system, the defining feature of a system in Stage 4 of the development cycle. However, at this level of development, the bypass will occur only at a local level. In a city center covered by overlapping hot spots, messages from one Wi-Fi transmitter could be sent to another located many blocks away without ever touching the wireline network, neighboring transmitters relaying the message forward all the way. However, if the message has to be sent from a transmitter in one city to another, say from New York to Los Angeles, the entire dynamic changes. Although theoretically one could imagine a chain of overlapping hot spots connecting New York to Los Angeles, the obvious advantages of using the existing wireline network would make the construction of a transcontinental Wi-Fi link impractical, to say the least. In other words, even with the emergence of Wi-Fi, inter-modal compatibility remains a factor that could potentially deflect the evolution of the wireline–wireless relationship away from the IDM pattern. Similarly, spectrum scarcity continues to be a constraining factor that reinforces the continuation of the existing complementary relationship. Yet, even if the present framework endures, there are likely to be significant changes in the wireline–wireless equation as wireless grows deeper into the network and the wireline recedes into the background as a long-haul transport vehicle. We are already seeing the deployment of Wi-Fi as a backhaul link, as opposed to the end-user link that has been the focus of the discussion so far, which connects unwired collection point towers to the wireline network. The advantage of Wi-Fi over microwave as a backhaul link is that it is not as severely constrained with regard to line of sight issues (Wilson, 2003).

Arthur C. Clarke classifies forecasting failures into two categories: "failures of nerve" and "failures of imagination". In the case of the former, the forecaster fails to see the obvious even when all the relevant facts are in front of him or her. The latter occur when the forecaster is unable to make the leaps of imagination necessary to grasp a new phenomenon (Clarke, 1962). The above analysis of the potential development of Wi-Fi technology in Stage 4 is reasonable and that is probably its greatest weakness. As we have seen time and again, future is rarely a logical extension of the present. IDM goads us to think differently.

4.6 IDM-based Projections

It is foolhardy to make projections about the future. Yet, in the realm of telecommunications, future is a pesky problem that can not be wished away. We are constantly faced with decisions about the future even though we do not know what it will entail. What technology should a company deploy to meet future demand? What regulatory frameworks should the regulators create that can cope with the pace of technological change? We have to act today and in order to do so we have to have some notion of the future, however tentative and ill formed. We would like to devise computer models that print out a detailed roadmap, but the processes driving technological change are too complex to model. We can fall back on raw hunch and intuition, but that can lead to impressionistic and historically uninformed decisions. Or, as this paper suggests, we can use heuristic models such as IDM to expand the range within which our imagination roams.

One of the primary reasons for the shortfall in our imagination is that our thinking is unable to go past the prevailing logic of the day. We seek answers within the confines of the entrenched structures. In the case of Wi-Fi, our mental energies are directed towards somehow fitting Wi-Fi within the prevailing scheme of things instead of considering new network architectures. IDM provides a springboard for breaking out of this mindset.

With that objective in mind, one can venture to make two projections. The first one challenges the fundamental assumption underlying the prevailing thinking about Wi-Fi. The second highlights a blind spot.

1. Even if Wi-Fi and 3G end up as complementary technologies, long-distance–short-haul alignment will not be the basis of such a relationship.

The seeming snugness of fit between a long-distance and a short-haul technology is often deceptive if the past is anything to go by. The canal companies invested in the early railroads enamoured by their potential to draw traffic into the canal system from difficult to reach areas. The railroads were seen as short-haul technologies that increased the canal system's catchment area by penetrating terrain into which the canals could not be extended. In fact, it was a canal company, Delaware & Hudson, which brought the first locomotive to America. The company imported "Stourbridge Lion", an English-made engine, to transport coal over a 16-mile stretch from Honesdale to Carbondale (Thompson, 1925). However, the engine weighed 7 tons instead of the specified 3 tons (Ringwalt, 1888). "The locomotive when tried out proved so heavy that they were afraid the track wouldn't sustain it, so it was discarded – but the damage had been done. The canal had taken to its bosom the serpent which later was to sting it to death" (Harlow, 1926, p. 78). The eventual development of long-distance railroads disrupted the cosy complementary relationship between the canals and railroads, leading to the near total decline of the former. Of course, all this was difficult to imagine when the railroads nicely fit in as little pieces of a larger canal-centric jigsaw puzzle. As discussed earlier, a similar pattern was repeated in the case of railroads and automobiles and telegraph and telephones.

There are notable differences between the 3G–Wi-Fi relationship and those mentioned above. Wi-Fi was at first seen as a threat by 3G interests and only later on came to be viewed as a complementary technology. This order is reverse of that of canal–railroad, railroad–automobile and telegraph–telephone relationships where the new technology was at first seen as complementary and only later on became a threat. Furthermore, the earlier relationships were between a backbone technology and a feeder. In the case of 3G and Wi-Fi, both of them are feeder technologies that draw traffic into a primarily wireline network. Yet, in spite of all such differences, IDM cautions us against lulling ourselves into the belief that today's long-distance–short-haul complementary relationship will continue into the future.

Already there are signs that Wi-Fi will not forever remain a short-haul technology. While the maximum range of Wi-Fi is generally considered to be 300 feet, community-networking activists have attained transmission of about 4 miles with home-made antennae using empty boxes of Pringle potato chips. According to the Guinness Book of World Records, the longest Wi-Fi link is 192 miles, created by the Swedish Space Corporation and Alvarion, an Israeli company (Talacko, 2003). The FCC has limited the power of devices such as remote controls, garage door openers and cordless phones that make use of the unlicensed spectrum in the Industrial, Scientific, and Medical Band to 1 watt so as to minimize interference and maximize the reuse of the spectrum. It is this 1-watt constraint that limits the reach of a Wi-Fi transmitter to a 300-foot radius. But then, a Wi-Fi transmitter need not be omni-directional. Directional transmitters that transmit in straight lines, as opposed to circles, can attain greater distances as the available power is channelled along a single path and not dissipated in all directions (Johnston and Snider, 2003). The Wi-Fi enthusiasts are exploiting this potential to tease out greater and greater transmitting range from a 1-watt transmitter.

There are now new terms such as Wi-Max and Wider-Fi for the newer Wi-Fi technologies with greater ranges. Correspondingly, the visions of what can be achieved with Wi-Fi are expanding. According to Johnston and Snider (2003), "Starbuck's 'hot spot' is a grossly constrained vision of the future of wireless networking" (p. 8). The talk is now of hot spots, hot zones, hot pathways and hot regions (Levy and Stone, 2002; Johnston and Snider, 2003). For example, two community groups, SFLAN and Bay Area Research Wireless Network, have set up 12 public nodes in San Francisco that are linked to each other in a simple mesh configuration (Markoff, 2003b). Similarly, on a commercial basis, WiFi Metro has positioned antennas to cover a six-block area in Palo Alto and San Jose. Furthermore, companies such as SkyPilot are looking to cover wider areas by "hopscotching bandwidth" from computer to computer (Levy and Stone, 2002). People have even started thinking of a future cooperative wherein users create ad hoc networks themselves instead of relying on service providers (Curry, 2001; Schrage, 2003).

> The concept is to create a Wi-Fi cooperative that turns individual laptops into potential nodes, routers and hubs of a global network analogous to the wireless-mesh networks being pursued by Intel, among others.
>
> So treat every laptop as a voluntary Wi-Fi hot spot. People could go online to retrieve software that effectively turns their machines into Wi-Fi access points. Instead of paying for broadband Internet subscriptions, individuals – and organizations – would agree to make their machines accessible to other machines, creating relays that eventually reach the Net (Schrage, 2003, p. 20).

If the cooperative idea develops legs, it will start to snowball as every new member added to the network will increase its overall reach as opposed to increasing costs in the case of cellular data networks (Markoff, 2002a, p. C4).

In sum, the past experience with railroads and other technologies suggests that the complementarity between 3G and Wi-Fi on the basis of long–distance–short-haul alignment is unlikely to be a lasting arrangement. This projection, however, does not preclude the potential co-existence of 3G and Wi-Fi. It is likely that the future wireless environment will be a heterogeneous one with multiple technologies co-existing (Lehr and McKnight, 2003). In this milieu, both 3G and Wi-Fi may co-exist. However, the basis of their co-existence is unlikely to be long-distance–short-haul complementarity. It may be mobility, less likely with the development of Wi-Fi handoff technologies, or some other basis that is difficult to imagine right now.

2. While our current thinking is dominated by the relationship between Wi-Fi and 3G, IDM predicts that Wi-Fi's main impact will be on the network core.

Both Wi-Fi and 3G are what Lehr and McKnight (2003) call "edge-network" technologies. In other words, "they offer alternatives to the last-kilometer wireline network. Beyond the last kilometer, both rely on similar network connections and transmission support infrastructure" (Lehr and McKnight, 2003, p. 357). Our focus to date has almost entirely been on the two edge-network technologies. We have given very little thought to the impact they may have on the core wireline network itself.

When some thought has been given to the relationship between Wi-Fi and wireline network, the focus has been on the backhaul links connecting hotspots to the IP backbone. According to Deutsche Bank analysts Viktor Shvets and colleagues, "the natural advantage of owning the backhaul connection, confers upon ILECs (Incumbent Local Exchange Carriers) a status as the best-placed operators to take advantage of Wi-Fi penetration growth" (quoted in Olavsrud, 2003). On the other hand, Wi-Fi networks, especially unsecured ones that open up a DSL line or cable modem connection to neighbors and passers by, have also been characterized as parasitic networks (Curry, 2001). The cable and telephone companies contend that subscribers who leave their transmitters unsecured are violating their service agreements and possibly breaking the law. They equate allowing neighbors and strangers to tap into one's Internet connection to cable theft. Wi-Fi enthusiasts argue that satellite broadcasts provide a more appropriate analogy. It is not a crime to pick up unscrambled satellite signals. However, it is against the law to decode the encryption of scrambled signals. In other words, securing connections is the responsibility of the network owner and not the users. These arguments have not yet been tested in the courts and it is difficult to predict how the justices will rule (Harmon, 2002). Today, the parasitic traffic is on a lightly loaded network. It is likely to become a problem with the growth of Wi-Fi networks (Lehr and McKnight, 2003).

When Wi-Fi and wireline technologies are considered as potential competitors, the frame is limited to the last mile. Both Wi-Fi and 3G proponents seek to attain data speeds comparable to that of current broadband wireline service and hence could emerge as competitors to DSL and cable modem service providers. However, Gartner, a market research company, believes that wireless technologies will simply not be able to keep pace with wireline ones as we move into the 1-gigabit world.

To survive – and thrive – Gartner believes that wireline carriers must use the current investment opportunity to deploy an unassailable competitive advantage. "Now is the time to make an investment wireless providers cannot match because of wireless technology limitations … Clearly we believe enabling integrated next-generation broadband services is the key to survival for wireline providers" (Gartner, quoted in Johnston and Snider, 2003, p. 16).

Interestingly, this purported "battle for survival" is actually a skirmish at the edges – broadband wireline (last mile) edge-technology vs broadband wireless edge-technology – rather than a contest for the heart of the network. IDM, on the other hand, predicts that a descendant of Wi-Fi, at some point in the future, will drive a stake through the heart of the network. How it will fundamentally reshape the core network will become clear only with the passage of time.

4.7 References

Anderson, O.W. (1985) *Health Services in the United States*. Health Administration Press, Ann Arbor, MI.

Christensen, C.M. (1997) *The Innovator's Dilemma*. Harvard Business School Press, Boston.

Clarke, A.C. (1962) *Profiles of the Future*. Victor Gollancz, London.

Curry, A. (2001) WiFi peering. *Stories.Curry.Com*, 6 September. http://stories.curry.com/stories/storyReader$12 (accessed on 27 August 2003).

Giese, M. (1996) From ARPAnet to the Internet: a cultural clash and its implications in framing the debate on the information superhighway. In Strate, L., Jacobson, R., and Gibson, S. (eds), *Communication and cyberspace*. Hampton, New York, pp. 123–141.

Harlow, A.F. (1926) *Old Towpaths*. Appleton, New York.

Harmon, A. (2002) Good (or unwitting) neighbors make for good Internet access. *The New York Times*, 4 March, pp. C1 and C4.

Johnston, J. and Snider, J. (2003) *Breaking the Chains: Unlicensed Spectrum as a Last-mile Broadband Solution*. Spectrum Series Working Paper No 7. New America Foundation, Washington, DC.

Lehr, W. and McKnight, L. (2003) Wireless Internet access: 3G vs. Wi-Fi? *Telecommunications Policy* **27**: 351–370.

Levy, S. and Stone, B. (2002) The Wi-Fi wave. *Newsweek*, 10 June, retrieved from Academic Search Premier database.

Markoff, J. (2002a) The corner Internet network vs. the cellular giants. *The New York Times*, 4 March, pp. C1 and C4.

Markoff, J. (2002b) High-speed wireless Internet Network is planned. *The New York Times*, 6 December, pp. C1 and C4.

Markoff, J. (2003a) More cities set up wireless networks. *The New York Times*, 6 January, p. C7.

Markoff, J. (2003b) Led by Intel, true believers in Wi-Fi say it will endure. *The New York Times*, 14 July, pp. C1 and C4.

Marvin, C. (1988) *When Old Technologies Were New*. Oxford University Press, New York.

Negroponte, N. (2002) Being wireless. *Wired*, October. http://www.wired.com/wired/archive/10.10/wireless.html (accessed on 29 June 2004).

Olavsrud, T. (2003) Analysts: Wi-Fi complementary to telecoms. Internetnews.com. http://www.internetnews.com/wireless/print.php/2222271 (accessed on 27 August 2003).

Ringwalt, J.L. (1888) *Development of Transportation System in the United States*. Published by the author, Philadelphia (Reprint by Johnson Reprinting, New York, 1966)

Sawhney, H. (1992) Public telephone network: stages in infrastructure development. *Telecommunications Policy*, 16: 538–552.

Sawhney, H. (1993) Circumventing the center: the realities of creating a telecommunications infrastructure in the USA. *Telecommunications Policy*, 17: 504–516.

Sawhney, H. (1999) Patterns of infrastructure development in the U.S. and Canada. In Sawhney, H. and Barnett, G. (eds), *Progress in Communication Science*, Vol. XV. Ablex, Stamford, CT, pp. 71–91.

Sawhney, H. (2003) Wi-Fi networks and the return of the cycle. *Info*, 5(b), 25–33.

Schrage, M. (2003) Wi-Fi, Li-Fi, and Mi-Fi. *Technology Review*, July/August, 20.

Stone, A. (2002) Wi-Fi: it's fast, it's here – and it works. *Business Week Online*, 1 April. http://www.businessweek.com/technology/content/apr2002/tc2002041_1823.htm (accessed on 27 August 2003).

Talacko, P. (2003) Going the distance. *The Financial Times*, 12 March, retrieved from LexisNexis Academic database.

Thompson, S. (1925) *A Short History of American Railways*. Bureau of Railway News and Statistics, Chicago.

Wilson, S. (2003) No strings attached: introduction to unlicensed wireless data technologies. Presentation at the 7th Annual Florida Communications Policy Symposium, 3 April, Tallahassee, FL.

Zuboff, S. (1988) *In the Age of the Smart Machine*. Basic, New York.

Mobile Phones as Fashion Statements: The Co-creation of Mobile Communication's Public Meaning

James E. Katz and Satomi Sugiyama

5.1 Introduction

This chapter explores public mobile communication technology as front-stage and back-stage phenomena. We explore the design aesthetics of the mobile phone from the standpoint of its commercial origins and public re-interpretation, emphasizing fashion and identity in the co-creation and consumption of mobile communication technology. The mobile phone in this context is analyzed as both a physical icon and an item of decorative display related to fashion and design. We begin by noting how the early telephone, because it enabled people to communicate efficiently over distance, served as a status symbol. We then highlight the role of fashion and display to show how the symbolic meaning of telecommunication has been evolving. In terms of fashion, we look at the way in which fashion and style have been used to promote the mobile phone by industry. In terms of display, we look at the collateral promotion of other products by reference to the mobile phone and body–technology relationship. Finally, we examine co-constructions that extend beyond the narrow, utilitarian purposes for which the mobile phone was originally designed to show how novel links are forged to deeper psychological and existential processes. That is, the mobile phone is strongly connected with ingrained human perceptions of distance, power, status and identity.

A few words concerning formal theory may be in order. Among the most prominent and influential sub-perspectives of the functionalist school are the "domestication" and "uses and gratifications" perspectives. They have been frequently employed by earlier researchers on mobile communication (e.g. Leung and Wei, 1999, 2000; Haddon, 2003; Wei and Lo, 2003). Indeed, they continue to exert substantial influence despite some criticism of their logical clarity. In this chapter, by contrast, we emphasize the expressive and symbolic dimensions of technology. These dimensions seem critical in understanding the reception and use of mobile communication technology. Indeed, in some cases they may actually supersede utilitarian motives in their importance (Pedersen, 2005). Certainly this view has been argued by Fortunati and her associates generally (Fortunati et al., 2003) and more narrowly in what has been dubbed the Apparatgeist theoretical perspective (Katz and Aakhus, 2002). From the expressive perspective, many mobile phone users are engaging in the same impulse that led to cosmetics and jewelry at civilization's dawn. The mobile phone thus becomes a device that is not merely a tool but as well a miniature aesthetic statement about its owner. In an edited volume, *Machines that Become Us: the Social Context of Personal Communication* (Katz, 2003), the sense of "machines becoming us" was used to capture the idea that the mobile phone could be our personal miniature representative. It was also used in the sense of "becoming" as in complementing and enhancing one's appearance. Fashion, Simmel (1957 [1904]) argued, encourages modification and adoption to individual needs. As Veblen (1934 [1899]) and others have pointed out, wearing of fashionable attire enables individuals to separate themselves from their family, to develop a more distinct identity and a more unique sense of self, and yet to maintain an affiliation with the prestigious aggregate (Fortunati, 1993; Lobet-Maris, 2003). Individuals can use fashion to tailor the social response they desire (Steele, 1997). Fashion then, is a form of communication which includes messages of status and power. Taken together, these various senses of the word "becoming," can be thought of as links between technologies of communication and aesthetic traditions, which in turn are part of the cultural and hierarchy-producing processes.

5.2 Communication over Distance as a Status Marker

Until fairly recently, communication over great distance was in all societies extraordinarily expensive if not downright impossible. Only supernatural beings or an occasional shaman were considered capable of such feats. Greek mythology did not even attribute this power to most of its pantheon. Against this backdrop, it is little wonder that those with access to modern communication innovations become imbued with status. This was true of the early telephone itself when it was first deployed more than 100 years ago. It is also true for the early era of the mobile phone. The devices were described in terms that suggested they were a "rich man's toy" (Marvin, 1987; Katz, 1999), which also alludes to the gendered as well as socio-economic aspects of early adopters.

From another perspective, though, early adopters were of interest since they were also the cutting edge showing the rest of society what was likely to become an everyday technology. As is true for many other technologies, such as air conditioning, automobiles and computers, early adopters were in a sense living in the near future – they were already experiencing, to a greater or lesser extent, what life would be like for subsequent adopters. Early adopters were by their experiences and behavior also helping shape and drive the future. How they chose to use the technology and how they re-configured their own interactional repertoires in light of their experiences and choices were also creating (and sometimes foreclosing) norms and practices that would be available eventually to the bulk of subsequent users.

Since communication over distance and mobility were seen as status symbols, it might be expected that possession of the power of mobile communication technology would be something seen as desirable. This would be especially the case when the technology bestowed upon the possessor: (1) the ability to reach others at the convenience of the possessor and (2) that aid or assistance could be summoned by the possessor. To the extent both of these conditions obtain, maximally "anywhere, anytime," the status of the possessor increases. The rationale for this claim is detailed in Katz (1999), but may be summarized here in terms accreting in the hands of the possessor power and security. Both of these attributes are highly desirable both psychologically and interpersonally [as argued by Maslow (1954) among many others].

5.3 Telephone as Aesthetic Expression

Although powers of distant communication and control of tools of communication technology are status symbols, it is also the case that the telephone itself has developed within a context of the machine age. In particular, certain aesthetic ideas, such as Modernism, have become incorporated the telephone as part of its design thesis. In fact, the phone is often portrayed and understood as part of the future (advanced, streamlined) world. Modernism [see Everdell (1997) for a review of Modernism], which became increasingly powerful throughout the first two-thirds of the 20th century, represents a sharp break with the adherence to traditional design motifs and modalities. That is, it wanted to replace tradition, which was most typically drawn from Classical and Medieval worlds. This old world would be replaced by a new one that emphasized spare stylization, streamlining and a design ethos dictating that form should follow function, and nothing more.

When Modernism as an aesthetic idea was formulated at the end of the 19th century, the telephone was a significant element in considerations of how the industrial and scientific ages were affecting perceptions of history, art and experience. The power of instantaneous command, which the telephone conferred, often figured in the leitmotif of the modern perspective. Early telephones initially embodied the fancy embroidery of the Victorian and Edwardian eras – not only was this fanciness a leitmotif of the era, but it also worked to disguise the proletarian principles of a machine in what should have been an elegant and tasteful setting such as one's home or office (Marvin, 1986). However, like all other sophisticated 20th and 21st century technological objects, it was captured (and largely dominated) in terms of its design by the Modernist impulse.

An extreme form of Modernism, known as Futurism, influenced many perceptions of the telephone and its meaning. As an ideological stance, Futurism emphasized speed, streamlining and rapid motion. Excessively embroidered artifacts and handicraft products were to be dispensed with and, if possible, destroyed. Futurism's prime exponent was Filippo Tommaso Marinetti (1876–1944), who propounded his views in a 1909 manifesto, *Le Futurisme*. His credo included that "the world's magnificence has been enriched by a new beauty: the beauty of speed." Futurism and related movements made themselves quickly felt through much of the industrialized world, and remain starkly visible and timely, and is especially reflected in the design of telecommunication technology.

It is worth recalling that the base of Modernism and Futurism is based on the idea of the machine. This dialectic leads to tension between humans (with thoughts and feelings) and machine (perceived as having power and endurance). As one scholar has written, "Futurism was the first attempt in the 20th century to reinvent life as it was being transfixed by new technologies and conceive of a new race in the form of machine-extended man" (Carey, 1993). There is indeed a long-standing theme in intellectual life of robots merging with humans (Katz, 2003) (a trend that appears a step closer to realization when one considers the headset wearing mobile phone user increasingly prevalent on city streets).

Nowadays, of course, the telephone has become mobile – it is taken out of our "back stage" areas of the home to the "front stage" of public life, where many onlookers can observe the self-presentation of others. The design of the mobile phone, lead by the Modernist impulse, has become part of its possessor's fashion and personal expression. In the following sections, we examine how the industry has been marketing this device as a modernistic item to the public, as well as how the public perceives it, focusing on the role of fashion and display.

5.4 Futuristic and Modern: Industry's Presentation of the Mobile Phone to the Public

Much of the public presentation of the mobile phone has been carefully crafted. This, obviously, is also typical of other commercialized products

(Himmelstein, 1994). In this section, we seek to add to the understanding of the themes that have been highlighted in the design of the mobile phone and in mobile phone ad campaigns.

When a usable mobile phone burst upon the scene in the late 1980s, it appeared to the public as a highly futuristic and sophisticated technology. It was an emblem of the rich and important, though not yet the famous. Interestingly, it appears that designers of the early Motorola StarTac clamshell mobile phone were inspired by the communicator of the TV series *Star Trek*; certainly the name chosen for the line, "StarTac", reinforces this belief when juxtaposed with the name "Start Trek" (Bormanis, undated).

In terms of handsets themselves, we believe that manufacturers from the outset were aware that they wanted to have an explicit futuristic and high-status design stance for the mobile telephone. Although this cannot be demonstrated directly because the archives of the design processes of mobile phone manufacturers remain proprietary, certainly there is scattered evidence to support this view. For instance, there have been various public discussions by leaders of design teams. Alastair Curtis, director of a Nokia design group, has said, "Design has been one of the key elements in the products from Day 1" (Swartz, 2003). Frank Nuovo, Nokia's chief designer, indicated that his design ethos is based on elegant simplicity, relying on high-end accessories as inspiration (Hafner, 1999). Of a specific handset device – Motorola's StarTac, the wildly popular clamshell design of the 1990s – a Motorola designer said, "We wanted a phone that would be visible enough to express something about you … It started as a couture product" (Oehmke, 1999).

A modern, futuristic design impulse has been strongly articulated in the advertising campaigns for mobile communication, according to several scholarly studies of mobile phone ads. For instance, based on semiotic analysis, Pajnik and Lesjak-Tušek (2002) suggest that the image of what they define interchangeably as "Modernity" and "Western values" are important themes in mobile phone ads deployed in Slovenia. In another study on China, conducted by Zhang and Harwood (2004), it appears that household appliances, including telephones and computers, are often associated with "modern" in advertising themes. According to Zhang and Harwood, although the TV advertising of these new technologies also employs traditional cultural values [e.g. family (and societal identity)] in conjunction with "modern" theme, it is noteworthy that "modernity" was one of the most frequently used value themes in recent Chinese TV advertising. In an analysis of the famous "1984" Macintosh advertisement, Stein (2002) argues that the ad employed the "freedom" and "revolution" rhetoric, which set a tone of later advertising for other new technologies such as for Microsoft and mobile phones.

These studies underscore the fact that the mobile is presented to the public via this medium as an embodiment of youth, modernism and futurism. They indicate a consistent presence of "modern", "cutting-edge" and "futuristic" themes in the way sophisticated consumer technologies are represented to various national cultures. Were these themes lacking in appeal to consumers, it is implausible that marketers would continue to spend lavishly on them. In other words, modernity resonates in the minds of the consumer and therefore sells. What continues to be sold is what is continued to be designed. Contrarily, despite ad blitzes, if something does not catch on with the public, it disappears. (This situation is exemplified by aggressively merchandized products which, despite being highly promoted, have failed. Examples include "new Coca-Cola" and the thick diet beverage "Metrical".) Indeed, a significant part of the argument is that the ads do not consistently trick or force people into thinking they want something as much as they do as capitalize upon latent desires or suggest new ones. However, the motive has to appeal to the consumer and serves the consumer interests. This line of reasoning is contrary to those who say that such ads are hegemonic moves by media/advertising industries to foster consumerism and certain social values. People seem to like these themes on their own.

Nonetheless, the industry has also sought to be sure that the public would understand the technology to be of high status and socially desirable. Nokia provides a good example of this process as it participates in the production of numerous Hollywood films and programs. The futuristic image was intentionally supplemented with one of status. Thus, as part of a product placement scheme, Nokia phones appeared on popular TV shows (*Beverly Hills 90210*; *Friends*) and were distributed gratis to Hollywood stars in an effort to build cachet (Oehmke,1999). Nokia paid handsomely to have a special model shown off in a James Bond film, and also in the first *The Matrix* film (which featured the Nokia 8110). Additionally, Nokia has paid for product placements in the films *Charlie's Angels* and *Minority Report* and also TV shows such as *The X-Files* and *The Sopranos* (Snellman, 2003). Said a Nokia marketing vice president, "If Tom Cruise transmits crucial evidence with his Nokia phone, many viewers may realize they need a camera phone too" (Snellman, 2003).

Nokia is not the only mobile phone manufacturer which is aware of the importance of influencing public images of the mobile phone. Observing Nokia's success, Sony Ericsson was able to displace Nokia in a subsequent Bond film, *Die Another Day*, while Samsung was the mobile phone star in the sequel *Matrix Reloaded*. For one cycle of the Emmy's, Nokia sent their newest mobile phone to all the Emmy award winners. These efforts have paid off in brand status and recognition. Interbrand, a consulting firm, says that its research shows that Nokia is the 11th most recognized brand, and that it even ranks ahead of Mercedes-Benz. Said one Interbrand official: Nokia is "selling an image, not technology ... They're very good at technology, but image is the key" (Hafner, 1999).

So, up to this point, we have argued that the mobile phone has been presented to the public as a modern and image enhancing technology. It is not only presented as modern and cutting-edge, however. As we argue in the next section, the mobile phone has also been marketed as a high-fashion item.

5.5 Fashion Imaging in Contemporary Mobile Phone Ads

The whole issue of jewelry and fashion is, of course, heavily laden with gender-specific connotations. The topic is worthy of separate treatment and we can only refer the reader to some relevant publications at this juncture (Rakow, 1992; Katz, 1999; David, 2004). However, here we can note that the active but feminine and always youthful girl is an iconic image for the promotion of mobile phones.

The emphasis on stylish design, elite status and fashion appears to have been a central part of marketing. There is no sign of the trend abating in terms of marketing mobiles as exclusive fashion and trend-setting style items. A press release of 1 July 2004 from Motorola claims that "New York City Elite Flip Over the New Motorola A630". The release bubbles on: "Staged on the rooftop of New York City's exclusive Hotel, top celebrities and style leaders were among the first to witness the unveiling of the next 'must have' mobile device from Motorola …" (Motorola, 2004).

Three recent mobile phone ads also underscore the continuing importance of fashion in the public presentation of the technology. These ads do not constitute a representative sample, but rather are chosen as illustrations; nonetheless, in our subjective judgment they are typical of the ads we have encountered.

The first photograph (Figure 5.1) is set in Budapest airport, taken in June 2004. It shows a pedestrian passing in front of a lit billboard-style ad. The ad depicts two high-fashion items: a luxury watch and a designer purse, accompanied by a Siemens phone. The equivalence is clear. As the photograph was being taken, a stylish lady, engaged with her own mobile phone, passed in front of the billboard ad and is captured in the foreground.

Another image in that emphasizes the fashion display of the mobile phone is shown in Figure 5.2. This figure includes an ad that depicts a mobile phone handset displayed literally as a piece of body jewelry. In this case the mobile phone could be interpreted as a belly-button pendant or piercing. It would certainly seem that the advertiser (in this case Motorola) is seeking to give its product an edgy and ultra-hip image consonant with current fashion. The possible appeal of body piercing as a form of risk and youthful status, which in this case is associated with mobile phone use, should not go unnoticed.

Figure 5.1 Mobile phone handset ad in Budapest airport, 2004. Source: collection of the author.

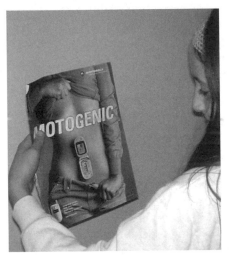

Figure 5.2 An individual examining a mobile phone handset ad, 2004. Source: collection of the author.

The larger visual environment can be substantially saturated by mobile phone ads. This seems particularly true in terms of airports, as suggested previously in Figure 5.1. However, the lateral space of an airport can provide a unique environment within which to promote mobile phones. For instance, the long corridors can be used to project extended images of mobile phones as statements of fashion and lifestyle. This is depicted in Figure 5.3, taken in July 2004. It shows a series of ads in Schiphol airport (The Netherlands). The ad depicts a group of girls on a beach sun

Figure 5.3 Mobile phone network ad displayed at Schiphol airport (The Netherlands), July 2004.

bathing; they would seem to be active, hip, daring and hedonistic. The linking of their phones and bodies emphasizes the social and clique nature of a mobile-leisure lifestyle. The close collocation of the bodies suggests far greater intimacy than is generally observed in Western Europe and North America (Goffman, 1974, 1979).

Although by no means definitive evidence, nonetheless taken together these recent images provide reasonable grounds to believe that fashion and body display are important aspects in the depictions by advertisers to sell mobile phones.

Fashion remains, then, an important aspect to mobile phone design and merchandizing. Yet not all attempts lead to success. To illustrate, we can cite IBM's effort in early 2000 wherein it created a line of digital jewelry including at one point earrings containing tiny mobile phones; despite the attractiveness of the fashion jewelry, it did not function well and was abandoned (Kharif, 2002). This example suggests that consumers need more than commercially driven appeals to fashion alone to be persuaded to buy a mobile phone.

Next we turn to the reception of the mobile phone by consumers, highlighting the perceptions of users and the way they modify the technology to serve as identity objects.

5.6 Consumer Perception and Reception

How does the public perceive mobile communication technology? There is some evidence that the public does evaluate it in terms of fashion and status, i.e. in the same way that the mobile phone industry also thinks the public perceives it. (The reader is reminded that it is the case that there is a reciprocal process of negotiating meanings between an industry that is seeking to frame the technology and the public that responds to, adopts, and modifies further the technology.) The importance of fashion in the consumer mind is suggested by some indicators based on our focus group interview and some of our surveys.

In 2001, we conducted a focus group interview with college students at a US university. Most of the participants said that they did not think of the mobile phone as a fashion item at first. However, a college student from Korea said that, "For our culture, cell phone is a part of the fashion thing. Yes, especially for younger generation, it kind of tells your personality, and it tells many things." Interestingly, this statement prompted other participants to reconsider their original response, and they started talking about how young people use the mobile phone for fashion. A female participant said, "for younger generation, they don't need cell phone, because they are not in college. High school, middle school … because I see younger kids having cell phone. I think they think it fashion." And another participant followed by saying, "It's changing. People are switching their phones … I don't' know, I think some people are trying to be cool about it, I think."

We observed some ambivalent feeling about how to understand the mobile phone and fashion in this focus group. They sense some association between the growingly ubiquitous mobile phone around them and fashion, but the association did not seem to be so clear in their mind. There seemed to be a kind of "third-person effect" operating. So that whereas they see themselves using the mobile phone for necessity, they see others as having a style dimension to their evaluations. This suggests that the mobile phone, in young people's discourse, takes on the role of a fashion accessory that is in great demand as a status symbol. A national survey taken in 2004 of youngsters in the USA concluded that for 8–10-year-olds the mobile phone is "as much a status symbol as a communications device". (Selingo, 2004).

In 2002, we explored the relationship between the timing of the mobile phone adoption and the importance of the aesthetic dimension of the phone (Katz and Sugiyama, to be published). The results indicated that American youths who adopted mobile phones earlier were more likely to think that the style of the mobile phone would be an important factor in selecting their own mobile phone. A similar trend was found in the sample group of Japanese youths (Katz and Sugiyama, to be published). In addition, our research showed that both US and Japanese heavy mobile phone users valued style more relative to non/light users. Moreover, Japanese heavy users even preferred style over battery life. If we consider battery life of the mobile phone as a functional aspect and style as an aesthetic aspect, our research seems to suggest that many buyers, especially early adopters and heavy users among youth, trade off functionality willingly for attractive styling. (Of course the mobile phone is not unique in this regard, since many people choose aesthetic or fashion appeal over functionality in many areas, including in their personal relationships, and even political candidate selection, but the point is that there is an affective as well as instrumental motive at work here.)

In a 2004 poll of our own, we surveyed a class of Rutgers undergraduate students about their attitudes towards mobile communication and fashion. The class was for non-technologists in the subject of information and technology and most students were about 20 years old; fortuitously, exactly 100 students (out of about 114 students attending that day) completed the survey, so percentages and number of respondents align exactly. Although, like the above surveys, the poll is not representative of all Rutgers students, it may shed some light on the relative frequency and intensity of attitudes towards mobile phone fashion. As indicated in Figure 5.4, over half of the students agreed with the statement that "my mobile phone should look cool" and of those about half (i.e. 25) also indicated that they notice the fashionableness of the mobile phones of their friends. (The category of "agree" and "strongly agree" were collapsed for the purposes of this analysis.) In other words, about one-quarter of the

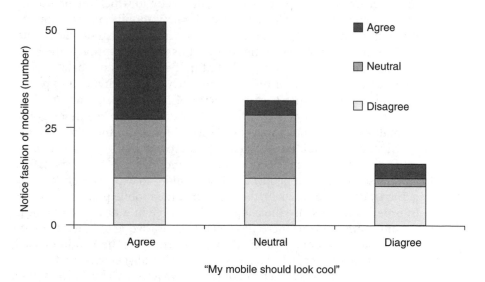

Figure 5.4 Fashion attitudes towards mobiles. $N = 100$ Rutgers students, April 2004.

students surveyed appear to be actively engaged in a fashion assessment of mobile phones. This would seem to be a striking figure, although it bears repeating that this is not a representative sample and cannot tell us to what extent, if any, the campaigns of the mobile phone industry has influence on the perception and attitudes of the students. Nonetheless, it is noteworthy that there is a similarity between the way in which the industry has been promoting the mobile phone and the way in which young users perceive it. At the other end of the spectrum, however, we must mention that 10 students disagreed with the statements, apparently rejecting interest in the fashionableness of the mobile phones of both themselves and their friends. Of course, we never meant to suggest that all users view the mobile phone as a fashion or status symbols, which they clearly do not. Rather, it is the case that many do, and that this perception is important for the commercial success of companies and also the use of public space and the condition of individual lives. We turn our attention to this topic next.

5.7 Promotion of Luxury and User Enhancement

Buyers of mobile phones may be clustered into two broad categories. First are those who purchase one simply as a communication tool, claiming to care little about its appearance or symbolism. Although immensely important to the overall market, this cluster is not the focus of this chapter. Second are those who buy one in part because of the status that a design, logo or brand imparts. They are the group to which the preceding discussion concerning style is addressed. Many mobile phone adopters seek to individualize them, personalize them and integrate them into their own local cultural meaning. Within this second approach, the concept of having the device be a symbol of individuality is important. "Our primary concern is to tailor products as much to the individual as possible", according to Alastair Curtis, director of Nokia's design group. "The phone is an extension of your identity" (Swartz, 2003). People respond, often making a conscious choice of the style of their mobile phone (Ling, 2003; Oksman and Rautiainen, 2003). This process has been investigated in other domains in terms of consumerism and identity (Massaris, 1997). Just in the same way that people employ "fashion" to express their identity (e.g. Davis, 1992; Crane, 2000), they consume the mobile phone.

In order to underscore the element of prestige in a technology that is becoming omnipresent, there has been a figurative arms race towards ever more lavish mobile phones. One approach, from the London-based boutique company Vertu, has been the marketing of specially made high-end phones, which are also a form of expensive jewelry. This fashion item has a platinum casing and a sapphire crystal screen. (It is also designed so that the phone's internal technology can be easily updated.) The price is about $26,000. It is worth noting in this context that Vertu was formed by Nokia in 2002, clearly a part of its fashion–luxury initiative. Motorola recently launched a clamshell V600 model that offers interchangeable covers studded with clear Swarovski crystals, and offers a variety of fashion handbags in which to carry its phones. Nokia's recently introduced 7200 model offers fabric covers that have analysts calling it the Louis Vuitton phone. While the emphasis on luxury is not a guarantee of success, it nonetheless shows that at least the handset manufacturers are persuaded that such an approach will appeal to an important market segment.

The phenomenon of transforming the mobile phone into high-fashion jewelry is also observable in other parts of the world. Gem-encrusted handsets have become extremely popular in China, which is experiencing rapid growth of mobile phone use. In 2000, TCL Mobile began offering diamond-studded mobile phones. TCL Mobile sold more than 12 million jeweled phones between 2001 and mid-2003 (most of which had fake gems) (Reuters Singapore, 2004).

"In Asia, phones are much more of an aspirational statement about who you are and who you want to be," Scott Durchslag, a Motorola corporate vice president said. TCL Mobile's managing director, Wan Ming Jian, asserts that in Asia, "attaching jewelry on the phone adds a cultural and spiritual dimension to the product." In further describing the popularity of jeweled phones, Mr Wan said that, "to many Chinese, precious stones symbolize esteem, good fortune, peace and love. So jeweled mobile phones are not just communication tools, they also act as lucky charms" (Reuters Singapore, 2004). This reinforces the importance of the cultural setting in considering the meaning of the mobile phone.

Although certain styles of the mobile phone are associated with high-fashion and prestigious brands, of course anything done by style leaders is subject to co-optation by the hoi polloi, thus diluting the brand's status value (which is always a challenge to fashion as Veblen noted). Nowadays it is common in department stores to see jewelry for adorning a mobile phone. (Notably, these items are offered not in the electronics section but in the jewelry section.) Street fairs too are a venue for mobile phone enhancements. As popular as these items are in the USA, they are even more so in Korea and other Asian countries.

Clearly, a growing segment of the public purchases these ready-made, futuristic-looking devices but then personalizes them. That is, they alter the appearance of the device to make it more individually meaningful and symbolic within their cultural contexts. Pasted-on photos, colored plates, dangling antenna ornaments and fake jewels are common "after-market" enhancements added by users. "With the colors, symbols, patterns and brands people choose, they are associating themselves with the meaning conveyed by them," according to Chris Conley, an assistant professor of design at the Illinois Institute of Technology (Swett, 2002).

Again, it is worthwhile reminding the reader that the mobile phone is not unique in this regard, since other consumer items are also enhanced and customized by users, including bicycles, jeans and backpacks. Still, the extent and meaningfulness of personalization, and also the substantial outlays of money and time to create the mobile enhancement culture, are potentially a worthwhile topic of investigation.

On the one hand, artisans and crafts people create their own enhancements. For example, one of our students brought from her trip to Namibia a hand-carved Sony-brand mobile phone which was purchased at an outdoor market in Namibia – it was a replica of a cell phone hand-carved from wood (and made before the Sony–Ericsson partnership, so there was no Sony branded mobile phone yet in existence). Was it a toy? Perhaps. Certainly other commercial enterprises have used the mobile phone as toy for babies and young children, promoting its "cool" image.

There is much collateral exploitation of the mobile phone from other quarters. For instance, a fast-selling fashion item in the USA among early teenaged girls is a make-up kit styled to look like a sleek, small mobile phone. It is realistic looking, and without careful examination could readily fool any casual observer. The kit is shown in Figure 5.5.

In order to use the mobile phone as enhancement of the self-image, how the mobile phone is carried and displayed becomes an issue. As a result, new opportunities are created in terms of fashion and display to carry and put mobile phones into operation. These include phone devices that flash brightly when in use to signal to ambient others that the device owner is important and/or connected enough to be talking to a distant person.

The dependence and ease of use exigencies concerning the mobile phone encourage creative solutions. Clothes and backpacks have been designed to integrate conveniently the mobile phone into what we wear. Digithongs (www.digithongs.com) offers around-the-body carrier straps and describes itself as an "innovative line of cellular phone accessories created to meet the needs of women". More massive than a thong is the "Scott eVest", which is designed to hold electronic equipment in numer-

(a)

(b)

Figure 5.5 Eye-shadow makeup kit made to look like a small mobile phone, USA, May 2004. (a) Kit with Motorola V60 phone; (b) kit in open position. Source: collection of the author.

ous zip, flap and Velcro pockets. It even has additional pockets inside some of the pockets. For headphone wires there are hidden tunnels within the clothing. Another firm, Dockers, makes the "Mobile Pant" (Newman and Wendland, 2002).

As users load themselves down with more gadgets, specialized harnesses have been created. An example of one of these, which seems to combine geek and chic, is shown in Figure 5.6. One of the authors has observed more than a half-dozen of these harnesses in the New York metropolitan area. Yet despite the plentitude of clothing enhancements and accessories, many users rely on their own creativity rather than commercial products to park their phones. Indeed, as the mobile phone becomes more commonplace, users are finding ever more convenient places on the body to park their mobile devices. A rather casual approach to mobile phone placement is illustrated in Figure 5.7, which is a photograph taken on a public tram in New Orleans (USA), May 2004.

Just like other fashion items, the mobile phone has become an aesthetic object that people adopt and modify according to their sense of self and group affiliation. Like other fashion items, they use the device to project a sense of identity and self into public arenas. However, in order to perform self-presentation with the mobile phone appropriate to the particular culture or social group, users need to be keen to the cultural meanings of the mobile phone, especially in an age that some marketers have dubbed "brand-morphing." ("Brand-morphing" refers to the way in which meanings of brands change across various social or cultural groups; Kates and Goh, 2003). The data we presented here suggest that not only meanings of certain brands but also the meaning of certain mobile phone designs are

Figure 5.6   E-belt wearing customer in office supply store, May 2004. Source: collection of the author.

Figure 5.7   Example of casual display of mobile phone. Source: collection of the author.

in the process of "morphing". At least to a group of people who are conscious about the style of the mobile phone, the phone is not a mere tool for convenience, but an expression of identity (Katz and Aakhus, 2002). Interestingly, consumers of the mobile phone do not seem to be happy with merely adopting the culturally appropriated meanings presented by the industry. They themselves attempt to "morph" the meanings of the mobile phone in various creative manners. Phones with expensive jewelries or the models that appear in popular movies are not the only "fashionable" phones. Pasted-on photographs, antenna ornaments, and an "ordinary" phone displayed on the body in a certain way could also become cool and fashionable. "Mobile phone morphing" occurs both by the well-calculated strategies of the industry and by the creative mind of consumers. It also occurs by the resourceful intervention of third parties or "after-market" manufacturers. Many examples may be seen in shops and markets around the world. Among the many examples in our own collection is a mobile phone earphone/microphone combination featuring a smiling kitty reflecting cuteness reminiscent of the "Hello Kitty" phenomenon (Belson and Bremner, 2003).

So, taken as a whole, we believe the evidence is strong that the creation and consumption of mobile phones becomes a multi-party process.

However, it should be abundantly clear that the style dimension is enormously important in the way the mobile phone is understood by users and audiences alike.

5.8 Conclusion

In this chapter, we have discussed the fashion forms and personalization of the image of the mobile phone. What we have sought to add with this analysis is a more precise connection between an industrial ethos (marketing a futuristic status symbol) and the popular reception and co-creation of a communication technology. Moreover, we have also sought to highlight the folk artistic aesthetic and uses of what has become an exceedingly important and socially significant device. We have shown, from a folk culture dimension, a variety of artistic endeavors to co-create a device. In this way it becomes not only a communication tool but also (depending on circumstances) a status symbol and individual value statement. It is every bit as much of a fashion statement as the choice of one's clothes. And equally as the choice of clothes could include secular as well as sacred garb, so too the "fashion" statement of a mobile phone can be used for purposes ranging from the temporal to the transcendental.

This analysis is at variance with theories of mass society, and of cultural studies of oppression, both of which traditionally tend to suggest that the consumer is a passive cow, milked by large corporations. We find instead a rich weaving of adoption, modification and embroidery.

Of course we certainly understand that the mobile phone *qua* decorative endeavor is not a universal impulse; many users just accept the mobile phone as an "off-the-shelf" item. They find the phone of neither inherent interest nor intrinsic beauty. They also refrain from use and display when others are about. However, although these types of people do exist, there also appear to be vast numbers of mobile phone users who are heavily fashion conscious. Many users invest the communication object with myriad personal decorations and also personal significance.

We do not have broad-based statistical evidence as to the prevalence and consequence of these outlooks, although we hope that these data will be forthcoming. Rather, here we have presented initial and exploratory evidence to suggest another image: consumers find an exquisite technology, which fits extremely well with their values and interests, and that they adopt in droves. Nevertheless, users are more than mere consumers. They are also co-creators. They achieve this status by, after purchase, further manipulating these devices to reflect personal tastes and to represent themselves to the outside world. As in the case of audience reception theory, mobile phone users are like audiences of various mass media texts, creating, interpreting, appropriating material, to develop meaningful, personalized and culturally appropriate new texts. As such, the technology is present in both front-stage and back-stage social processes. The mobile phone then may be seen not only as a "necessary accessory" to the body, as Fortunati (2002, p. 58) has argued. It also becomes a communication device that reflects and embodies the user, and is used to communicate effectively with the physically present audience, passive though it may be, as much as the distant interlocutor, as ethereal as it may be.

5.9 Acknowledgments

We thank Ronald E. Rice, Dafna Lemish and Richard S. Ling for their helpful comments on earlier drafts.

5.10 References

Belson, K. and Bremner, B. (2003) *Hello Kitty: the Remarkable Story of Sanrio and the Billion Dollar Feline Phenomenon*. Wiley, New York.

Bormanis, A. (undated) Features: an interview with Andre Bormanis. Startrek.com. Retrieved on 10 July 2004 from www.startrek.com/startrek/view/ features/firstperson/article/3840.html

Carey, J.W. (1993) Everything that rises must diverge. In Guard, P. (ed.), *Agendas: New Directions in Communications Research*. Greenwood Press, Westport, CT, pp. 171–184.

Crane, D. (2000) *Fashion and its Social Agendas: Class, Gender, and Identity in Clothing*. University of Chicago Press, Chicago.

David, K. (2004) Gender and mobile phones: a cultural investigation. Presented at the New Jersey Communication Association Annual Meeting, April, New Brunswick, NJ.

Davis F (1992) *Fashion, Culture, and Identity*. University of Chicago Press, Chicago.

Everdell, W.R. (1997). *The First Moderns*. University of Chicago Press, Chicago.

Fortunati, L. (1993) *Gli Italiani al Telefono*. Angeli, Milan.

Fortunati, L. (2002) Italy: stereotypes, true and false. In Katz, J.E. and Aakhus, M.A. (eds), *Perpetual Contact: Mobile Communication, Private Talk, Public Performance*. Cambridge University Press, Cambridge.

Fortunati, L., Katz, J.E. and Riccini, R. (eds) (2003) *Mediating the Human Body, Technology, Communication, and Fashion*. Lawrence Erlbaum, Mahwah, NJ.

Goffman, E. (1974) *Frame Analysis: an Essay on the Organization of Experience*. Harper, New York.

Goffman, E. (1979) *Gender Advertisements*. Harper, New York.

Haddon, L. (2003) Domestication and mobile telephony, in Katz, J. (ed.), *Machines that Become Us: the Social Context of Personal Communication Technology*. Transaction Publishers, New Brunswick, NJ, pp. 43–56.

Hafner, K. (1999) A designer who sets a worldwide standard for technophiles, *New York Times*, 9 December, p. G.1. Retrieved on 29 April 2004 from ProQuest electronic retrieval service, document ID 46995192.

Himmelstein, H. (1994) *Television Myth and the American Mind*, 2nd edn. Praeger, Westport, CT.

Kates, S.M. and Goh, C. (2003) Brand morphing: implications for advertising theory and practice. *Journal of Advertising*, 32: 59–69.

Katz, J.E. (1999) *Connections: Social and Cultural Studies of the Telephone in American Life*. Transaction Publishers, New Brunswick, NJ.

Katz, J.E. (ed.) (2003) *Machines that Become Us: the Social Context of Personal Communication*. Transaction Publishers, New Brunswick, NJ.

Katz, J.E. and Aakhus, M. (eds) (2002). *Perpetual Contact: Mobile Communication, Private Talk, Public Performance*. Cambridge University Press, Cambridge.

Katz, J.E. and Sugiyama, S. (to be published) Mobile phones as fashion statements: evidence from student surveys in the US and Japan. *New Media and Society*.

Kharif, O. (2002) *Business Week Online*, 10 April 2002. Ebsco Accession Number 6469884. Retrieved on 4 May 2004 from Ebsco database.

Leung, L. and Wei, R. (1999) The gratifications of pager use: sociability, information-seeking, entertainment, utility, and fashion and status. *Telematics and Informatics*, 15, 253–264.

Leung, L. and Wei, R. (2000) More than just talk on the move: uses and gratification of cellular phone. *Journalism and Mass Communication Quarterly*, 77, 308–320.

Ling, R.S. (2003) Fashion and vulgarity in the adoption of the mobile telephone among teens in Norway. In Fortunati, L., Katz, J.E. and Riccini, R. (eds), *Mediating the Human Body: Technology, Communication and Fashion*. Lawrence Erlbaum, Mahwah, NJ, pp. 93–102.

Lobet-Maris, C. (2003) Mobile phone tribes: youth and social identity. In Fortunati, L., Katz, J.E. and Riccini, R. (eds), *Mediating the Human Body: Technology, Communication and Fashion*. Lawrence Erlbaum, Mahwah, NJ, pp. 87–92.

Marvin, C. (1986) *When Old Technologies Were New*. Oxford University Press, New York.

Maslow, A.H. (1954). *Personality and Motivation*. Harper, New York.

Massaris, P. (1997) *Visual Persuasion. The Role of Images in Advertising*, Sage, London.

Motorola (2004). New York City elite flip over the new Motorola A630. Retrieved on 30 September 2004 from www.motorola.com/mediacenter/ news/detail/0,,4413_3747_23,00.html

Newman, H. and Wendland, M. (2002) Las Vegas Electronics Show makes gadgets, gear fashion statement. Detroit Free Press, 10 January 2002. Retrieved 30 April 2004 from Ebsco, Accession Number 2W61527420999.

Oehmke, T. (1999) The hot phone. *New York Times*, 14 November 1999, p. SM74. Retrieved 30 April 2004 from ProQuest database.

Oksman, V. and Rautiainen, P. (2003) Extension of the hand: children's and teenager's relationship with the mobile phone in Finland. In Fortunati, L., Katz, J.E. and Riccini, R. (eds), *Mediating the Human Body: Technology, Communication and Fashion*. Lawrence Erlbaum, Mahwah, NJ, pp. 103–112.

Pajnik, M. and Lesjak-Tušek, P. (2002) Observing discourses of advertising: Mobitel's interpellation of potential consumers. *Journal of Communication Inquiry*, 26: 277–299.

Pedersen, P.E. (2005) Instrumentality challenged: the adoption of a mobile parking service. This volume, Chapter 25.

Rakow, L. (1992) *Gender on the Line: Women, the Telephone, and Community Life*. University of Illinois Press, Champaign, IL.

Reuters Singapore (2004) Diamante mobile phones anyone? *Taipei Times*, 20 April 2004, p 16. Retrieved on 27 April 2004 from http://www.taipeitimes.com/News/feat/archives/2004/04/20/2003137447

Selingo, J. (2004) The cellphone: a new joy of childhood. *International Herald Tribune*, 20 March 2004, p 18. Retrieved on 10 April 2004 from Gale InfoTrac.

Simmel, G. (1957) [1904] Fashion. *International Quarterly*, 10(1), October, 130–155; reprinted in *American Journal of Sociology*, 1957, 62(6), 541–558.

Snellman, S. (2003) A movie can be the best commercial. *Helsingin Sanomat, Business and Finance*, 4 February 2003. Retrieved on 17 October 2003 from http://www.helsinki-hs.net/news.asp?id=20030204IE2

Steele, V. (1997) *Fifty Years of Fashion: New Look to Now*. Yale University Press, New Haven, CT.

Stein, S.R. (2002) The "1984" Macintosh ad: cinematic icons and constitutive rhetoric in the launch of a new machine. *Quarterly Journal of Speech*, 88, 169–192.

Swartz, K. (2003) Style ranks high in cellphone design. *Knight Ridder/Tribune Business News*, 10 November 2003. Retrieved on 27 April 2004 from InfoTrac database, ITEM03314009.

Swett, C. (2002) Cell-phone accessories market booms as devices become fashion statements. *The Sacramento Bee (CA)*, 1 September 2002. Retrieved on 29 April 2004 from EBSCO host, Accession Number 2W62983392961.

Veblen, T. (1934) [1899] *The Theory of the Leisure Class*. Modern Library, Random House, New York.

Wei, R. and Lo, V. (2003) Staying connected while on the move: cell phone use and social connectedness. Paper presented at the 53rd annual conference of International Communication Association, San Diego, CA, 25–28 May.

Zhang, Y.B. and Harwood, J. (2004) Modernization and tradition in an age of globalization: Cultural values in Chinese television commercials. *Journal of Communication*, 54, 156–172.

Part 2

The Public and Private Spaces

Introduction

Alex Taylor

This part of the book presents seven chapters that address the interplay between space and mobile communications. Viewed in detail, each of the chapters deals with the notion of space differently, sometimes implicitly and sometimes explicitly. In some respects, however, they might be seen broadly to tackle their subject matter from a similar perspective, namely in terms of the public and private delineations of space and the phone's role in (re)negotiating this two-part division. This area of study has already received a good deal of attention in mobile phone-related research (e.g. Cooper, 2001; de Gournay, 2002; Green, 2002) and also in the broader sociology and sociology of technology literatures (e.g. Urry, 1995; Harrison and Dourish, 1996; Curry, 2002), so it is of no surprise to see this as a salient topic in the work presented in this volume.

The chapters from Tom Julsrud and Fernando Paragas, for example, deal directly with the distinctions between public and private space and how they interrelate with mobile phone use. In his chapter on the mobile workplace, Julsrud considers how changes in modern office settings have transformed the way private and public spaces, or "territories", are constituted, managed and coordinated. Whilst highlighting the importance of organizational arrangements such as the movement towards open-plan offices and "hot-desking", he emphasizes the crucial role mobile and wireless forms of communication have had in such transformations. Although in a very different context, Paragas deals with a number of similar issues in his fascinating account of mobile phone use on public transport in the Philippines. He argues that the various forms of transport available in urban Manila, varying in safety and overcrowding, shape the "private" or "public" methods for handling phone calls and messages.

The five remaining chapters in this part handle the public and private character of space in more subtle ways. In both Mimi Ito's and my own chapters, distinctions are made between the "private" spaces that are mutually produced through young people's collective communications and the "public" spaces that either explicitly or implicitly come about through the exclusion of others – be they peers, parents, bystanders, etc. The physical setting, the mobile as a material object and young people's social arrangements are all shown to imbricate and to "occasion" public and/or private senses of space. Addressing the timely and hitherto under-examined role of picture messaging, Carole Rivière's chapter is yet further removed from any immediate analysis of space. However, her reading of the post-modern corpus and the work she does to relate it to the ephemeral character of phone-based, picture messaging casts the phone as a resource that enables temporary bonds, uniting distributed parties into an apparent "third" space of privacy when in public.

Dafna Lemish and Akiba Cohen and Leopoldina Fortunati have a different take on space, although both groups remain interested in the Goffmanesque sense of public performances. Lemish and Cohen examine the relationship between the mobile phone and identity in Israel. Key to their analysis is the consideration of the ways in which the mobile phone is enlisted in the "performance" of identity. They describe the demonstrable qualities associated with the phone and how Israelis construct various discourses around phone use in public – framing their identities in terms of gender, age and ethnicity. Fortunati is likewise interested in the enactment of one's social character through the mobile phone. She suggests that public phone-talk can alter people's efforts at self-presentation, offering something of their *back-stage* characters whilst located on the publicly accountable *front-stage*. She also refers to the tension between the presentation of self made in public physical space and that made through talk over the phone.

6.1 Making Space

Given that the following chapters are related – in so far as they deal with various interpretations and representations of private and public space – it might seem reasonable in this Introduction to present a more detailed overview of how the public and private are rendered as discernible in the studied contexts, e.g. in offices, on public transport and among gatherings of young people. To take an alternative and, it is hoped, more interesting route, however, I have chosen to examine the ways in which each of the authors has carved out their own versions of public/private space through their choice of analytical method, as much as through their examination of data. In taking this route, the implication is that by drawing on particular analytical orientations, the chapters' authors have rendered space in such a way that it is amenable not simply to the empirical data to which they refer, but also to the analytical resources they bring to bear. I tread carefully here, to imply not that the following chapters have imposed some theoretical constructs that are removed from real-world practice. Rather, I wish to raise an awareness of the interpretative character of the work that follows and to reflect on this interpretive work in the tradition of reflexivity that is central to modern sociological modes of inquiry.

Let me begin this examination by reflecting on my own chapter in this section, where I present a study of teenagers and how their phone is used in the management and organization of ordinary forms of talk. Through my use of particular analytical resources, drawing specifically on an ethnomethodologically inspired orientation, space is revealed to be something that is practically accomplished on a moment-by-moment basis. Taking a reflective position here, it becomes apparent that I have configured the distinctions in space by deciding where the action is (and is not). Although I do not use these terms, my "private space" is where the mobile phone is being used and the public space is assigned to the onlookers and bystanders to that action. I argue that the mobile phone, through systematic and coordinated action, can be used to draw a division between those who are party to *phone-talk* and those outside of it. Indeed, I go as far as to suggest the phone allows for the *subversion* of the recognized order of a conversation, or what is referred to as a localized system of order. My position thus orients space around specific centers of mobile phone-related activity, marking out public and private through my own observations of where phone talk gets done.

The central role that a work's analytical motivations has on formulations of space is further revealed in three chapters in this part that tackle the distinction between public and private in more concrete terms. All three chapters orient their analysis around the distinction that the sociologist Irving Goffman makes between public and private, *front-stage* and *back-stage*, and then, individually, go on to extend their work by drawing on others' theories.

Paragas, for example, enrols Goffman's theories on the presentation of self to consider how space is managed and negotiated on public and private transport in Manila. He suggests that those traveling in the city employ particular methods of phone use to demonstrate their engagement (or lack of it) with fellow passengers. His argument hinges on the premise that these methods are shaped by the different modes of transport available and, specifically, how safe they are felt to be. On a different tack, he also suggests that the potential for lying about one's location, when in earshot of other passengers, is moderated because talk on a mobile phone is accountable not only to the interlocutor, but also to bystanders who can overhear the conversation. As a result of these foci, we find that Paragas has viewed both phone users' and bystanders' behaviors in terms of their public performances. Space, for Paragas, is thus characterized by the extent to which safety and the level of accountability are demonstrably occasioned. Indeed, Shimanoff's work on rules is briefly invoked to suggest that there are nascent rules of etiquette and decorum, governing the ways in which these performances are orderly accomplished.

In her chapter on public phone use in Italy, Fortunati borrows on Goffman's theatrical metaphor and Simmel's accounts of the performativity of everyday action. Once again, the analytical frame becomes a useful resource to examine how her respondents have divided phone use into public performances, open to all, and those behaviors that are intended to remain private. As readers, though, we must be mindful of Fortunati's own participation in a separation business; her analytical work separates out her respondents' excerpts into a classification scheme to reinforce the combined Simmel–Goffman theory. Owing to the mobile's dual role in supporting both publicly available and private talk, tensions are shown to exist between the front- and back-stage presentations of self. Fortunati is to be credited for her recognition of and reflection on applying what she refers to as "the analytical categories of Simmel and Goffman". However, her brief epilogue calling for a re-articulation of the front-stage–back-stage metaphor does little more than shift and arguably blur the line drawn between public and private "stages". Fortunati's work thus serves to recast the two-part theatrical model, extending its empirical application whilst orienting her data to confirm and reinforce its theoretical basis.

The chapter from Julsrud similarly draws on Goffman's theories of self-presentation and dramaturgy. Julsrud's focus is grounded in the specific use of office space, where he suggests that the assembly of new workplace regimes, architectural styles and communication technologies have precipitated particular "information environments". He considers these environments in terms of the spaces that have been made available because of mobile communication technologies and, like Fortunati, separates space into front and back stages or "territories". Drawing on Meyrowitz's view of technology's role in (re)defining space, Julsrud further elucidates a taxonomy of socially constituted territories that are fluidly defined through a space's changing patterns of human relations and uses of information. This produces a parallel with Rivière's "third", communicative space, where the new forms of mobile communication are formative in establishing a "space" that sits between the purely physical, on the one hand, and the immaterial/interactional, on the other.

What is particularly interesting about Julsrud's analysis is the link he makes between territories and identity. He suggests that as the locations for communication in the workplace have evolved, so too has our means of self-presentation and thus our active production of identity. Again, we are presented with Goffman's theorizing of public performances in social space. Of interest here, however, is the manner in which identity is graphed on to such performative displays. Julsrud equates identity with self-presentation so that when an employee temporarily lays claim to a territory in an office, by using it to talk on their mobile phone, for example, they are unavoidably engaged in an expression of identity.

For the purposes of reflection, it is important to acknowledge Julsrud's choice to link performativity and identity because, as with the chapters discussed so far, his focus motivates one analytical stance over others. Let me illustrate this by suggesting that by paying heed to public expressions of identity, Julsrud – perhaps intentionally – produces a gloss that is one step removed from an understanding of the resources and methods people bring to bear to *do* "office life". What I would argue is lost sight of is just *how* people's ordinary, taken for granted methods for moving around in offices, talking on phones and so on, come to occasion a known about order to a situation. That is, by framing his analysis around identity and concerning himself with the demonstrations of rank, status, etc., Julsrud cannot help but see what people do *as* "performing" identity. What is lost in his analysis is how it is they matter-of-factly go about their business and are seen to do so regardless of any abstract notions of identity. To reiterate, my argument here is not with any failing in Justrud's analysis but, rather, one of recognition that his work privileges a particular analytical orientation that relies on inbuilt assumptions about the relations between human behavior, space, technology and identity.

Remaining focused on mobile phone use and its relations to space, but lessening the emphasis on Goffman's work, are the chapters from Rivère, Ito and Lemish and Cohen. These three chapters attempt to tackle what Button (1993) and others have referred to as the *grand themes* in sociological inquiry through their studies of mobile telephony. Their analytical orientations lead them to articulate interpretations of phone use that recapitulate a number of heavily theorized themes in the social science literatures.

For example, Rivère engages with the social constructivist discourse and post-modernism. She writes of how the "photographic act", performed with the mobile phone, allows for the transformation of one's usually personal and mundane experiences into a means of public self-expression. Her analysis focuses on the functional qualities of the phone, and specifically picture messaging. She views the integration of the camera with the phone as providing a new social function, where the transitory image – captured and stored for the moment – is immediately available as a resource to communicate and cement feelings of togetherness with remote others. Interleaved with this functional perspective is an attention to the symbolic character of the picture message and how this is constituted via the photographic act; the spontaneous act imbues the image with the affective qualities of playfulness, intimacy and so on. This all amounts to the mobile phone's role in constituting a new, "third" communicative space where the temporal and physical are fused and relocated, albeit temporarily, on to a phone's memory.

What is relevant, in Rivère's chapter, to my own argument is how her analytical position hinges on a post-modern discourse. She depicts space as something to be socially constructed through locally situated acts; the photographic act furnishes the re-articulation of space, suggesting that space is open to (re)negotiation – continually up for grabs. This position provides for her own playful dialogue with space, a dialogue in which her

89

use of the post-modern corpus and own data serve as tools to furnish one reading over another. Thus, Rivière sets herself a site for the articulation of an ontological proof. The mobile phone is drawn into a line of reasoning that is more than about its practical and routine use and comes, also, to stand in is as an anchor point in the post-modernist project.

Ito overlays the 'lens' of power-geometries to interpret how young people in urban Japan manage and indeed contest conventional constructions of public and private space. Relevant to my argument here is that Ito has inserted *power* into the readings of her data. This allows her to render the phone as a resource through which young Japanese can distinguish themselves from their peers, and others such as adults. Moreover, the phone is seen to provide a repertoire through which the category of 'being young' can be actively accomplished and used to challenge the authority in-built into what Ito refers to as "adult-controlled institutions and spaces of activity". Consequently, Ito's work provides an examination of the features of the phone and its relations to institutionalized space, and how these relations can enable such negotiations of power.

By adopting her orientation of power-geometries, Ito is able to present her findings in terms of the intergenerational hierarchies that are upheld and subverted through the interleaving of the mobile phone with the movement through space–time. Ito's chapter is a revealing account of how power relations may come into play, laid bare through the incorporation of a particular analytical orientation. This articulation of power is, however, not of her participants' making, but rather an old and well-rehearsed sociological (and anthropological) 'problem'. Whether the negotiations of power are of immediate relevance to young people's uses of the phone is debatable, but what is certain is that power is in no way a readily observable feature of routine phone use. The chapter is, in this sense, a reworking of one of sociology's *grand themes*, as much as it is about phone use amongst Japanese youth.

Lemish and Cohen are similarly drawn to such grand themes. In their study of phone use in Israel, they, like Ito, rearticulate some familiar sociological themes – this time, those of gender, age and ethnicity. Their treatment of the data brings these topics to the fore, by examining how Israelis visibly express their identities in their practical uses of mobile phones (and more particularly through their descriptions of others' phone-related behaviors). By eliciting the discourses of gender, age and ethnicity, however, their end product detaches itself from what is said to be the work's central concern: the practices of everyday life. Lemish and Cohen render the demonstrably features of people's everyday actions as somehow bound up with these glossed features of identity and, in doing so, de-situate action from what is made visibly and witnessably available. What their work does do is reveal the ways in which Israelis construct explanations of their identities vis-à-vis the mobile phone, suggesting, specifically, that their informants' post-hoc rationalizations incorporate the themes of gender, age and ethnicity. Adopting a reflective position allows us to be sensitive to the fact that there is no inherent property of phone use that privileges such explanations over and above the alternatives we see in this part of the book or elsewhere in mobile phone research.

In sum, I wish to reiterate that what I have sought to foreground are the ways in which the authors of each of the following chapters have chosen analytical orientations that shape the way in which we are able to see space being produced through mobile phone use. From the many possibilities available, the authors have intersected with and dissected their data sources in specific ways and thus, intentionally or not, favored one reading over others. My hope is that this Introduction stands as a reminder of the importance of reflexively interpreting the ways in which research is undertaken, encouraging a critical examination of the methods and procedures that we bring to bear in examining where the action is.

6.2 References

Button, G. (1993) The curious case of the vanishing technology. In Button, G. (ed.), *Technology in Working Order: Studies of Work, Interaction, and Technology*. Routledge, London, pp. 10–28.

Cooper, G. (2001) The mutable mobile: social theory in the wireless world. In Brown, B., Green, N. and Harper, R. (eds), *Wireless World: Social and Interactional Aspects of the Mobile Age*. Springer, New York, pp. 19–31.

Curry, M.R. (2002) Discursive displacement and the seminal abiguity of space and place. In Lievrouw, L. and Linvingstone, S. (eds), *The Handbook of New Media: Social Shaping and Consequences of ICT*. Sage, London, pp. 502–517.

de Gournay, C. (2002) Pretense of intimacy in France. In Katz, J.E. and Askhus, M.A. (eds), *Perpetual Contact: Mobile Communication, Private Talk, Public Performance*. Cambridge University Press, Cambridge, pp. 192–205.

Green, N. (2002) On the move: technology, mobility, and the mediation of social time and space. *The Information Society*, 18, 135–152.

Harrison, S. and Dourish, P. (1996) Re-place-ing space: the roles of place and space in collaborative systems. In *Proceedings of the ACM conference on Computer Supported Collaborative Work (CSCW)*, Boston, MA. ACM Press, pp. 67–76.

Urry, J. (1995) *Consuming Places*. Routledge, London.

7

Behavioral Changes at the Mobile Workplace: A Symbolic Interactionistic Approach

Tom Erik Julsrud

7.1 Introduction

> The next generation of cordless telephones may give everyone their own portable telephone to be used anywhere at affordable places. [...] More interestingly, a telephone will then belong to a person not a place. We will call a person and not know where they are. (Handy, 1991, p. 15)

During the last 10 years, mobile communication media have diffused into almost every corner of society. The professional users have, in this period, been eager to take advantage of the new technology to develop productivity or gain other strategic advances. In the initial phase of this development, the mobile media (including mobile telephones, portable PCs, pagers and Personal Digital Assistants or PDAs) functioned as a *supplement* to the traditional communication equipment, giving the user a new option for receiving and making calls or messages outside the office or the main workplace. The mobile phone was an extra phone mostly used in the car or on holidays and often shared by a family or a work group.

Today, we are entering a new phase in the utilization of mobile communication media among professional business users. Increasingly, modern organizations are exploiting mobile media and their infrastructure in a more strategic manner, developing work styles and office designs that are evolving around the possibilities of the new technology. The introduction of the *mobile workplace* is today the most distinct evidence of this change. In these new workplaces, the mobile telephone – like Charles Handy forecast 10 years ago – has largely outmoded the fixed telephone. For many office workers today their only communication tools are a mobile telephone, a laptop and an e-mail address.

The object of this chapter is to explore how this new mode of working contributes to changes in behavior in work style among office workers. In particular, it will focus on how changes in the use of individual territories might have consequences for how the daily work is conducted. Empirically, the study is based on a series of qualitative interviews and direct observations of managers and employees in an insurance company and in a telecoms company. Hence, the scope of this chapter is limited to discussing changes at the "office", including typical white–collar, knowledge workers. A general theoretical framework based on the work of sociologist Erving Goffman will be the lens through which these changes will be examined.

The chapter will start by taking a closer look at how the mobile telephone has penetrated work life during the last decade. It describes how a new kind of "territorial setting" has been introduced to knowledge workers, to a large degree based on mobile communication tools. It then goes on to describe, in a general way, the concepts of territories and roles at the workplace, and how new communication media challenge these. It will be argued that the new office architecture and communication technology might be studied as a general "information environment." It is then discussed how changes in the use of territories at the workplace can trigger new behavior patterns, presenting selected observations of knowledge workers in two companies. Finally, these is a discussion of how the mobile workplace might challenge general motives associated with the mobile workplace, such as efficiency, information sharing and knowledge development.

7.2 The Advent of the Mobile Workplace

The mobile telephone has, for a relatively long period, been popular among professionals in many business areas. The diffusion of professional mobile telephone users is indicated with a look at the number of employees who have their mobile telephone bills paid by their company. In businesses such as IT/Telecoms and building and construction in Norway this is today a common practice.

The interesting change that has happened during the last 2–3 years is that the mobile telephone seems to have *replaced* the traditional telephone (POT). This means that office workers in a number of businesses are equipped with only a mobile phone. This shift is in itself dramatic, since it implies that the office worker now has a sharply *increased availability*. A clerk working in a public services department equipped with a mobile telephone is in principle available whenever he or she has it switched on and is not occupied with another call. The situation might offer the individual employee a higher level of flexibility in his or her choice of workplace. However, this new availability might also result in

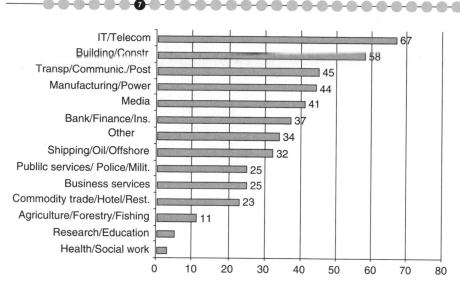

Figure 7.1 Percentage of Norwegian employees in various business segments who get their mobile telephone expenses covered by the company. (Estimates made by Telenor Mobile 2002.)

higher expected availability by the public and the employee's managers. The switch from fixed to mobile telephones is, however, made even more radical by the development of *new workplace regimes* evolving around the mobile worker. During the 1990s, there were discussions among architects and organizational developers about the mobile office. For instance, Harrison (1991, p. 76), in the early 1990s, held that:

> In the next decade communications technology will have its most dramatic impact in freeing many workers completely from their constraints. The free address workstation will become a technical reality with truly portable voice and data links ...

Relating it to the development of a network society, Castells (1997, p. 247) argued that the mobile office was emerging as a common feature of the networked economy:

> There might be a fourth phase of the office automation brewing up in the technological cauldrons of the last years of the century: the mobile office, performed by individual workers provided with portable, powerful information processing/transmitting devices. If it does develop, as seems likely, it will enhance the organizational logic I have described under the concept of the network enterprise ...

Today we have a clearer picture of what this actually means, as many companies have now implemented concepts of mobile workplaces. In

Norway, mobile workplaces have been introduced in many companies. Large numbers of companies in insurance, consultancy and technology development have transformed themselves – or parts of themselves – into mobile workplaces. Actually, it is hard to find any new building project in the business sector that *is not* implementing offices that include some form of mobile work practice. Indeed, a recent survey of managers in four major business segments indicates that open-plan offices are the most common type of workplace in 35% of Norwegian companies. The study also found that every third manager plans to make rearrangements at their workplace during the coming 2 years (DEKAR 2004).

However, there is no single concept that is being utilized to describe these different solutions. Some common characteristics are easy to spot, however:

1. Widespread use of open work-zones: most workers conduct their tasks in "open offices".
2. No fixed desks: the use of desk sharing is common in the open work-zones. Use of individual offices is an exception.
3. Increased use of functionally focused areas: small rooms for "quiet work" or private phone calls are common, and also special rooms for project work, brainstorming, etc.
4. Use of mobile technology: portable computers and telephones are used in tandem with wireless LANs and Bluetooth solutions.
5. Flexible work schedules: there is openness towards work at home, at the customers, in public places, etc.
6. Egalitarianism: managers use the same workplaces and technologies as the others.

These points do not, of course, cover every mobile office, but they reflect the most common characteristics, at least in a Norwegian and Nordic setting. In addition to the mere office plan, one should also note that the "workplace" includes several private and/or public places such as a home office, an airport lounge or a restaurant table. I will argue that this is a new way of designing the white-collar workplace that gives a new *workplace setting.*

The transformation of the white-collar workplace from a fixed to a mobile workplace is, of course, due to factors other than the diffusion of mobile telephones. The mobile workplace regime is motivated by a potential reduction of costs for housing and overheads, better contacts with customers and clients and enhanced methods for collaboration. However, the mobile telephone, together with advances in wireless infrastructure for data communication (local WLANs), is at the core of these changes, making the vision of a mobile office a reality. One should note that the mobile office is also a *virtual office*. Mobile communication media develop in interaction with the Internet and with other media. Hence the mobile workplace might have a "media environment" that is more or less tuned towards mobile or fixed media.

At a company policy level, these changes in the setting of the workplace are introduced under the guise of creating a more innovative work practice, developing knowledge work, etc. (Leaman, 1991; Becker and Steele, 1994; Duffy, 1997; Bjerrum and Bødtker, 2003). I will not hesitate to admit that these types of changes may occur. There is, however, a significant challenge for researchers to investigate if and document how these ambitious goals are fulfilled. This is a task that is often overlooked in the period after a new workplace has been introduced.

Taking up this thread, this chapter will discuss how the new workplace setting affects *behavior at the office*. I will not, however, focus directly on the parameters mentioned above, but rather explore the more unnoticed changes in expressive behavior following the introduction of a mobile office. To understand these changes, it is necessary to have a perspective on "place", space" and "territories" at the workplace.

7.3 Territories and Human Behavior

The core of the new mobile office is the increased mobility, inside and outside the office building. The reason for making the workers more mobile are, from an organizational point of view, two-sided: on the one hand, it is a goal to "speed up" the communication so that the circulation of information goes faster and new ideas are conceived. On the other hand, it is a goal to reduce the use of space to save costs and handle temporary employment in a more efficient manner (Leaman and Borden, 1993). In most cases these motives are intertwined in a general "policy" encouraging the mobile workplace as well as mobile work styles. An experienced freedom in the choice of where and when to work thus frequently goes hand in hand with reduced access to permanent individual rooms and possessions. This is not something that simply is stated as a new rule in the organization; it is manifested in the architecture, where there simply is less room for individual work Thus, one implication of the mobile office is that behavior that used to take place in a private setting – doing individual work tasks, talking with customers or family members on the phone, etc. – now must be conducted in "public areas". However, special rooms for private calls, project work, etc., are often implemented to counterbalance the lack of individual space. Still, these are not always available and it is necessary to plan these situations slightly in advance. Two important transformations in organizations introducing a mobile workplace should be noted:

1. It is a transformation from work in individual offices towards work in a larger group of employees.
2. It is a transformation from one preferred place of work to multiple possible locations, inside and outside the building.

At the core of this is a reorganization of the workers' expected use of territories. The reduction of private space is most often the point where the mobile office interferes with user expectations in many organizations (van Meel et al., 1997). The understanding of territories at a workplace, and how this is linked to behavior, is therefore critical.

7.3.1 Territories of the Self

The importance of physical space in social settings has been recognized by social scientists for sometime. For instance, the anthropologist E.T. Hall has distinguished between four individual spatial zones, which are changing between different cultures (Hall and Hall, 1990).

The connection between space and behavior is in sociology most elaborated by the so-called "symbolic interactionists", and in particular by Ervin Goffman (1956, 1961, 1971). From this perspective, the social situations are crucial factors to understanding social behavior. Moreover, understanding the situation is also crucial for the individuals to understand their own conception of themselves. Following Goffman, the frame of action for human behavior is the social setting, where people can play out their different *roles*. The analogy of a "stage" is used to describe the place where individuals have the chance to put into action their own character. Underlying the dramaturgical metaphor is the more general point that the self should be recognized as a social product. The self of an individual is what comes to life on the stage in different social interactions. The self depends on the making of a performance that can be socially supported in the context of a given status hierarchy (see Lemert and Branaman, 1977).

Territories are the setting where action takes place; they are the frames for the social situations that put constraints on how the individuals play their roles. However, they have an even more profound function in Goffman's work: territories make it possible for individuals to move between different settings and play out different roles. As pointed out by Fortunati in this volume (Chapter 13), many role performances rely on the segregation of a front and a back region. Whereas the *front region* is where a particular performance takes place, the *back region* is where the performance itself is rehearsed or discussed by members of the same social group or team. Front and back regions are thus different territories where one can express different roles.

According to the perspective outlined by Goffman, social life is a constant staging of social interactions – alone or together – aimed at giving a certain impression on an audience, which in turn has the option of accepting or rejecting what they see. The separation of social situations is crucial for the segregation between on-stage and off-stage, but also between "different stages".

However, for Goffman the individual also uses the territories in a more *ritualistic* way. In later work, his interest turns from the dramaturgical side of social interactions to studying how individuals constantly seem to be seeking to balance in social situations by using different kinds of "ritual behavior". The individuals are not only – as implied in the dramaturgical metaphor – striving to present themselves in a positive manner for a public, they are also striving to fit in with a particular social system. When people meet, there appears to be a common goal of affirming social order and individuals are marking out their own territories as well as respecting those of others in a manner consistent with their social position or rank. In *Relations in Public*, Goffman (1971) argues that "claims of territories" are one crucial way individuals use to maintain social order.[1] The following eight "territories of the self" are discussed: (1) personal space, (2) the stall, (3) use space, (4) the turn, (5) the sheath, (6) possessional territory, (7) information preserve and (8) conversational preserve.

Personal space resembles much of what has been written by Hall and others. This is "the space surrounding an individual, anywhere within which an entering other causes the individual to feel encroached upon, leading him to show displeasure or sometimes withdraw". *The stall* is a physical space where the individual can lay temporary claim. Often this is a scarce good, such as a table with a view in a restaurant. In a mobile workplace this would, of course, typically be the shared desk. *Use space* refers to space that is temporarily claimed by an individual, such as a space between a picture and a viewer at an exhibition. *The turn* refers to the right to gain access to a good in a particular sequential order, for instance in a queue. This is more the right to access a good than a physical space. *The sheath* refers to control of the body, but also the clothes. *Possessional territories* refer to "any set of objects that can be identified with the self and arrayed around the body whatever it is". This includes personal effects such as handbags, matches, jackets and gadgets (for instance, mobile telephones). *Information preserve* refers to control over information about the individual person accessible while in presence of others. Finally, *conversational preserve* refers to the possibilities to control who to talk to, and who to include in the conversation.

It is easy to see that this list of territories has been influenced by the work Goffman conducted at different institutions (mental hospital, jails, etc.). He was here concerned with how individuals under extreme

---

[1] Claims of territories are not the only practices described by Goffman in *Relations in Public*. He also describes remedial interchanges, supportive interchanges, "Tie-signs" and normal appearances.

circumstances were able to preserve a feeling of self-identity. When all the territories mentioned above were removed, the self was gradually "mortified". The general point Goffman makes, however, is that claim of territories is as central to social life as handshaking, small talk, interchange of glances and other kinds of "social rituals". Moreover, he stresses that these territories are socially determined. They may vary widely according to situation and people involved. But particularly the individuals' place in the social hierarchy is important: the higher the rank, the greater is the size of all territories and the greater the control over their boundaries.

For our purposes it is first interesting to see the difference that Goffman makes between *physical and abstract territories*. Individual territories are related not only to place, but also to personal gadgets and symbols in addition to access to information. As I will discuss later in the paper, it seems that the physical territorial claims may be on the decline, but other more abstract territorial claims may pop up instead. Goffman's taxonomy of territories offers an interesting framework to understanding these variations. Second, the way he treats territories as an integrated part of

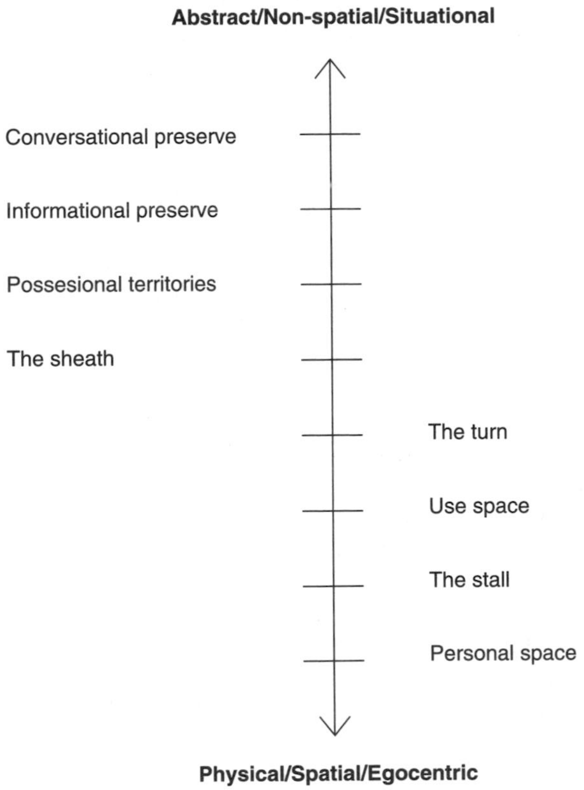

Figure 7.2 Goffman's eight territories of the self.

the social dynamics balancing a social organization suggests that if we manipulate these territories we will probably also intervene with basic social functions.

7.3.2 Physical and Virtual Environments

Goffmans perspective on the territories of the self gives us an idea of why change in territories at work might conflict with social roles and identity. Claims of physical territories – such as an individual office – is traditionally linked to self-identity and to status in an organizational hierarchy. The same goes for size of the office desk, keys to different areas of the building, etc. And people also use territories strategically to support the social order and to receive support for their own status. Following Goffman's theories, then, changing territorial practices might lead to individual uncertainty about place in the social system.

However, the office has not been left untouched by the "communication revolution". Since Goffman wrote his theories, a general theme in studies of organization and in communication has been how media change the understanding of place or territories (see, for instance, Giddens, 1990; Castells, 1997; Baumann, 2000). Joshua Meyrowitz (1986) and others have argued that electronic media influence the social interactions in leisure and work. Building explicitly on Goffman's situational analysis, Meyrowitz holds that in particular the television tears down the boundaries that surround traditional social settings or "situations". Basing his analysis on situational theory, and also medium theorists[2] such as Marshall McLuhan, Harold Innis and others, he argues that new electronic media have gradually changed the meaning of social situations. Electronic media increasingly destroy or alter the barriers between social situations and between front regions and back regions, he argues. The result is first of all a general *homogenization* of roles and behavior in many different areas of society. Many roles that used to be acted out in isolated front and back regions are currently being replaced by new *middle-region roles* that integrate former separate role behavior. However, the media might also divide existing social situations in different situations and develop *deeper back regions or more forward onstage behavior*[3] (Meyrowitz, 1986, p. 51).

Meyrowitz thus recognizes the importance of social situations for behavior. However, to get a clearer picture of the effect of the media, he

---

[2] In later works Meyrowitz positions his contributions as a prolongation of the older medium theories (see Meyrowitz, 1994).

[3] The electronic media tend to integrate informational spheres according to Meyrowitz, while older printed media tended to segregate them. However, he also suggests that electronic media might have this effect.

argues that the social situations should be understood as *information systems*. Territories and spaces are really "barriers of perceptions" that hamper or enhance the distribution of information to others. Following Meyrowitz, it is not the physical setting that determines the nature of the interaction, but "the patterns of information flow" (Meyrowitz, 1986, p. 36). Roles, then, can be thought of as "fluid information networks" that are susceptible to restructuring through changes in information-flow patterns. The explosion of electronic media seriously changes the information flows that influence our roles and our self-identity.

The information system point of view is in many ways a re-interpretation of Goffman's analysis of territories. As we saw earlier, his understanding of territories included also abstract concepts. Meyrowitz's contribution is to enhance the more immaterial dimensions of Goffman's theory. The information system perspective on the social situation, however, illustrates how physical space and communication media are very closely related. Meyrowitz frequently uses architectural metaphors in his works, to stress this point:

> ... different media are like different types of rooms – rooms that include and exclude people in different ways. The introduction of new media into a culture restructures the social world in the same way as a building or removing of walls may either isolate people into different groups or unite them into the same environment. (Meyrowitz, 1994, p. 62)

A "mobile worker" stands with one leg in the physical workplace with immediate contact with his co-workers and the other leg in the virtual world of electronic communication networks. A view on the new workplace as a blending of information from face-to-face and mediated communication seems to be reasonable. However, what seems to be lost from Meyrowitz's "renewed" perspective is the symbolic dimension that was so evident in Goffman's work. Social roles seem to be reduced to a passive process of "information receiving". Applying an information system approach on the workplace, we should remember that identity development is after all an active, and also a creative process.[4] An even more critical problem with Meyrowitz's contribution is that he focuses so much on the mass media that he tends to overlook other media, such as the telephone. Much of the described consequences of mobile telephony do not seem to be directly comparable to the effects of television.[5] Further, Meyrowitz does not describe how the new mobility of the individual influences the blurring of social boundaries. In reality, the mobile work-

---

[4] A problem with the information approach to roles is also that much of the focus is removed from the active controlling of space. When Goffman talks about "stalls", possessions and visual space as territories he stresses how these are embedded in social systems of power.

[5] For a review of studies on the social consequences of mobile telephony see Haddon (1997).

place represents a blurring of boundaries in another way than the mass media; it is basically caused by the movement of people, not (only) on the distribution of information.

Still, the mediated or virtual environment must be recognized in an analysis of the mobile workplace. The so-called "knowledge workers" are increasingly drawn to social relationships outside the physical office. These relationships or networks will in many cases compete with the face-to-face relationships (Castells, 1997; Zuboff, 1988; Nardi et al., 2000). Changing the media in addition to changing the physical territories has implications for social behavior, but the question of how these are integrated is, of course, a complex one. Certainly, the new media at the workplace create a new social landscape, which goes beyond the mere physical interaction. In itself the development of a more virtual workplace means that the individual has become less visible from a single point of view. The "open office" does not necessarily show the people you work with, and it is not necessarily the place where you do most of your work. It is a paradox that the virtualization of knowledge work also has made it more visible. To cover both the mix of mediated (virtual) and face-to-face social networks, I will use the term *information environment* in the remaining parts of the chapter.[6]

7.4 New Behavioral Patterns at the Workplace

The discussion so far has argued – with Goffman and Meyrowitz – that territory has a significant impact on how we interact in social settings. We claim territories to assure others and ourselves of social position, and it is a setting for staging of a certain performance. An information system approach argues that the core issue is the information to which we have access in a social situation, and that electronic media challenge this with a wider access to electronic networks. As we have seen, Goffman suggests two important ways that territories are used in everyday (work-)life. First, it is a setting where we perform or role-play, or where we can withdraw from this "public" setting. Second, the territories are part of the "symbolic environment" that we need to signal and negotiate our place in a particular social system. There is a deliberate tension here, between the individual as a manipulator of a social interaction and the individual as a contributor to a social structure, evident in much of Goffman's writing (Branaman, 1997). Figure 7.3 sums up the two dimensions of territoriality. The information environment may on the one hand impinge on the role-play when the distinctions between front and back regions are made

---

[6] This is a term coined by Lundby (1994). The information environment approach on the office has similarities with a conception of the office as a *hybrid infrastructure* (Bakke and Yttri, 2003).

the open office as a front stage. If she had decided to stay at her place, she would have had to find a right balance between the two roles involved. Some may find this more difficult than others.[7]

The mobile media are a continuous threat for everyone working to keep up a front-stage image. For some reason some individuals want to keep a strict line between front and back regions, whereas others tend to prefer a "homogenization of roles".

7.4.2 Territories as Social Rituals

In the traditional office it was fairly easy to read the individual's position in the organization by his or her use of space. The higher the rank, the larger is the office. In many cases there were even rules that were written down in job descriptions and company policies (Becker and Steele, 1994). The mobile workplaces have largely rejected the use of space as status symbols for the employees. In the literature arguing for extended use of flexible offices, this is often described as one of the serious drawbacks in the old regime (Becker and Steele, 1994; Duffy, 1997).

However, the need for assurance of position in a social system is not easy to get away from. According to Goffman and others, this is a very basic human action that goes on everywhere where there is a social institution. What we might see is that the use of fixed physical territories such as an office yields more egocentric or situational territories. Consider the following three examples:

Example E: *A young man enters the mobile office in the morning. He spends some time walking around, greeting colleagues with his voice and face. After a while he puts the PC bag on a free desk, close to another colleague. When connecting his PC to the network he engages in small talk with his "new neighbor".*

Example F: *A woman suddenly stops the computer work, and walks across the floor to a male colleague. She starts to discuss how to finance a particular project. They talk quietly, but it is possible to hear what they talk about. He gives her advice on how to do things, and after a while she returns to her desk.*

Example G: *A man goes over to a colleague and starts talking. After a while they both go over to a third person – apparently a manager – and talk to him about a project. The conversation is a sort of "briefing" where they inform the manager about the latest actions taken in a certain project. After 3–4 minutes they both return to their desks.*

---

[7] Ling (1997) has described how these kinds of situations might imply that the speaker must handle the management of many "parallel front stages."

Example E shows how a man searches out the most appropriate place to work. The desk is similar to what has been labeled a "stall", a physical space on which the individual can lay temporary claim. It is not a fixed territory, but something one may claim for a period. He chooses, after several minutes of considerations, a place near to a co-worker, perhaps of equal position. This will probably strengthen his bonds and position towards this person. Sitting down next to the manager, however, would be a more risky choice, since this would be a direct signal of contact upwards in the social hierarchy. This could also weaken the bonds towards his colleague.

Example F shows clearly how the woman is of "lower rank" than the male (informally, if not formally). She is dependent on his view on a particular matter, and this is exposed to everyone. The territory the male manager has control over is not physical space, however, but information about a particular matter. He seems to be more central in an information network then she is, so she is dependent on his benevolence. Thus he controls an informational territory that gives him authority and control over the situation.

In example G, we observe a similar situation. Two persons move over to the manager to get his opinion. It is clear that these two people have permission to disturb the manager when needed. They are members of the same "team" and they know he has an interest in what they are doing. We can speculate that there are certain rules regarding who has access to disturbing managers in this information environment. A newcomer would probably not have access to disturbing the manager so abruptly. This illustrates that the right to interrupt is essential as part of the social interaction. It is, however, much more invisible than in a large office. There are also more diffuse rules involved here, with no doors to close or to knock on. A general pattern in both examples F and G, however, is that it is those with a lower rank that move towards the one of the higher rank to get information. The initiative is the responsiblity of the "weaker" part.

The examples above indicate how the new territorial regime in the mobile workplace may lead to new types of social behavior patterns. I would like to stress that these behavior patterns are not "right or wrong". I stick to the general view held by Goffman that the performing of a role is not immoral or suspect in any way. It is how we all strive to achieve self-identity in a social community, even if our attitude towards this role-playing varies. Neither should it be perceived as negative when people use energy to mark out social position in one way or another. This is social dynamics that will occur one way or another in most social systems. The interesting point is that the roles change as a result of changes in the context in which the roles are played out.

7.5 Discussion: Implications

I have so far discussed how the mobile workplace has become more like a front region, in the theoretical perspectives of Irving Goffman. This indicates that the office worker has become open for social interaction in a new way. The difference is due to the new visibility (or information access) experienced in an open office area. The newness here is related to the general lack of private spaces or back regions. At the same time, much of the traditional ways of using territories as a way of assuring social position has been changed. This might cause uncertainty, and a situation where new sets of behavioral rules must be worked out in an organization.

7.5.1 The Role of the Mobile Media

The mobile media play a central part in the changes that take place: first, they are a constant intruder, introducing new situations that challenge the performance of a certain role. This is particularly challenging for people receiving private calls when in a "public" work setting. Outgoing calls are easier since the worker can move to a suitable place – in the corridor or outside the building – before they dial the number. Second, the media are also tools that are used actively in "performances". Loud conversations in the open office are a way to make an impression of a particular kind towards managers and others. Communication equipment can also enhance symbols that are strategically used to give an impression of a certain identity. Third, the media provide access to alternative relations and networks outside the building. The mobile media connect the worker to a virtual network of co-workers and contacts that is not visible from observations, but crucial for the status of the worker in the social system.

7.5.2 Behavioral Change

The observations described above are selections from larger studies involving observations and interviews with office workers at two mobile workplaces (Julsrud et al., 2002a, b). These studies indicate that the mobile workplace regime may result in certain consequences for behavior. A drawback with these observations is that they focus only on behavior at the central workplace. As mobile workers also conduct work on complementary places outside, the cases do not give a complete picture of their workplace setting. Still, it gives glimpses of typical behavior at their main place of work. I would here like to highlight four trends or movements related to the new way of using territories, using the theoretical terms exploited in this chapter:

From fixed to individual and situational territories. First it appears that physical territories are giving way to more situational and personal territories. This means that whereas the use of physical space becomes less important, other ways seem to become more important. In particular, people are making use of possessions, clothes and gadgets as symbols of social position and rank. This is in accordance with other studies reporting on changes in the use of symbols in organizations that tries to downplay traditional status symbols (Hatch, 1997, p. 255).

From visible to invisible territories. Second, more of the territorial claims move from the visible to the invisible. Invisible access to informational sources and their position in a social network is more important than actual access to physical space or privileges at the workplace. Much of the knowledge worker's social capital may be connected to his or her social networks outside the organization (Nardi et al., 2000).

From back regions to front regions. Third, it appears that knowledge workers in the mobile office have to cope with an almost constant participation in a front region. This new setting, where anyone can be seen and heard, encourages multiple role-plays. In a group where there is always someone new, the need to play out work-related roles will be a more prominent part of the behavior at work.

From back region to alternative back regions. It should be noted that most of the workers in our cases had access to alternative places to do their work, such as home offices and other places. Frequently these were used as alternative back regions in our cases. It may be that the pressure of working in a front stage most of the working hours enhances the interest for exploiting alternative back regions.

A common goal for the introduction of mobile offices is – as mentioned at the beginning of this chapter – to get direct business advantages connected to more efficient internal information, knowledge building in teams and reduction of real-estate costs. The underlying idea seems to be that if people get the chance to conduct more work together in common areas, this will enhance the circulation of ideas and the building of knowledge. The social consequences I have discussed in this chapter suggest that other effects may also occur. Some of these changes may trigger a development in a more negative direction: workers may experience stress from being in a "front region"; workers may withdraw to alternative back regions so that the social group falls apart; workers may feel uncertain of their position in a social system based on invisible territories; newcomers without any network or contacts may feel isolated. Hence the intended goals of the introduction of a new workplace setting may experience problems that come out of a lack of understanding of social interactions. This is not to say that the mobile workplace is doomed to fail: social systems are extremely adaptable, and work can be done successfully in a number of settings. However, from a "situational perspective" such as that suggested by Goffman, the new regime of mobile workplaces clearly has several pitfalls. Knowledge about where these pitfalls are is the first step to avoid them.

7.6 References

Bakke, J.W. and Yttri, B. (2003) Hybrid Infrastructures for knowledge work. Paper presented at the 4th International Space Syntax Symposium, London, 17–19 June.

Baumann, Z. (2000) *Liquid Modernity*. Polity Press, London.

Becker, F. and Steele, F. (1994) *Workplace by Design*. Jossey-Bass, San Francisco.

Bjerrum, E. and Bødker, S. (2003) Knowledge sharing in the "new office" – possibility or problem? Paper presented at the 8th European Conference of Computer-supported Cooperative Work, Helsinki, 14–18 September.

Branaman, A. (1997) Goffman's social theory. In Lemert, C. and Branaman, A. (eds), *The Goffman Reader*. Blackwell, Oxford, pp. xlvi–lxxxii.

Castells, M. (1997) *The Rise of the Network Society*. Blackwell, London.

DEKAR (2004) Status and strategies for the knowledge intensive Nordic workplace. Newsletter No 1-04. See www.dekar.org

Duffy, F. (1997) *The New Workplace*. Conran Octopus, London.

Giddens, A. (1990) *The Consequences of Modernity*. Polity Press, Oxford.

Goffman, I. (1956) *The Presentation of Self in Everyday Life*. Doubleday, Bantam Books, New York.

Goffman, I. (1961) *Encounters; Two Studies in the Sociology of Interaction*. Allyn and Bacon, New York.

Goffman, I. (1971) *Relations in Public: Microstudies of the Public Order*. Doubleday, Bantam Books, New York.

Hall, E.T. and Hall, M.R. (1990) *Understanding Differences*. Intercultural Press, London.

Handy, C. (1991) *The Age of Unreason*. Business Books, London.

Harrison, A. (1991) The workstation, the building and the world of communications. In Duffy, F. and Laing, A. (eds), *The Responsible Workplace*. Butterworth Architecture, Estate Gazette, London, pp. 65–77.

Hatch, M.J. (1997) *Organizational Theory. Modern Symbolic and Postmodern Perspectives*. Oxford University Press, Oxford.

Julsrud, T., Yttri, B., Mjøvik, E. and Nielsen, P. (2002a) *Nomadiske Brukere*. Report No. 40/2002. Telenor RandD, Fornebu.

Julsrud, T., Yttri, B., Bakke, J., and Brunland, A. (2002b) *Validation in Telenor, Cycle I*. Report from the SANE project. See www.saneproject.com

Leaman, A. (1991) The responsible workplace: user expectations. In Duffy, F. and Laing, A. (eds), *The Responsible Workplace*. Butterworth Architecture, Estate Gazette, London, pp. 16–31.

Lemert, C. and Branaman, A. (eds) (1997) *The Goffman Reader*. Blackwell, Oxford.

Ling, R. (1997) One can talk about common manners! The use of mobile telephones in inappropriate situations. In Haddon, L. (ed.), *Communications on the Move: the Experience of Mobile Telephony in the 1990s*. Report from COST 248. Norsteds Tryckeri, Stockholm, pp. 73–96.

Lundby, K. (1993) *Mediekultur*. Universitetsforlaget, Oslo.

Manning, P.K. (2000) Information technology in the police context. In Yates, J. and van Maanen, J. (eds), *Information Technology and Organizational Transformation. History, Rhetoric, and Practice*. Sage, London, pp. 205–222.

Meyrowitz, J. (1986) *No Sense of Place*. Oxford University Press, New York.

Meyrowitz, J. (1994) Medium theory. In Crowley, D. and Mitchell, D. (eds), *Communication Theory Today*. Polity Press, London, pp. 50–77.

Nardi, B., Whittaker, S. and Schwarz, H. (2000) It's not what you know, it's who you know: work in the information age. *First Monday*, 5 (5), May.
http://firstmonday.org/issues/issue5_5/nardi/index.html

Silverstone, R. and Haddon, L. (1996) Design and domestication of information and communication technologies. In Mansell, R. and Silverston, R. (eds), *Communication by Design: the Politics of Information and Communication Technologies*. Routledge, London.

Van Meel, J., de Jonge, H. and Dewulf, G. (1997) Workplace design: global or tribal? In Worthington, J. (ed.), *Reinventing the Workplace*. Institute of Advanced Architectural Studies, Architectural Press, University of York, pp. 50–63.

Zuboff, S. (1988) *In the Age of the Smart Machine. The Future of Work and Power*. Basic Books, New York.

8

Being Mobile with the Mobile: Cellular Telephony and Renegotiations of Public Transport as Public Sphere

Fernando Paragas

8.1 Overview

As mobile phone users travel in public spaces, their mobile telephony raises issues of etiquette, privacy and regulation. This confluence of mobile telephony and public spaces has been discussed from the perspectives of simultaneous communication within real and cyber/cellular places (Ling, 1997; Fortunati, 2000; Geser, 2002), nuanced coordination (Haddon, 2000; Ling and Yttri, 2002) and societal/parental control (Haddon, 2000; Ling and Helmersen, 2000; Green, 2001). More specifically, mobile telephony and mobility per se have been studied alongside the development of cities (Ling and Yttri, 2002), gender, travel and location (Ling and Haddon, 2001; Green, 2002).

That mobile telephony now affords the public the opportunity to travel and engage in mediated communication at the same time has indeed resulted in interesting conceptions of communicative activities in public transport as a public sphere. This chapter explores these conceptions as regards the experiences of commuters in Metro Manila,[1] the capital region of the Philippines, by looking at

1. How mobile telephony is regulated by the users' perceived personal security and social norms depending upon their mode of transport.

---

[1] Metro Manila and the National Capital Region are used interchangeably, as in most government documents.

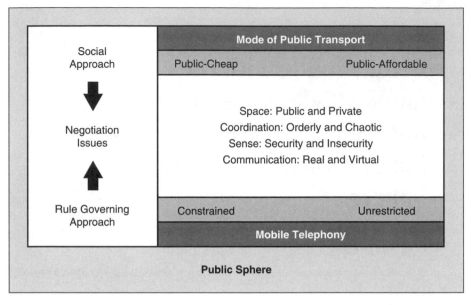

Figure 8.1 A framework on mobile telephony and public transport in the Philippines.

2. How mobile phones afford their users a sense of security, which, in itself, is threatened by the prospect and incidences of mobile phone theft.
3. How mobile phones help provide a semblance of order, just as they contribute, to the seeming chaos of commuting.

To organize the concepts in this chapter, I have constructed a model that depicts how negotiations in space, coordination, security and communication result from the convergence of public transport and mobile telephony in the public sphere (Figure 8.1).

As will be discussed in the next section, public transport in Metro Manila could be classified according to cost and space (top right section of the model). It is a continuum of cheap modes of public transport where space is generally public at one end (jeepneys and buses) and their expensive counterparts where space is private on the other (tricycles and taxis). In the middle are modes of transport where the fare is still affordable and space is generally neither expansive nor constrained (FX[2] shuttles and the

---

[2] The term FX, read as ef-ex, comes from the vehicle model Tamaraw FX, which was the forerunners of this mode of transport. An FX is a medium-sized Asian utility vehicle (AUV) that carries 10 passengers.

overhead railways). The choice of public transport to a great extent influences the use of mobile telephony (bottom right section of the model). Where the fare is cheap and space is limited, the use of mobile phone is constrained by choice and/or by context. Conversely, where the fare is expensive and space is exclusive to the passenger, mobile phone use is relatively unrestricted.

To say that only the mode of public transport influences mobile telephony in this regard, however, is simplistic. Between the two continuums are complex negotiation processes where passengers (1) consider their private nooks in the otherwise public sphere of public transport; (2) coordinate both covertly and overtly with people within and outside the vehicle; (3) take into account the security which their mobile phone affords them at the same time as the dangers that using it entails; and (4) manage their conversations across real and cyber/cellular realities within the greater public sphere that is the metropolis.

The discussion of these negotiations is anchored on Goffman's ideas on symbolic interactionism and the elements of Shimanoff's definitions of rules. Using Goffman's constructs, the focused and unfocused interaction between and among passengers in any given moment, for example, can be considered a frame where they all self-present themselves (Goffman, cited in Littlejohn, 1996). Meanwhile, the negotiations are "governed" by Shimanoff's followable, prescriptive and contextual rules, unwritten constructs that evolve from the interactions among the passengers themselves. This argument echoes Murtagh's (2001) assertion about situation-contextualized rules over mobile telephony.

Drawing on insights from cultural interpretation (Littlejohn, 1996, pp. 215–216), this chapter uses the data from 52 college students (mean age 21 years) on their mobile telephony during their everyday commute. Half of the students were enrolled in my senior-level undergraduate class on Data Interpretation and Reporting from June to October 2002. They submitted three-page reports of their fortnight's observation of their and other people's use of mobile phones in public transport. Further, each of them interviewed a student enrolled in a different university to locate the experiences included in this chapter beyond our campus. Likewise, the data from these interviewees were designed to serve as a point for comparison for the observation reports as I thought there was a risk that my students would be hypersensitive to their mobile telephony as a result of the assignment. That there were no significant variations in the answers among the students from within and beyond the university showed the universality of the students' experiences. For attribution purposes, the quotations included in this paper are those of my students.

The study's approach, although lacking the detail of a daily diary report, distills the experiences of the students inasmuch as it allows them to make sense critically of their mobile telephony while in public transport. Students-as-observers is an effective technique in what cultural interpreter Clifford Geertz (Littlejohn, 1996) refers to as thick description, or the discussion of cultural practices from an indigenous perspective.

8.1.1 Public Transport in Metro Manila

Public transport in Manilla is primarily composed of road-based vehicles (Hashimoto, 2001) with six basic modes that range from the mass-based (buses and the two transit railways), the intermediary (jeepneys and FX shuttles) to the intimate (taxis and tricycles). These vehicles account for 78% of all trips made in the metropolis (Bayan, 1995; Hashimoto, 2001).

A Filipino innovation, jeepneys, can accommodate as many as 20 people and account for 40% of all vehicles in Metro Manila (Bayan, 1995). Originating from the US service jeeps during World War II, jeepneys are now the king of the roads in the metropolis as they ply the main (line haul) and neighborhood (feeder trip) road networks. Jeepneys and buses, which ply main highways, serve the lower to middle income commuters with fares pegged at around 4 pesos for the first 4 km and 25 centavos for every succeeding kilometer.[3] Buses and jeepneys are noted for their cramped spaces, especially during peak hours. Commutes by jeepney and bus are not on a fixed schedule as they stop randomly to load or unload passengers (Labastilla, 1999).

The two mass transit railways,[4] meanwhile, have too limited a network to serve a significant section of the commuting public, although they are noted for speed and convenience. Using these railways almost always means having to take the other modes of transport as well. The average train fare is 12 pesos.

Relatively more comfortable ways of travel come with a price. The air-conditioned FX shuttles serve passengers between the suburbs and inner-city destinations. Commutes in these shuttles are comparatively faster because of their smaller load and less frequent stops. Fares for FX rides start at a minimum of 10 pesos, or twice that of jeepneys and buses. A survey on FX passengers reveals that they are mostly young middle-income professionals who cannot afford to buy their own vehicle (Kamid, 1999). As Jasmin, one of my students, observed:

> Riding an FX is more convenient than riding a jeepney because it is less congested and it is air-conditioned. FX vehicles also make fewer stops because they have fewer passengers. However, riding a jeepney is more practical financially since the fare costs less.

---

[3] To an idea of costs, the foreign exchange rate currently floats around 55 Philippine pesos = US$1. The minimum wage is pegged at approximately 250 pesos per day (US$4.50).

[4] A third railway system opened in Manila in April 2003, at the time of the latest revision of this chapter.

Meanwhile, taxis can be hired for a flag-down rate of 25 pesos, and 50 centavos for each of the succeeding 200 m, although cab drivers are notorious for rejecting passengers bound for traffic-jammed suburbs. Finally, tricycles,[5] the poor man's cab, ply tertiary roads (village streets, for example) where jeepneys are scarce or unavailable. Hired for a fixed rate of around 15 pesos from the terminal to the exact destination, tricycles provide relative comfort, except for their noisy engine.

These vehicles navigate the city's ever-busy streets. Traffic crawls at 15 kph during rush hour in the main avenues (World Bank, 2002) since the city only has 2900 km of public roads for its over three million vehicles (Elefante, 2002).

Many issues confound the confusion that the congestion in the city brings. Beyond sitting through traffic jams, commuters have to endure noise pollution from trucks, buses and jeepneys (Fajardo, 1999) and also encounter numerous vehicular accidents (Bautista, 1995). Further, commuting in Metro Manila follows no regular trip schedules. As trips depend on passenger volume (Hosomi, 1997) and vehicle availability (Labastilla, 1999), waiting times at public transport terminals are variable (Kamid, 1999). This collective frenzy obviously brings about some concerns among the passengers, including Elinor:

> The word that best describes the general feel of space in public utility vehicles (PUVs) is intimidating. Unless I get preoccupied with my own thoughts, or get caught up in my own world, I feel vulnerable by the bombardment of sights, sounds, smell and touch.

Still, despite the seeming frenetic system of public transport, there remains a general feeling of safety, as Rhea noted:

> They are crowded, but, at the same time, there seems to be enough (space) for passengers. I feel quite safe in PUVs; I am more afraid to venture in the streets at late hours. I still take precautions though. In all PUVs, except in the MRT, I keep my hands on my bag so that it won't get snatched.

8.1.2 Telecommunications in Metro Manila

Telecommunications in Metro Manila, even with its origins dating as far back as 1872 (Aquino, 1994), has been slow to develop. Efforts to improve the country's telecom system even before World War II only led to a monopoly that did not improve interconnectivity within the country (Paragas, 2002a). It was only with the implementation of the National Telecommunications Plan in 1989 that the industry was deregulated and invigorated.

---

[5] A tricycle is a motorcycle with a sidecar. It can carry two or three people.

Coinciding with the development of the country's landline[6] infrastructure at the turn of the century was the entry of the mobile phone networks. This meant that Filipinos, while long deprived of telecom facilities but lacking the financial resources to subscribe to both mobile and landline services, had to choose between these two options (Paragas, 2002a). Industry data reveal that mobile seems to have won over landline. The strong preference for mobile phones accounts for the estimated 11 million mobile phones in the country, up by 70% from 2000 figures (Paragas 2002a). Indeed, data from the International Telecommunication Union showed that in 2002 there were 4.2 landline subscriptions for every 100 persons in the Philippines compared with 19.1 mobile subscriptions. Mobile phones are popular because they are relatively affordable to maintain considering the features that they provide. While the monthly maintenance fee for a basic landline service is pegged at over 700 pesos (or 10% of minimum wage), owning a mobile phone depends primarily on use (10 pesos for a 1-minute call) because there are no monthly charges.

Further, the use of short message services (SMS) significantly decreases telecom expenses. At 1 peso per message, SMS is so affordable that the country's texting population sends around 120 million messages per day (Paragas, 2002b). This means that each subscriber sends approximately eight SMS messages per day. This compares with about three for a country such as Norway. Perhaps because it is convenient, fast and cheap, texting has also become "in the Philippines an alternate media and communication tool that can rally as well as entertain" (Yu, 2001). Although lower costs have made mobile telephony accessible to lower socio-economic groups, there remains a class-based distinction in mobile phone ownership and use in terms of being updated with the hottest brands and the latest models.

The popularity of mobile phones, however, is not unique to the Philippines. The same is true in Cambodia, Japan, Korea and Singapore (Paragas, 2002b). However, whereas "in richer countries, these may be complementary, in low income ones, these are substitutes" (Fink et al., 2001, cited in Paragas, 2002b).

8.2 Public Transport and Mobile Telephony

The convergence of public transport and mobile phone communications has technological and societal implications. In terms of technical services, mobile phones can now help track traffic areas in the Philippines (Cabagnot, 2001; Casiraya, 2002), pay for parking in Estonia (Turism, 2002) and fares in Finland (Turku, 2002) or obtain information about transport schedules in Spain (TMB, 2002).

---

[6] Landline phones are called POT in other countries.

Beyond these services, however, the convergence is raising etiquette, privacy and security issues, among others. For instance, making a phone call while on public transport – which perhaps compares with performing a private activity in a public place – is a riddle in decorum that seems to be more confounding in the West (ESRC, 2002; Greenspan, 2002; GSM Association 2002; TNS, 2002) than in Asia (Yu, 2001). Attempts to answer this riddle, in fact, have ranged from providing "mobile phone train carriages" for phone users (Baird, 2002) to offering "mobile-free cars" for those who do not like to hear mobile conversations (GSM Association, 2002).

In Metro Manila, the convergence of the "high technology" that is mobile telephony and the "old technology" that is the transport system is having profound effects not only on how the mobile phone is used in particular but also on social dynamics in general. In its congested streets, mobile phones connect individuals as they navigate through traffic, a situation that is similarly observed in Asia's other busy cities (Yu, 2001). Further, mobile phones, while serving as security blankets on the one hand, are also becoming threats to one's security on the other, as seen in the incidence of mobile phone-snatching incidents (Ho, 2001; Tubeza, 2001; Felipe, 2002). At least on one occasion this has led to murder (Tubeza, 2001). Further, a mobile phone is said to have detonated the bomb that exploded in a bus in central Metro Manila (Oca, 2002).

8.2.1 Public and Private Spaces

The dynamic between public and private spaces as venues for mobile telephony has been explored as regards the technology's social intrusiveness (Love and Perry, 2004) and its user's handling of the conversation (Weilenmann, 2003). In Philippine PUVs, that tension is compounded by congestion. Except for cabs, PUVs afford commuters limited space such that bus aisles and jeepney backdoors are converted to the private spaces of standing passengers. Commuters thus manage their transient, constricted territories by defending these using non-verbal communication. Rarely is there a direct verbal request for a passenger to decrease his/her personal space to benefit another.

> To make others aware of my own space I usually occupy the space that I think I can get. When others intrude my space, I don't usually tell them directly, but I try to "fight" for my space by bumping on them. Arlen

> Private space in PUVs is often violated (and) becomes limited to where you are seated. Even the space in the floor becomes more public as the passengers take turns in resting their feet further than their boundaries. Jan

Further, the definition of private from the public space in public vehicles entails an understanding of unwritten norms that have evolved from the shared difficulty in commuting.

> Space inside the PUVs is about owning a "territory" and a tacit agreement of carefully following the rules that go with that agreement. Elinor

> In a PUV, I believe there has to be some sort of compromise. You have to consider it as a public space where you naturally share it along with many other sorts of folks. Pamela

8.2.2 Real and Virtual Communication

It is easy to imagine that in such cramped public spaces, communication either in person (real) or by mobile (virtual) is taboo. In the UK, for instance, the collective public holds sway over the actions of a private individual in public places (ESRC, 2002; Greenspan, 2002; O'Sullivan, 2002; TNS, 2002). Further, even with their own private nooks in public transport, passengers are urged not to perform private acts such as using the mobile because of expectations of quiet in, say, the train (GSM Association, 2002). However, in Metro Manila and in many parts of Asia where there is rarely an expectation of quiet in public streets (Fajardo, 1999; Yu, 2001), users of mobile phones do not break the "silence" of the public place but fight to be heard above din of the street. Thus, in such congested surroundings, commutes entail not so much total silence but simple consideration for the public sphere of the vehicle.

> I respect that they have the right to talk on the phone since PUV is a public space. Marah

> Since PUVs are a public space, people ought to show enough consideration to others by at least being polite when they talk. By polite conversation, I mean toned down voice, no outright naming of people, no sordid details, no giggling or laughing out loud. Kabzeel

The student-informants say they believe the same rules apply for mobile phone calls (Murtagh, 2001), although they are also more aware of the volume of their phone and their voice when making such conversations:

> If the call is urgent, I accept it but with much caution regarding the loudness of my voice. I also make sure that my message alert and ring volumes are not loud so as not to attract attention. Shella

> Both virtual and real communications are usually limited to phatic conversations in most PUVs. However, there is more caution in virtual space communication because aside from worrying about the voice volume, I also worry about snatching and hold-up incidents. Marah

Similarly, they say that mobile phone calls are generally acceptable so long as the phone calls are relatively modulated and short as a concession to the shared public space.

> It is OK for people to talk over the phone or in real life. It is only irritating if the ring tone and speaking voice are too loud that it seems as if he owns the vehicle. Romila

> Phone calls in PUVs are OK with me but I really get pissed when their voice is too loud or everybody in the PUV hears him/her and it was a long call. But if it was really short then it is OK even if his/her voice is a little bit loud. Shella

> If the topic is important, it is OK to speak more loudly if the other party cannot hear you properly. If it is silent, like in the FX, it is only proper to lower your voice out of courtesy to the other passengers. Mary Grace

However, estimating the correct volume of conversations is more complex than ensuring the voice is not loud, as even this is deemed acceptable in some instances.

> It should not appear as if you are whispering to each other because the other passengers may be offended. Romila

> You can't keep people from having a conversation, and to do that, people have to raise their voices up some notches in decibels in competition with roaring engines. Pamela

Still, callers who exceed the concession for modulated phone conversations are sometimes concerned that they are irritating the other passengers.

> I think it is impolite to use the mobile phone in PUV because you usually have to raise your voice to be heard over a mobile phone. Marcus

> I accept mobile calls in jeepneys but I don't make them. In tricycles, I have a harder time receiving calls because of the loud noise emanating from the engine. Vyx

Interestingly, even when someone tilts the precarious balance between private and public spaces by talking a tad too loudly and for too long on their mobiles, rarely do people speak their minds. The student-informants believe a strong glance is already an appropriate sanction for these trespassers (see also Love and Perry, 2004).

> I give them a dirty look if their mobile phones or their voices are too loud because these are an annoyance. Rhea Claire

> There was one time last summer when (someone) answered (her phone) with the loudest voice I've ever heard. I stole a glance at her with my meanest stare. I was really annoyed. Iris

> Sometimes when the voice is super loud and it is not necessary, I really stare, especially when there is too much body movement. When I am in a good mood, I just shrug it off. Sarah Jane

Perhaps a stronger sanction for making private calls in public spaces is the knowledge that the other passengers are listening, even if they do so quite discreetly. Similarly, Love and Perry (2004) found this disinterested eavesdropping in their exploratory in the UK.

> Of course people will look at you when your ring tone is very loud. Others listen when they have nothing to do. It's OK since it is a public vehicle. Deborah

> I really don't like answering phone calls because I know people are going to look at me. Kathleen

Conversely, there seems to be gradations of guilt among the passengers in overhearing otherwise private conversations.

> Sometimes I listen in on other people's conversations. When I know that the topic is a little sensitive, I act as if I am not listening ... It is a form of entertainment. Iris

> When one passenger accepts/makes mobile phone calls in PUVs, the other passengers usually look at that person first and then it appears as if they seem not to listen at all. Personally, just like in normal conversations, I tend to do selective listening. Arlen

Considering the din of the city streets and the prospect of eavesdroppers, texting then becomes a practical means of contact via the mobile phone.

> I find it difficult to use my mobile phone in tricycles because of the noise. I just use text instead. Romila

> When it is too noisy and I cannot understand the party at the other end of the line, I just tell him to call back or text. Sarah Jane

8.2.3 Security and Insecurity

However, while a strong glare and the prospect of eavesdroppers can wilt the ardor of a mobile phone caller, a greater sanction against using mobile phones is the fear that the gadget may invite harm to their owner. The increasing number of mobile phone thefts in the city (Ho, 2001; Tubeza, 2001; Felipe, 2002) has given this new facet to the mobile phone, which is otherwise considered as a tool for improving one's sense of security.

Protecting the mobile phone, rather than the mobile phone helping to protect its owner, seems to be a preoccupation among the student-informants. This entails acting as if they do not own a phone, not using it in public transport or not bringing it to public spaces at all. This concern for the mobile phone defeats the sense of perpetual contact (Katz and Aakhus, 2002) that mobile phones supposedly afford their users.

> I pretend not to own a mobile phone because one was snatched from me before. Leslie

> To be safe, I hide my mobile phone in my bag whenever I commute and I rarely put it out. Every now and then, I peek at it to see if there is an important call or text. Rafael

> As much as possible, I leave my phone in the dorm because I am quite paranoid that somebody might snatch my phone. Mariel Andrea

Mobile phones are used only when the other passengers in the vehicle seem to fit stereotypes of upright citizens or after a keen surveillance of people within and outside the vehicle.

> When I have to receive a phone call, I am aware of the people around me and behind me (those outside the vehicle). When I have to text, I put my hand in my bag to reach for the phone instead of taking it out. Vyx

> In buses, I am more cautious because there are more passengers, who are all potential snatchers. I don't know who to trust because they are all strangers. Marah

Further, the student-informants weigh the mobile phone friendliness of the places where they travel before they bring out their phone. The ease of using mobile phones in a particular place thus serves as an indicator of its safety from thugs compared with other locations.

> I am more lax (in checking my phone) while in the MRT and buses than in jeeps especially those that do not ply the university route. Daniel

> I am particularly more comfortable taking out my phone when inside a university jeep inside the campus. When I go home, I don't take it out even if I really have to, because of a holdup experience a year ago. Iris

Central to my thesis, the mode of transport bears strongly upon mobile telephony. The informants believe that the worst place to use the mobile phone is the jeepney because its open structure makes it an ideal venue for snatchers to do their business. Buses are similarly suspect as venues for phone theft because their passengers come from all walks of life. Since jeepneys and buses are the most popular mode of transport in the city, the mobile telephony of the informants is thus significantly curtailed.

> I don't feel safe inside buses and jeepneys because I have a mobile phone. Because of the increasing number of phone snatching incidents in these PUVs, I feel insecure when I commute. I have a friend who lost a phone in a PUV. Rafael

> In jeeps, since there are people from the low-income groups there, I take my cell out if I am sure that I am in the company of students and professionals who probably own cells themselves. Rhea Claire

> In the jeepney I never use my mobile phone because I am afraid that somebody would grab it and steal it from me. Mary Grace

Similarly, more private modes of transport such as FX shuttles and taxis are considered to be relatively safe places for mobile telephony. Both modes are deemed exclusive (a function of cost) and "contained".

> Usually, the number of times I take my mobile phone bag out is proportionate to the fare of that certain PUV. Rhea Claire

> There is a higher tendency for me to use my mobile phone in FX and in buses because they are less open than in the jeepney, therefore less chances of my mobile phone being stolen. Elinor

> I also feel safe to take my mobile phone out of my bag when riding a taxi because I'm the only passenger or I'm with someone I know. Shella

The concern for the safety of the mobile phone, ergo also its user, extends not only to the person him/herself but also to his/her co-passengers.

> Taking your phone out in PUVs can be dangerous to you and other passengers so I avoid it. If a thief pulls a knife or gun because he wants your phone, it also endangers the other passengers. Marcus

> Whenever I see people using their mobile phone, regardless of whatever PUV they are riding, the first thing I think of is caution. Mobile phones are very prone to being snatched nowadays. Marah

Similarly, those who show disregard for the safety of the phone earn the ire of their co-passengers.

> When other people make or answer calls from their mobile phone, I usually glance over them and say something to myself, like, "I only wish your mobile phone is snatched". Elinor

> I get irritated with or find stupid those who use or display their mobiles in PUVs. I sometimes feel like saying, "I wish you are robbed so that you would know how to take care". Zara

Despite these fears for the security of the phone, perpetual contact using the mobile phone is achieved perhaps not by a voice conversation but by texting. Even then, this is done with extreme caution such as putting the phone in silent mode or texting while camouflaging the phone, so as not to invite the attention of snatchers. Hiding the phone from the view of the other passengers thus serves a dual purpose: it maintains the private nature of texting just as it protects the phone and its user from the snatchers.

> I usually don't accept calls inside PUVs, if the call was really important I would text that person calling me with my mobile phone inside my bag. And I always set my mobile phone in silent mode. Elinor

> I text while on tricycles but not on jeepneys. I only text on buses when I happen to be alone in a row [of seats]. Marcus

8.2.4 Order and Chaos

Mobile phones serve as a link between the student-informants and the people who are concerned for their safety, especially in reporting their location (Laurier, 2001; Weilenmann, 2003) to their parents (Green, 2001).

> When it's my parents, siblings or my best friend on the other line, I tell my real location because it is important that they know where I really am in case of an emergency. Dianne Kristine

> I always tell the other party where I am, mostly because that is the usual reason I get called in the first place; my caller wants to know my location, if I'm OK and on my way. Jemimah

However, that link is sometimes severed because of the fear that mobile telephony in public places can result in danger. In effect, mobile phones are not able to deliver the feeling of peace which mobile phones can bring by linking people anytime, anywhere.

> When I don't feel safe and comfortable to use my phone, I don't use it so as not to endanger myself. Arlen

In another respect, mobile phones still imbue the chaotic streets with some semblance of order by helping locate people through micro-coordination (Ling and Haddon, 2001; Ling and Ytrtri, 2002).

> When accepting phone calls, I have no qualms telling other people about my location. At least they know where I am, which is important if I am late for a meeting with the person who called. Vyx

> I tell other people my exact location so they'll know more or less when to expect me. Arlen

However, the fact that people could, and do, lie about their whereabouts, especially when they are late defeats their ability precisely to locate and time each other. Mobile telephony thus becomes a tool for the negotiation of time and space (Green, 2002).

While lying "protects" the tardy from the ire of the punctual party, it adds a dimension of chaos to what could be an orderly coordination, since people may think someone is nearby already when he or she is, in effect, some distance away.[7] Tardiness thus complicates both the content of the mobile phone conversation (Laurier, 2001; Weilenmann, 2003) and the use of the phone itself in the PUV.

> I usually tell the person I'll be meeting that I'm somewhere close to the meeting place in order not to get a lecture for being late. M

> When asked about my "real" location, I sometimes lie because the people waiting for me may get bored. R

> Sometimes I have to lie. When I have a meeting and I am running late, and the person I am supposed to meet calls me up, I tell him I am in this place already even when I have just left the house. S

A sanction against lying, as it is with engaging in mobile phone calls in public transport, is the prospect that people are listening to the conversation.

> When someone asks me over a mobile phone call where I am, I give my location, "In an FX" but I don't say exactly where I am – when I am late. For example, I will say I am in this road, but I won't specify that I am actually at the farther end of it from my actual destination. It's not totally lying; I can't lie either because I'm concerned that other people hear. P

> I do not lie about my location during phone calls because I know that my co-passengers will know that I am lying; I just text the lie instead. L

8.3 Summary and Conclusion

Contact via mobile phones is far from being perpetual, not so much because of etiquette, but because of more practical considerations of security.

Despite the congestion in public transport in Metro Manila, conversations, either real or mobile, are generally accepted since passengers recognize that they have private, if only transient, territories within the public sphere. However, activities within these personal bubbles are governed by

---

[7] For this section, it was decided to keep the student-informants anonymous because of the sensitive nature of lying.

rules in terms of topic, length and tone, the last two being particularly important for mobile phone conversations given the noise of the metropolis. Sanctions against those who do not acknowledge these rules (which have evolved from the collective difficulty of the commute) include a stare or a bump, but rarely a confrontation.

What seems to be the strongest sanction against mobile telephony in public transport in the Philippines is the fear of theft, given that mobile phones (being hot commodities) have become a target for snatchers. The prospect that their phones may be stolen from them prevents passengers from using their mobile phones at length while in public. This fear is strongest in PUVs, whose structures are open (jeepneys), whose passengers come from all walks of life (jeepneys and buses) and which ply routes that are thought to be unsafe. That these PUVs account for the most trips in the metropolis in any given day means that mobile phone owners are not able to access or use their phones for long periods of time during their commute. This situation transforms the mobile phone from being a security blanket to being a threat to one's security in public transport.

Still, amid this frenzy, mobile phones are able to help locate people in the metropolitan sprawl. This is important in timing their arrivals in meetings given the congestion and the traffic in the city streets. However, the fact that people could, and do, lie about their whereabouts – particularly when they are late – defeats this ability of the phone to help them locate and time the arrival of each other properly. The knowledge that people listen to conversations helps moderates this lying.

However, while passengers may have difficulty using their mobile phones to call while commuting, they still try to keep in touch by texting. Texting enables them to swap messages without having to worry about being loud or exposing the phone to possible snatchers.

Ultimately, mobile telephony in public transport reflects the precarious balance between the private activities of the passengers in their transient territories and the unwritten rules of decorum in the public sphere. Further, it indicates socio-economic dynamics as observed in public transport given the perceived safety of the more exclusive forms of transport, the chaotic coordination using mobile phones because of lying, and the value of mobile phones that makes them a target for snatchers. Mobile telephony in public transport thus entails a Goffmanian sense of self-presentation and interactionism in a setting governed by Shimanoffian rules.

For such are the travails of travel while being mobile with the mobile.

8.4 References

Aquino, T. (1994) The Philippines. In Noam, E., Komatsuzaki, S. and Conn, D. (eds), *Telecommunications in the Pacific Basin: an Evolutionary Approach.* Oxford University Press, New York.

Baird, J. (2002) Mobile-mad world shatters the comfort zone. Retrieved 2 January 2002 from <http://www.smh.com/au>

Bautista, D. (1995) *Serious Road Accidents in the City of Manila for Traffic Safety Planning.* University of the Philippines School of Urban and Regional Planning (UP SURP), Quezon City.

Bayan, J. (1995) *Cost Characteristics of Bus and Jeepney Transport Systems in Metro Manila.* UP SURP, Quezon City.

Cabagnot, R.J. (2001) Lost in Metro Manila? *IT Matters*, 31 January. Retrieved 7 January 2002 from <http://itmattters.com/ph/>

Casiraya, L. (2002) Smart launches vehicle tracking service. *IT Matters*, 18 April. Retrieved 7 January 2002 from <http://itmattters.com/ph/>

Economic and Social Research Council (ESRC) (2002) Simple may be best in mobile phone sales this Christmas as youngsters embrace the ritual qualities of texting. Retrieved 5 January 2002 from <http://www.esrc.ac.uk/>

Elefante, F. (2002) 2002: The year Metro Manila woke up. *The Manila Times*, 31 December. Retrieved 25 January 2002 from <http://www.manilatimes.net>

Fajardo, B. (1999) *A Study on Individual Perceptions of Road Traffic and Noise.* UP SURP, Quezon City.

Felipe, C. (2002) Manila renews drive vs cell phone snatchers. *The Philippine Star*, 7 November. Retrieved 5 January 2002 from <http://www.philstar.com>

Fortunati, L. (2000) The Mobile phone: new social categories and relations. *Information, Communication and Society*, 4, 513–528.

Geser, H. (2002) Towards a sociological theory of the mobile phone release. Retrieved 5 January 2002 from <http://socio.ch/mobile/t_geser1.htm>

Green, N. (2001) Who's watching whom: monitoring and accountability in mobile relations. In Brown, B., Green, N. and Harper, R. (eds), *Wireless World: Social and Interactional Aspects of the Mobile Age.* Springer, Godalming.

Green, N. (2002) On the move: technology, mobility, and the mediation of social time and space. *Information Society*, 18, 281–292.

Greenspan, R. (2002) Almost half of Brits want quiet. Retrieved 5 January 2002 from <http://cyberatlas.internet.com>

GSM Association (2002) Etiquette. Retrieved 2 January 2002 from <http://www.gsmworld.com>

Haddon, L. (2000) The social consequences of mobile telephony: framing questions. Retrieved 25 June 2004 from <http://www.mot.chalmers.se/dept/tso/haddon/Framing.pdf>

Hashimoto, T. (2001) *Transferability of the '87 LRT Jeepney Mode Choice to the MRT and Passengers' Demand Forecasting.* UP SURP, Quezon City.

Ho, A. (2001) Cell phone firms sign deal blocking use of stolen units. *Inquirer News Service*, 23 July Retrieved 2 January 2002 from <http://www.inq7.net>

Hosomi, A. (1997) *An Analysis of the Locational and Functional Characteristics of Terminal Areas in Metro Manila.* UP SURP, Quezon City.

Kamid, S. (1999) *Estimating the Passengers' Mode Switching Behavior: the Case of the Tamaraw FX.* UP SURP, Quezon City.

Katz, J. and Aakhus, M. (eds) (2002) *Perpetual Contact: Mobile Communication, Private Talk, Public Performance.* Cambridge University Press, Cambridge.

Labastilla, G. (1999) *Empirical Analysis of Route Operating Characteristics. Level of Service and Profitability: the Case of Jeepneys in the City of Manila.* UP SURP, Quezon City.

Laurier, E. (2001) Why people say where they are during mobile phone calls. *Environment and Planning D: Society and Space*, 4, 485–504.

Ling, R. (1997) One can talk about common manners!: the use of mobile phones in inappropriate situations. In Haddon, L. (ed.), *Themes in Mobile Telephony: Final Report of the COST 248 Home and Workgroup.* Retrieved 5 January 2002 from <http://www.telenor.no/fou/program/nomadiske/articles/09.pdf>

Ling, R. and Haddon, L. (2001) Mobile telephony, mobility and the coordination of everyday life. In Katz, J. (ed.), *Machines That Become Us.* Transactions Publishers, New Brunswick, NJ.

Ling, R. and Helmersen, P. (2000) It must be necessary, it has to cover a need: the adoption of mobile telephony among pre-adolescents and adolescents. In *The Social Consequences of Mobile Telephony: the Proceedings from a Seminar about Society, Mobile Telephony and Children Conference*, Norway, 16 June. Retrieved from <http://www.telenor.no/fou/program/nomadiske/articles/06.pdf>

Ling, R. and Yttri, B. (2002) Hypercoordination via mobile phones in Norway. In Katz, J. and Aakhus, M. (eds), *Perpetual Contact: Mobile Communication, Private Talk, Public Performance*. Cambridge University Press, Cambridge.

Littlejohn, S. (1996) *Theories of Human Communication*. Wadsworth, Belmont.

Love, S. and Perry, M. (2004) *Dealing with Mobile Conversations in Public Places*. CHI, Vienna.

Murtagh, G. (2001) Seeing the "rules": preliminary observation of action, interaction and mobile phone use. In Brown, B., Green, N. and Harper, R. (eds) *Wireless World: Social and Interactional Aspects of the Mobile Age*. Springer, Godalming.

Oca, M. (2002) Kapag ang cell phone ang tumunog, may sasabog … (when the cell phone rings, something will explode). *The Philippine Star*, 25 October, p. G5. Retrieved 5 January 2002 from <http://www.philstar.com>

O'Sullivan, C. (2002) 93% of Britons support mobile phone driving ban. Retrieved 5 January 2002 from <http://www.europemedia.net>

Paragas, F. (2002a) A case study on the continuum of landline and mobile phone services in the Philippines. Presented at the Social and Cultural Impact/Meaning of Mobile Communication Conference, Korea.

Paragas, F. (2002b) Policy, phones and progress: peculiarities and perspectives in the Philippine telecom industry. Presented at the 2003 International Association of Media and Communication Research Conference, Spain.

Taylor Nelson Sofres (TNS) (2002) Strong support in Britain for ban on mobile phones in public places. Retrieved 5 January 2002 from <http://www.tnsofres.com>

TMB (2002). TMB on your mobile phone. Retrieved 2 January 2002 from <http://www.tmb.net>

Tubeza, P. (2001) Another cellular phone owner shot dead. *Inquirer News Service*, 31 August. Retrieved 2 January 2002 from <http://www.inq7.net>

Turism (2002) Public transport ja parking. Retrieved 2 January 2002 from <http://turism.tartumaa.ee>

Turku (2002) A 24-hour visitors ticket by mobile phone! Retrieved 2 January 2002 from <http://www.bussit.turku.fi>

Weilenmann, A. (2003) I can't talk now, I'm in a fitting room: an initial investigation of the ways in which location features in mobile phone conversations. *Environment and Planning A*, 35, 1589–1605.

World Bank (2002) Improving traffic problems in the Philippines. Retrieved 25 January 2002 from <http://web.worldbank.org>

Yu, F. (2001) *The Wireless Effect*. Asia Internet Report, 9 November. Retrieved 7 January 2002 from <http://www.asianinternetreport.com/>

9

Mobile Phones, Japanese Youth, and the Re-placement of Social Contact

Mizuko Ito

9.1 Introduction

The mobile phone is often perceived as an emblematic technology of space–time compression, touted as a tool for anytime, anywhere connectivity. Discussions of young people's mobile phone use, in particular, often stress the liberatory effects of mobile media, and how it enables young people to escape the demands of existing social structures and parental surveillance. This chapter argues that the mobile phone can indeed enable communication that crosses prior social boundaries, but this does not necessarily mean that the devices erode the integrity of existing places or social identities. While Japanese youth actively use mobile phones to overcome limitations inherent in their weak social status, their usage is highly deferential to institutions of home and school and the integrity of existing places. Taking up the case of how Japanese teens' mobile phone use is structured by the power geometries of place, this chapter argues that characteristics of mobile phones and mobile communication are not inherent in the device, but are determined by social and cultural context and power relations. After first presenting the methodological and conceptual framework for this chapter, I present ethnographic material in relation to the power dynamics and regulation of different kinds of places: the private space of the home, the classroom, the public spaces of the street and public transportation, and the virtual space of peer connectivity enabled by mobile communications.

9.2 Method and Conceptual Framework

9.2.1 Research

This chapter draws from ongoing ethnographic research on mobile phone use and location, centered at Keio Shonan Fujisawa Campus near Tokyo. I draw primarily from two different sets of data. One is a set of ethnographic interviews I conducted in the winter of 2000 with 24 high school and college students about their use of media, including mobile phones. The central body of data behind this paper is a set of 24 "communication diaries" and interviews collected between July and February 2003 with Daisuke Okabe. For this study, our intent was to capture the usage patterns of particular individuals. We adapted data collection methods piloted by Rebecca Grinter and Margery Eldridge (Grinter and Eldridge, 2001), where they asked 10 teenagers to record the time, content, length, location and recipient (or sender) of all text messages for seven days. As with interviews, this data collection method still relies on second-hand accounting, but has the advantage of providing much more detail on usage than can be recalled in a stand-alone interview.

We expanded the communication log to include voice calls and mobile Internet, and more details about the location and context of use. Participants were asked to keep records of every instance of mobile phone use, including voice, short text messages, e-mail and web use, for a period of two days. They noted the time of the usage, who they were in contact with, whether they received or initiated the contact, where they were, what kind of communication type was used, why they chose that form of communication, who was in the vicinity at the time, if there were any problems associated with the usage and the content of the communication. After completion of the diaries, we conducted in-depth interviews that covered general attitudes and background information relevant to mobile phone use and detailed explication of key instances of usage recorded in the diaries. Our study involved seven high school students (aged 16–18), six college students (aged 18–21), two housewives with teenage children (in their forties) and nine professionals (aged 21–51). The gender split was roughly equal, with 11 males and 13 females. A total of 594 instances of communication were collected for the high school and college students and 229 for the adults. The majority of users were in the Tokyo Kanto region. Seven were recruited in the Osaka area in southern Japan to provide some geographic variation. This paper focuses on the communications of the high school and college students. I turn now to the theoretical and conceptual framework for the analysis.

9.2.2 Mobile Phones and Youth

In countries where there is widespread adoption, there are cross-cultural similarities in the intersection of youth and mobile phones. For example, Ling and Yttri (2002) describe adolescence as a unique time in the lifecycle, how peers play a central role during this period, and how the mobile phone becomes a tool to "define a sense of group membership, particular vis-à-vis the older generation (2002, p. 162). In a more recent paper, Ling and Yttri (to be published) extended this developmental perspective to examine how mobile phone use is located in the power relations of family and peer group. A growing body of work with teens in locations such as the UK (Grinter and Eldridge, 2001; Green, 2003, to be published; Taylor and Harper, 2003), Finland (Kasesniemi and Rautianinen 2002; Kasesniemi, 2003), Norway (Skog, 2002) and Sweden (Weilenmann and Larsson, 2002) finds similar patterns in other countries. Text messaging appears as a uniquely teen-inflected form of mobile communication, in that is lightweight, less intrusive, less subject to peripheral monitoring, inexpensive and allows easy contact with a spatially distributed peer group (Grinter and Eldridge, 2001; Kasesniemi and Rautianinen, 2002; Ling and Yttri, 2002).

Our data and other material on Japan (Matsuda, 2000, 2005b; Okada and Matsuda, 2002; Yoshii et al., 2002; Habuchi, 2005) also support these general findings. My analysis here, however, focuses less on the distinctive qualities of youth communication and more on the institutional and material conditions in which this distinctiveness is produced. Rather than locating the affinity between messaging and youth in the developmental imperatives of teens, I take a context-driven approach. I argue that the practices and cultures of youth are not solely outcomes of a certain state of developmental maturity, or even of interpersonal relations, but are also conditioned by the regulative and normative force of places. I shift the center of attention from the practices and identities of youth themselves to their institutional and cross-generational surrounds.

Behind our approach is the "new paradigm" in childhood studies that has argued that "youth" and "childhood" are categories constructed and consumed by people of all ages, and produced in particular power-geometries (James and Prout, 1997; James et al., 1998). In other words, youth practices need to be analyzed in relation to adult social structures that limit and regulate youth activity as well as discourses and research frameworks that often construct youth as frivolous and not fully "socialized". Modern teens, despite their physical and psychological maturity, do not yet have access to a full repertoire of adult rights, responsibilities and resources, such as their own homes where they can meet friends, and lovers, or a workplace where they are considered productive members of society (as opposed to "consumers" and "learners"). Teens are considered legitimate objects of external regulation, control and redirection in a way that even young adults are not. While enjoying mobile phone use to stay in touch with friends and current technology and fashion trends, young people also use these devices to push back at their own disenfranchised position within adult-controlled institutions and spaces of activity. I cut our data along these lines as well. I apply the category of youth to those institutionalized as such – high school and college students who are financially dependent on adults.

Among the many contextual factors that drive youth patterns of mobile media usage (Ito and Okabe, to be published), our focus here is on the structuring force of place. For a theory of place, I look to cultural geographical conversations on new media. I see place as a hybrid of the social, cultural, and material (including technology, architecture and geography). Massey's (1993) insistence that hierarchical relations are key components of place-making are particularly important in the analysis. Critiquing "easy and excited notions of generalized and undifferentiated space-time compression", Massey argues that

> different social groups are placed in very distinct ways in relation to late modern flows of media, people, and capital. This point concerns not merely the issue of who moves and who doesn't, although that is an important element of it; it is also about power in relation to the flows and the movement. Different social groups have distinct relationships to this anyway-differentiated mobility; some are more in charge of it than others; some initiate flows and movement, others don't; some are more on the receiving end of it than others; some are effectively imprisoned by it (Massey, 1993, p. 61).

I draw from Massey's framing, but also work further to specify the particularities of these power geometries of space–time compression by looking at the particularities of intergenerational power dynamics. Just as social theory has interrogated race, class and gender, generational dynamics need to be analyzed with a similar social structural lens (Alanen, 2001). Further, it is not only that certain people are differentially located within power geometries. The same person can be alternatively in control or lacking control of communicative and cultural flows depending on lifecycle stage, different spatial and temporal locations and their access to new technology. Youth communications are regulated by peers or adults depending on place and time of day, and that access to mobile media takes a central role in managing and inflecting that control. Conceptually, our approach shares much in common with Green's (2002) and Ling and Ytrri's (to be published) analysis of the role of mobile phones in surveillance and monitoring between adults and teens and among teens. Mobile phones are embedded in existing power geometries and create new social disciplines and accountabilities. After an overview of issues in Japanese youth usage of mobile phones, I examine the institutionalization of mobile phone use in relation to the urban home, school, street, public transportation and online mobile space.

9.2.3 Japanese Youth and Communication Technology

In Japan, young people have led mobile media adoption since the early 1990s. Although both the pager and the mobile phone were originally designed and marketed as business-oriented devices, in both cases, teenage girls appropriated the technology for their personal communications and in turn, informed the design of subsequent devices. Mobile texting, in particular, was an innovation largely initiated by young people, with origins in pager cultures where girls sent numeric codes to pagers from home phones and payphones (Fujimoto, 2005; Okada, 2005). Short message services on mobile phones were developed in direct response to teenage pager texting. In the case of short message services and the mobile Internet, again, young people drove adoption. In his study of youth mobile media cultures, Fujimoto (2005) describes "the girls' pager revolution" as a technology-linked paradigm shift, where certain cultural values became embedded in mobile technologies that have now infiltrated the general population. Although mobile phones are not pervasive among people of all ages and occupations, young people continue to use their phones more, spend more on phones and engage in higher frequencies of text mobile e-mail[1] exchanges (VR, 2002; Yoshii et al., 2002).

As the stereotypical user shifted from the businessman to the teenage girl in the mid-1990s, the popular press reported on rising concern about young people's mobile media usage. In her review of popular and research discourse surrounding young people's mobile phone usage in Japan, Matsuda (2005a) describes how mobile phone use stirred anxieties about young people making indiscriminate social contact, devolving manners and unruly behavior in public places. In particular, public attention focused on *kogyaru* (high school girls), a label attached to the newly precocious and street-savvy high school students of the 1990s who displayed social freedoms previously reserved for college students. *Kogyaru* sported new school uniforms with extremely short skirts, congregating in city centers and disrupting prior social norms that young girls should be tightly regulated by the imperatives of home and school. Until recently, the popular assumption has been that heavy mobile phone users are low-achieving socialites. Even worse, ongoing reports suggest links between mobile phones and teen prostitution and crime, and a new term was

---

[1] I use the term mobile e-mail to refer to all types of textual and pictorial transmission via mobile phones. This includes what Japanese refer to as "short mail" and Europeans refer to as "short text messages" and also the wider variety of e-mail communications enabled by the mobile Internet. At the time of writing, this includes text, graphics, photographs and, just recently, video clips.

coined, *enjo kousai*, to refer to high school girls dating middle-aged men for cash (see Tomita, 2005).

In contrast to these widely publicized images of undisciplined, footloose, mobile phone-wielding youth, research has demonstrated that young people's usage is not particularly distinctive in term of manners and promiscuity of social contact (Matsuda, 2005a), and their communications are overwhelming developed within the frame of adult-run institutions of family and school. Matsuda (2005b) argues that most young people are becoming more selective rather than superficial in their social relationships, focusing on friends with whom they identify closely. Most of these friends are tied to institutions such as schools, family or local community. In her study of American and Japanese youth, White (1994) sees fewer conflicts between Japanese parents and youths than their American counterparts, and less pathologization of youth as a problematic life stage. Dependency has less social stigma than it does among Euro-American youths, and this is institutionalized in the protective functions of family that extend through college and often beyond. White (1984) also describes how youth are defined by marital and employment status rather than by age, and "such institutional definitions have more weight than social and psychological identities" (11).

The daily rhythms of young people are largely determined by school, and life rhythms of high school and college students are substantially different. High school students spend most of their free time in school, particularly if they have sports and other after-school activities. By contrast, college students have extremely flexible schedules, routinely miss classes and keep late hours and do not adhere to the household schedule for meals. Rather than a time of independence, however, where they leave the parental home, most urban college students live with their parents and are financially dependent. Our sample of college students for the communication diary part of our study is slightly skewed in this respect, as our student pool at the Keio campus was largely comprised of youth living on their own. As a suburban campus of an elite urban university, the situation at our campus is unique in attracting students from around the country in an area with a relatively low urban density. I turn now to our ethnographic material in order to explicate how these power geometries of place operate in the everyday practices of young people's mobile phone use.

9.3 Mobile Youth Culture and the Politics of Place

9.3.1 Mobile Phones in the Home

There are peculiarities to the urban Japanese case with respect to the politics of location, particularly the home context. Most notably, Japanese youths, through college, have less private space compared with their U.S. and European counterparts. The Japanese urban home is tiny by middle-class American standards, and teens and children generally share a room with a sibling or a parent. Most college students in Tokyo live with their parents, often even after they begin work, as the costs of renting an apartment in an urban area are prohibitively high. Unlike the U.S., there is no practice for teens to get their own landline at a certain age, or to have a private phone in their room. The costs of running a landline to a Japanese home are very high, from US$600 and up, about twice what it costs to get a mobile phone. It is therefore extremely rare for a home to have more than one landline.

Here is an excerpt from an interview with four high school girls who are close friends.

> *Interviewer:* You all live close to each other. Do you visit each other's homes?
>
> *Student 1:* We don't. It's not that we are uncomfortable, or our parents get on our case, but it's like they are too sweet and caring, and you worry about saying something rude, or talking too loud. You can't be too rowdy. So we don't meet in our homes.
>
> *Student 2:* Occasionally. Maybe once a year. Actually, that's not even occasional.
>
> *Student 1:* And if it happens, it is at a friend's house where they have their own room.

This stance was consistent across the youths that we interviewed. Meetings among friends almost always occurred in a third-party space run by indifferent adults, such as a fast food restaurant, karaoke spot or family restaurant. Even for college students living on their own, their space is generally so small and cramped that it is not appropriate for hanging out with groups of friends.

The phone has always provided a way of overcoming the spatial boundary of the home, for teens to talk with each other late at night and to shut out their parents and siblings. As noted in other studies (Green 2002; Ling and Yttri, 2002, to be published; Skog 2002), the mobile phone has further revolutionized the power geometry of space–time compression for teens in the home, enabling them to communicate without the surveillance of parents and siblings. This has freed youths to call each other without the embarrassment of revealing a possible romantic liaison, or at hours of the day when other family members are likely to be asleep. All those interviewed were consistent in stating a preference for calling a friend on a mobile rather than home phone, despite the higher cost. Youths now do not have the home phone numbers of any but their most intimate friends. A high school girl describes how she makes gender-based choices of what phone to call:

> If it is a boy, I will call their mobile. If they have one, I will call their mobile. If it is a girl, I will call their home. If it is a girl, well, I'm a girl right? So if I call they think I am just a regular friend. But if it is a boy, his family might tease him, and I've made a friend very uncomfortable in the past because of this. I've also been told some nasty things by a parent. I was totally pissed off when a parent of a boy told me off like I wasn't a proper girl. So since then, I don't use the home phone.

This girl describes how the home phone is tied to household collective identity, where it is appropriate for a teen to have childlike relationships with same-sex peers, but the more adult identity of romantic liaisons is an uneasy fit. The person-to-person mode of the mobile phone is a more comfortable alternative "place" to conduct these kinds of communications that exceed the young person's status of a "child" in the power geometries of the home.

The home phone once was a means for parents to monitor and regulate their children's relationships with their peers. With the mobile phone, the spatial boundaries of the home become highly porous to discretionary communication. The spatial dispersion of homes, coupled with the freedom of communication via the mobile phone, is an inversion of the dynamics of the classroom, where kids occupy the same physical space, but are not in control of their communications with each other. In addition to enabling communications with peers who would not call on the home phone, the mobile phone also supports frequent and lightweight communication that would not be appropriate with a household-identified communication device. Among close friends and couples, most youths (two out of three in our sample) maintained ongoing lightweight contact as they went about their daily routines, sending each other messages about their current status and thoughts, such as: "Just woke up with a hangover," "The episode (of the TV drama) really sucked didn't it," or "good night." They enjoy a sense of co-presence with peers that they are not able to realize physically because of their dependence on the parental home.

For teenage couples living with their parents, this means that the mobile phone can be the primary means for staying in touch. The logs of one teenage couple in our study represent a somewhat more intense version of couple communications that we saw in other instances. Their typical pattern is to begin sending a steady stream of e-mail messages to each other after parting at school. These messages will continue through homework, dinner, television shows and bath, and would culminate in voice contact in the late evening, lasting for an hour or more. A trail of messages might follow the voice call, ending in a good night exchange and revived again upon waking. On days when they were primarily at home in the evening, they sent 34 and 56 messages to each other. On days that they were out and about the numbers dwindled to six and nine. The content of the messages ranged from in-depth conversation about relational issues to coordination of when to make voice contact, to lightweight notification of their current activities and thoughts. Messaging became a means for experiencing a sense of private contact and co-presence with a loved one even in the face of their inability to share any private physical space.

Informants for our communication diary research included 10 cases of high school students living at home: seven high school students, three mothers and one father. We also had one example of a college student living at home. Parents appreciated how the device can extend the parameters of their own contact with and surveillance over their children, but they were less comfortable with the ways in which their children use their mobile phones to engage in peer communication. In communicating with their kids, parents exploited the fact that their kids were constantly checking their mobile e-mail. Without exception, parents with children at home would send messages telling them that it was time to come home, or coordinating details such as pickup and meeting times. We saw mild to acute tensions surrounding the extensive use of mobile phones by the children. Parents and children alike voiced a rule that e-mailing should not happen during mealtimes. All the parents we interviewed described a sense of unease and curiosity about their children's mobile communications. Conversely, all the children took measures to keep parents in the dark about the content of their e-mail and calls. Generally this was done by going to their bedroom or other location out of earshot when taking a voice call. One mother voices what we take to be a typical parental stance:

> *Interviewer.* Do you have a problem with her using her mobile phone during meals, or after meals in the living room, when you are together?
>
> *Mother.* I don't have a problem with it when we are just lounging around. But during meals or when she is studying, I try to tell her to tell the other person on the line.
>
> *Interviewer.* Are you curious or concerned about with who and what she is communicating?
>
> *Mother.* I am concerned about all of it … though I can usually guess who it is.
>
> *Interviewer.* When you tell her to stop, does she stop?
>
> *Mother.* She goes to her room … if I am strict about it.
>
> *Interviewer.* Do you ever ask her, like "What in the world are you talking about?!"
>
> *Mother.* I do ask sometimes. But I just get a vague reply.

This mother is curious and concerned about the content of her daughter's mobile communications, but is largely in the dark. She is able to regulate, to some extent, the degree to which mobile communications disrupt specific activities in the home (meals, studying), but largely fails to regulate the content, partners and overall frequency of her daughter's communications.

Young people's heavy reliance on mobile phone usage grows out of the constraints on communication imposed by the home and the home phone, but this usage does not substantively challenge or reshape the power geometries of the home. Rather, mobile phones become a tool for circumventing the normative structures of the home with minimal disruption to its institutional logic. Dobashi (2005) has described how Japanese housewives use mobile phones in ways that are consistent with and supportive of gendered household roles. Although these uses by housewives, focused on accomplishing household coordination, are completely different from their children's uses, in many ways, both forms of uses have the effect of maintaining the integrity of the place-based logics of the household. Rather than intrude on the family unit and household expectations with calls to the home phone, young people take their personal communication elsewhere, to their mobile phones.

9.3.2 Mobile Phones at School

In the school context, there is variability in how teachers deal with mobile phones. Almost all schools officially ban phones from the classrooms, but most students do use e-mail during class at least occasionally. It is not uncommon for students to leave their mobile phones out on their desks during class, claiming that they use the clock function. All students, in both high school and college, voiced the rule that they would not use voice communication in class, but almost all said that they would read and sometimes send messages. The mobile phone gets used most frequently during lunch time and immediately after school, as students scurry to hook up with their friends. We saw e-mails being sent during class in only two of our communication diary cases, but almost all students reported in their interviews that they would receive and send messages in class, hiding their phones under their desks. Here is the response from one of the high school students who we did see using her phone during class:

> *Okabe*: What sorts of places and situations do you use your phone a lot?
>
> *Student*: At school, during class … I leave my phone on my desk and it vibes.
>
> *Okabe*: Do you take voice calls during class?
>
> *Student*: No. That would be going to far.
>
> *Okabe*: Oh, so you wouldn't answer. What kinds of exchanges do you have over e-mail during class? Do you send e-mails to people sitting in the same classroom?
>
> *Student*: Yes, I do that too.
>
> *Okabe*: What do you say?

Student: "This is boring."

Okabe: And you get a reply?

Student: Yes.

Okabe: When you write your e-mail, do you hide it?

Student: Yes. When the teacher is facing the blackboard, I quickly type it in.

Like this student, three other students described conversations with students in the same classroom, making comments like "this sucks," "this is boring," or "check it out, the teacher buttoned his shirt wrong." More commonly, students reported that they conducted "necessary" communications during class, such as arranging a meeting or responding to an e-mail from somebody with a specific query. The communications in class that we saw in the diaries involved coordinating meetings after school or receiving e-mails from friends who were absent, asking for notes or other class information. In all these cases the mobile e-mail is being used to circumvent the communicative limitations of the classroom situation, much as passing notes and glances across the classroom did in an earlier era. Perhaps more uniquely, the mobile phone in the classroom is a way to challenge the communication hierarchy of the traditional lecture format that insists that students passively listen to an active teacher. Mobile e-mailing enables students to resist their role in this one-way communication and to make more productive use of their attentional "dead time" between jotting notes and waiting for teachers to finish writing theirs (Taylor, this volume, Chapter 10). Just as in the case of the home setting, the low-profile modality of mobile, particularly text communication, means that students can engage in personal chatter while remaining respectful to existing power geometries of the classroom.

9.4 Mobile Phones in Urban Space

Our research has focused on the greater Tokyo metropolitan area, which is an extremely dense urban setting well connected to its more suburban surrounds. This urban landscape is amenable to appropriation by youth because of the extensive public transportation system, and the fact that it is safe to be on public transportation and out on the street even for young women at night. Youth will take public transportation from city outskirts and congregate in city centers such as Shibuya and Ikebukuro, considered the epicenters of youth culture. Not surprisingly, Shibuya crossing has the highest density of mobile phone use in the world.

In the early years of mobile phone adoption, when business uses dominated, there were few efforts to regulate usage in public spaces and transportation. It was only after young people became prominent users of the mobile phone in the late 1990s that public institutions stepped up efforts to regulate usage on trains and buses, in the midst of a wave of articles in the popular press about poor mobile phone manners and the annoyance of having to listen to teenage chatter in public space (Matsuda, 2005a; Okabe and Ito, 2005). Now, most trains and buses display "no mobile phones" signs, and announcements about mobile phone usage are made after every train stop. These announcements have been evolving towards more specificity through the years. Currently, they say, "We make this request to our passengers. Please turn off your mobile phone in the vicinity of priority seating [for the elderly and disabled]. In other parts of the train, please keep your mobile phone on silent mode and refrain from voice calls. Thank you for your cooperation." In a separate study, we observed uses of mobile phones on trains and subways. In hundreds of cases observed, almost all involved text input. When voice calls are received, people will, generally cut the call quickly and speak in a low voice, often shielding their phone and mouth with a hand or magazine. When passengers do take a voice call, they are often subject to subtle social sanctioning by other passengers in the form of quick glances or even sustained glares (Okabe and Ito, 2005).

In contrast to voice calls, mobile e-mail is considered ideal for use in public spaces. Although recent fears of negative effects on pacemakers have initiated a trend towards blanket prohibition in some areas, the prevailing social norm is that non-voice mobile communication is permissible. Although bus drivers will prohibit someone speaking on a mobile phone from entering a bus, we have not observed any instances of regulation of silent mobile phone uses. Just as the power geometries of the home and classroom make e-mail a privileged, private form of communication, regulatory efforts on public transportation have also contributed to the rise in e-mail as a preferred form of mobile communication. Largely because of the risk that their interlocutor may be on public transit, a social norm has arisen among the younger generation that you should not initiate voice calls without first checking availability with a text message. Unless certain that their recipient is at home, most youths (there were two exceptions in our study) will send a message first asking if they can call.

During a physical gathering, youth will generally prioritize the co-present encounter, but there are instances when they are interrupted by a mobile e-mail or call. Unless in public transport or fancy restaurants, they will attend to the interruption. All interviewed voiced a general rule that family and fast food restaurants were acceptable for voice calls. E-mail will be attended to regardless of place. Those that require an immediate response, such as a mother asking when they are going to return, or a message from somebody they are planning to meet, will be responded to right away. When with friends, youth will almost always take the call, but will cut it short if it is a one-on-one gathering. When multiple parties are meeting up (we saw this in two of our documented cases), it is common for mobile communications to be used to contact those that have yet to appear on the scene, adding relevant information to the current co-present encounter. At other times, contact with distant others can be used to augment a particular gathering. In one observation I made on a bus, a group of high school students were discussing a gathering they were arranging. As they discussed who was or was not coming among themselves, they also sent numerous text messages to others not co-present, confirming whether they were coming or if they had information about others.

A feeling of urban anonymity is disappearing as youths stay in ongoing and lightweight contact though messages with their peers and loved ones. Out shopping, a lone girl sends a picture of the shoes she is buying to a friend. Another sends a message announcing that she just discovered a great sale. After a physical gathering, as friends disperse on trains, buses, cars and on foot, a trail of messages often continues the conversation, thanks somebody for a ride or announces that they forgot to return an object. Rather than fixing a meeting place, gatherings between youth are now almost always arranged in a fluid way, as people coordinate their motion through urban space, eventually converging on a shared point in time and space. At the same time, mobile communications are highly responsive to the power geometries of urban space and an emergent set of social and communicative norms. I present one example from our communication diaries of one female college student who carried on a text message conversation while moving between different forms of public transportation. She has just finished work, and makes contact with her boyfriend after she boards the bus.

22:30	(boards bus).
22:24	(send) Ugh. I just finished (>_<).[2] I'm wasted! It was so busy.
22:28	(receive) Whew. Good job (>_<).
22:30	(send) I was running around the whole time. Are you okay?
22:30	(Only other passenger leaves. Makes voice call. Hangs up after 2 minutes when other passengers board).
22:37	(send) Gee I wish I could go see fireworks (; _ ;).
22:39	(receive) So let's go together! I asked you!
22:40	(gets off bus and moves to train platform).
22:42	(send) Sniff, sniff, sniff (; _ ;). Can't if I have a meeting! I have to stay late!
22:43	(receive) You can't come if you have to stay late?

---

[2]Japanese emoticons are written "head-on" rather than sideways. (>_<) is a grimacing face, (; _ ;) is a crying face, and (^o^) is a happy face.

22:46	(send) Um, no … I really want to go … (; _ ;).
22:47	(receive) Can't you work it out so you can make it?
22:48	(boards train).
22:52	(send) Oh … I don't know. If I can finish preparing for my presentation the next day. I really want to see you (>_<). I am starting to feel bad again. My neck hurts and I feel like I am going to be sick (; _;). Urg.
22:57	(receive) I get to see you tomorrow so I guess I just have to hang in there! (^o^)
23:04	(gets off train)
23:05	(send) Right, right. I still have a lot of work tonight. I can't sleep!

In our interview, she describes how her messaging embeds subtle clues that indicate her status and availability for communication keyed to her physical location. She keeps the conversation going as she continues her ride on bus and train, even though she is not terribly interested in the content of their chat. As she prepares to get off the train, she initiates a change of topic (about feeling bad and her neck hurting) as an indicator that the conversation has come to an end. She has enlisted a companion on her solitary bus ride, successfully filling dead time with small talk, ending it at precisely the moment when she arrives at her destination. This is but one example of many that we have gathered that attests to the highly nuanced and place-sensitive nature of mobile communications made in transit. Approximately half of the students in our study engaged in this sort of chat-like sequence while in transit. The regulatory efforts of public transport operators in Japan have structured a set of emergent social practices coordinated to the rhythms of youths' motion through urban space and relying on the non-disruptive modality of text communication. We conclude with an analysis of the technical and social structuring of the online space of mobile connectivity.

9.4.1 Mobile Virtual Places

The use of e-mail, and the growing expectation that mobile phones define a space of persistent connectivity, point to an alternative sort of technosocial space being defined by new mobile technologies for Japanese youths. While mobile phones have become a vehicle for youths to circumvent the power geometries of places such as the home, the classroom and the street, they have also created new disciplines and power geometries, the need to be continuously available to friends and lovers and the need always to carry a functioning mobile device. These disciplines are accompanied by new sets of social expectations and manners.

When unable to return a message right away, there is a sense that a social expectation has been violated. When one girl did not notice a message sent in the evening until the next morning, she says that she felt terrible. Three of the students in our diary study reported that they did not feel similar pressure to reply right away. Yet even in these cases, they acknowledged that there was a social expectation that a message should be responded to within about 30 minutes unless one had a legitimate reason, such as being asleep. One describes how he knows he should respond right away, but doesn't really care. Another, who had an atypical pattern of responding with longer, more deliberate, messages hours later, said that her friends often chided her for being so slow. All students who were asked about responses delayed an hour or more said that they would generally make a quick apology or excuse upon sending the tardy response.

With couples living apart, there is an even greater sense of importance attached to the ongoing availability via messaging. The underside to the unobtrusive and ubiquitous nature of mobile e-mails is that there are few legitimate excuses for not responding, particularly in the evening hours when one is at home. Five of the 10 student couples in our study were in ongoing contact during the times when they were not at school, and all these couples had established practices for indicating their absence from the shared online space. They invariably send a good-night e-mail to signal unavailability, and would often send status checks during the day such as "are you awake?" or "are you done with work?" We saw a few cases when they would announce their intention to take a bath, a kind of virtual locking of the door. Just as mobile workers struggle to maintain boundaries between their work and personal lives, youths struggle to limit their availability to peers and intimates. Although the "place" constructed through the traffic of e-mail and voice calls on a mobile device does not have the same institutionalized weight as places such as home, classroom and train car, an emergent set of social expectations are defining the parameters of these new power geometries.

9.5 Conclusions

This chapter has described the institutions and places which condition Japanese youths' mobile phone use. We see place as a power geometry that integrates the social, material and cultural. We have argued that this perspective is a useful inflection of prior research that has examined the more personal and relational aspects of mobile phone use by teens. Another goal of this work has been to argue that far from destroying the integrity of place with unfettered communication, mobile phones are keyed to the norms attached to existing places, and participate in the structuring of new forms of place-based norms and disciplines.

Given this perspective, we can understand youths' penchant for text messaging as an outcome of a wide range of factors. These include the unique expressive functions and styles of this form of communication, in addition to certain economic and historical factors unique to this generation (Ito and Okabe, to be published). In this chapter, we have focused on factors that relate to regulation and surveillance in particular places. Japanese youth, particularly high-school students, move between the places of home, school and urban space that are all subject to a high degree of regulation and surveillance by adults. Unlike the institutions of family and school, youth peer groups and couples are "institutions" that lack ownership and control of place. The outcome of these power geometries is that couples and friends have few opportunities for private conversation. Although a limited form of contact, mobile e-mail has fulfilled a function akin to co-presence for people that lack the means to share the same private physical space. New technologies become infrastructure for new disciplines and institutional relations as much as they challenge old ones that they grow out of, and the mobile phone is no exception.

In addition to explicating some of the factors behind distinctive patterns of youth mobile phone usage, our analysis argues more generally for a socially and culturally contextualized vision of technological "effects". The mobile phone is not "inherently" a device that disrupts existing social norms and places. Rather, its form, functionality and use are keyed to specific social settings as well as specific social groups occupying those settings. In the case of Japanese youth, this has meant the prevalence of devices with mobile Internet and e-mail access, and uses that operate under the radar of adult institutions and surveillance. Mobile e-mailing on a handheld device is lightweight, low-profile, concealable and not disruptive to the normative structures of most places that young people find themselves in. As Alex Taylor (this volume, Chapter 10) suggests, these everyday maneuverings with a handheld device are not so much subversion in the macro-political sense, but are "concealed, locally assembled resistance against an established set of social structures or 'rules'." In the case of young people soon to graduate into a full repertoire of adult rights and responsibilities, circumventing rather than overthrow of existing power geometries seems an appropriate strategy. By contrast, other groups (most famously, in the Philippines) with more pressing macro-political agendas have mobilized similar features of the mobile phone to defy governments (Rheingold, 2002; Agar, 2003) or challenge their financial disenfranchisement (Ilahiane, 2004). Whether the mobile phone functions as a socially conservative or transformative tool is determined by its status as a socio-technical device, embedded in specific social contexts and power geometries.

9.6 Acknowledgments

The work described here was supported by NTT Docomo and "Docomo House" at Keio University Shonan Fujisawa Campus and by the Annenberg Center for Communication at the University of Southern California. I thank Kenji Kohiyama and Hiromi Odaguchi of Docomo House and our student research assistants Kunikazu Amagasa, Hiroshi Chihara and Joko Taniguchi. This work has also benefited from the comments of Rich Ling, Alex Taylor and reviewers and participants at the conference "Front Stage/Back Stage: Mobile Communications and the Re-Negotiation of the Public Sphere."

9.7 References

Agar, J. (2003) *Constant Touch: a Global History of the Mobile Phone.* Icon Books, Cambridge.

Alanen, L. (2001) Explorations in generational analysis, pp. 11–22. In Alanen, L., and Mayall, B. (eds), *Conceptualizing Child–Adult Relations.* Routledge Falmer, New York.

Dobashi, S. (2005) Gendered usage of keitai in domestic contexts. In Ito, M., Okabe, D. and Matsuda, M. (eds), *Personal, Portable, Pedestrian: Mobile Phones in Japanese Life.* MIT Press, Cambridge, MA.

Fujimoto, K. (2005) The anti-ubiquitous "territory machine" – The third period paradigm: from "girls' pager revolution" to "mobile aesthetics". In Ito, M., Okabe, D. and Matsuda, M. (eds), *Personal, Portable, Pedestrian: Mobile Phones in Japanese Life.* MIT Press, Cambridge, MA.

Green, N. (2002) Who's watching whom? Monitoring and accountability in mobile relations. Pp. 32-45 In Brown, B., Green, N. and Harper, R. (eds), *Wireless World: Social and Interactional Aspects of the Mobile Age, Computer Supported Cooperative Work.* Springer, London.

Green, N. (2003) Outwardly mobile: young people and mobile technologies. Pp. 201-218 In Katz, J.E. (ed.), *Machines that Become Us.* Transaction Publishers, New Brunswick, NJ.

Grinter, R.E. and Eldridge, M. (2001) y do tngrs luv 2 txt msg? Pp. 219-238 In Prinz, W., Jarke, M., Rogers, Y., Schmidt, K. and Wulf, V. (eds), *Seventh European Conference on Computer-Supported Cooperative Work,* Kluwer, Dordrecht.

Habuchi, I. (2005) Accelerating reflexivity. In Ito, M., Okabe, D. and Matsuda, M. (eds), *Personal, Portable, Pedestrian: Mobile Phones in Japanese Life.* MIT Press, Cambridge, MA.

Ilahiane, H. (2004) Mobile phones, globalization, and productivity in Morocco. In *Information, Technology, Globalization, and the Future.* Intel, Hillsboro, OR.

Ito, M. and Okabe, D. (to be published) Intimate connections: contextualizing Japanese youth and mobile messaging. In Harper, R., Palen, L. and Taylor, A. (eds), *Inside the Text: Social Perspectives on SMS in the Mobile Age.* Kluwer, Dordrecht.

James, A. and Prout, A. (1997) *Constructing and Reconstructing Childhood: Contemporary Issues in the Sociological Study of Childhood.* Routledge Farmer, Philadelphia.

James, A., Jenks, C. and Prout, A. (1998) *Theorizing Childhood.* Teachers College Press, New York.

Kasesniemi, E.-L. (2003) *Mobile Messages: Young People and a New Communication Culture.* Tampere University Press, Tampere.

Kasesniemi, E.-L. and Rautianinen, P. (2002) Mobile culture of children and teenagers in Finland pp. 170–192. In Katz, J.E. and Aakhus, M. (eds), *Perpetual Contact: Mobile Communication, Private Talk, Public Performance.* Cambridge University Press, Cambridge.

Ling, R. and Yttri, B. (2002) Hyper-coordination via mobile phones in Norway, pp. 139–169. In Katz, J.E. and Aakhus, M. (eds), *Perpetual Contact: Mobile Communication, Private Talk, Public Performance.* Cambridge University Press, Cambridge.

Ling, R. and Yttri, B. (to be published) Control, emancipation, and status: the mobile telephone in teen's parental and peer group control relationships. In Kraut, R., Brynin, M. and Kiesler, S. (eds), *New Information Technologies at Home: the Domestic Impact of Computing and Telecommunicatons.* Oxford University Press, Oxford.

Massey D (1994) *Space, Place, and Gender.* Minneapolis, MN: University of Minneapolis Press.

Matsuda, M. (2000) Friendship of young people and their usage of mobile phones: from the view of "superficial relation" to "selective relation". *Shakai Jouhougaku Kenkyuu*, 4, 111–122.

Matsuda, M. (2005a) Discourses of keitai in Japan. In Ito, M., Okabe, D. and Matsuda, M. (eds), *Personal, Portable, Pedestrian: Mobile Phones in Japanese Life*. MIT Press, Cambridge, MA.

Matsuda, M. (2005b) Mobile communications and selective sociality. In Ito, M., Okabe, D. and Matsuda, M. (eds), *Personal, Portable, Pedestrian: Mobile Phones in Japanese Life*. MIT Press, Cambridge, MA.

Okabe, D. and Ito, M. (2005) Keitai and public transportation. In Ito, M., Okabe, D. and Matsuda, M. (eds), *Personal, Portable, Pedestrian: Mobile Phones in Japanese Life*. MIT, Press, Cambridge, MA.

Okada, T. (2005) The social reception and construction of mobile media In Japan. In Ito, M., Okabe, D. and Matsuda, M. (eds), *Personal, Portable, Pedestrian: Mobile Phones in Japanese Life*. MIT Press, Cambridge, MA.

Okada, T. and Misa, M. (eds) (2002). *Keitai-Gaku Nyumon*. [*Understanding Mobile Media*]. Tokyo: Yuhikaku (in Japanese).

Rheingold, H. (2002) *Smart Mobs: The Next Social Revolution*. Perseus, Cambridge.

Skog, B. (2002) Mobiles and the Norweigian teen: identity, gender and class. In Katz, J.E. and Aakhus, M. (eds), *Perpetual Contact: Mobile Communications, Private Talk, Public Performance*. Cambridge University Press, Cambridge, pp. 255–273.

Taylor, A. and Harper, R. (2003) The gift of the gab?: a design oriented sociology of young people's use of mobiles. *Computer Supported Cooperative Work*, 12, 267–296.

Tomita, H. (2005) Keitai and the intimate stranger. In Ito, M., Okabe, D. and Matsuda, M. (eds), *Personal, Portable, Pedestrian: Mobile Phones in Japanese Life*. MIT Press, Cambridge, MA.

VR (2002) *Mobile Phone Usage Situation*. Video Research, Tokyo.

Weilenmann, A. and Larsson, C. (2002) Local use and sharing of mobile phones. Pp. 92-107 In Brown, B., Green, N. and Harper, R. (eds), *Wireless World: Social and Interactional Aspects of the Mobile Age, Computer Supported Cooperative Work*. Springer, London.

White, M. (1994) *The Material Child: Coming of Age in Japan and America*. Berkeley, CA: University of California Press.

Yoshii, H., Matsuda, M., Habuchi, C., Dobashi, S., Iwata, K. and Kin, K. (2002) *Keitai Denwa Riyou no Shinka to sono Eikyou*. Mobile Communications Kenkyuukai, Tokyo.

Phone Talk

10

Alex Taylor

10.1 Introduction

Ordinary talk is, of course, an orderly accomplishment achieved through the use of such resources as turn sequence, posture, gesture and gaze. In this chapter, I aim to demonstrate that teenagers sometimes use these discursive resources in conjunction with the mobile phone to produce a particular order to their face-to-face conversations. Specifically, I will show that, amongst teenagers, talk about the phone, or more generally, *phone talk*, is routinely used in the management of a conversation's *topic* and the organization of *participation status*. Of particular interest is the manner in which these features of ordinary talk are organized and managed through the mobile phone's physical presence and its material qualities, and not simply by means of utterance, gaze and bodily posture.

The chapter will go on to demonstrate that teenagers can sometimes use the phone and its features to *subvert* the orderly progress of talk – in short, how the phone is used to participate in practices that disrupt talk-in-progress, "subverting" what is known to be the right and proper conduct for an occasion. This form of "*localized subversion*" should be seen as distinct from the commonly referred to subversions that are framed politically; subversion, in this chapter, is seen as a subordinate move – in talk or action – against the recognized order of a specific social situation and not a dissenting movement against a political authority or system.[1]

This work draws on a limited number of excerpts from both interview transcripts and field notes. These data were collected over the course of a study investigating teenager's use of mobile phones. The field notes were made during observations at two suburban sixth-form colleges attended by students between 16 and 19 years of age. The observations, amounting

---

[1] I am aware that this use of the word subversion may be confusing. My reason for choosing the word is that it captures the nature of those covert but purposeful actions that aim to challenge the recognized social order of a situation. I have used the phrase "localized subversion" in an attempt to specify these forms of subversive actions as locally achieved and set against locally constituted systems of order.

to approximately 18 hours in the field, recorded phone-related activities in the college canteens, playgrounds and hallways. The transcripts were the product of nine video-recorded group interviews with six students from one sixth-form college. The interviews took place over a period of 10 weeks in various locations, including the college classrooms and common area, The Mound, a small park outside of college buildings, and in the food hall of a local shopping mall.

10.2 Topic

To begin, I have chosen to focus on a short extract from the interview transcripts. The extract is taken from an interview with two students, Jackie and Lauren. Lauren talks, uncomfortably, of her relationship with her new boyfriend and, because of her discomfort, feels the need to redirect the topic of the conversation. It is in this situation that the phone is shown to provide an acceptable means to manage a conversation's topic and bring closure to a topic-in-progress.

(1. Interview: local shopping mall)[2]

1	Lauren:	It's rea:lly annoying: cos Ali's really brown: an– he's my new
2		boyfriend.- an he's like so brown:: I'm just like (.) [oh my
		God!
3	Jackie:	That's
		like
4		Bob! It's not fa:ir:: .hh
5	Lauren:	I like go up to him an I'm like YEAH! look how brown I
		am(.)=
6	Alex:	huh [huh
7	Lauren:	=and then he's like- (.) His nat(.)tral colour is like
		way
8		[brown: er. hhh
9	Alex:	huh-huh-huh-ha You've got a new boyfriend already:?
10		Arr: [she works quickly doesn't she
11	Lauren:	yeah, well: it's been a month! GOD:!(.)=
12	Alex:	a month::? ((sarcastically))
13	Lauren:	=I've bin suff:er:ing:! [.hhh
14	Alex:	huh-huh-ha-ha
15	Jackie:	.hhh have-you-been-going-out-with-him-for-a
16		month?=
17	Lauren:	=No! I haven't been out with anyone for a month
18	Jackie:	Oh, °huh-huh°
19	Lauren:	It was aw::ful!
20	Jackie:	hhh DA:mn! .hh

---

[2] See Sacks et al. (1974) for an explanation of the notation system.

21	Alex:	khe-kheh-ha-ha-[ha,
22	Lauren:	.hhh I didn't like it at all::! It was horrible!
23	Alex:	Rear:lly?(.)
24	Lauren:	No. I can't handle it.
25	Alex:	Only a month? Lauren? khe-kheh-
26	Jackie:	°Awe::°
27	Lauren:	→ But I don't even talk to him that much- I don't think it's
28		going to work °out° (.)
29	Alex:	Ar [r::
30	Lauren:	→ But anyway, this isn't ab[out ( )
31	Alex:	Okay then.
32	Jackie:	You talk to Imie more:
		anyway.
33	Lauren:	→ °Yeah:: ° But he's great! I can talk to him about lots of °stuff:°
34		((reaches to pick up Jackie's mobile that is placed on the
35		table and balances it on one end))
36	Alex:	°Yeah°=
37	Jackie:	=Oo, Oo look! This is really cool! This is really sad actually.
38		((takes the phone from Lauren, leans back and starts to
39		press the keys with both hands))
40	Lauren:	.hhh he-her-her
41	Jackie:	°ok watch° (.) Look at my phone- it's really really corny- look
42		( ) ((leans towards Lauren holding the phone with one hand.
43		Lauren also leans in and the two huddle closely to view the display))
44	Lauren:	[[ha-ha-ha-ha
45	Jackie:	huh-huh-huh(hhhh)
46		((both lean back and look towards each other laughing. Jackie shows
47		Lauren her screen-saver that displays an animated heart))

My main interest in the above excerpt occurs when Lauren and Jackie shift topics by turning their attention to the mobile phone (lines 33–47). However, before attending to this, I wish to make some general comments about the beginning of the excerpt; despite the apparent simplicity of this initial sequence of turns, the interpretive work that is performed accomplishes a great deal and does much to set the scene for the subsequent topic change.

From the outset, it is evident that both girls are willing to cohere to the orderly undertaking of the conversation and that Jackie, importantly, is prepared to align herself with Lauren. This is apparent, for example, in the way the two girls "play out" what is termed a *paired sequence* (Schegloff and Sacks, 1973), in the first four lines. The first turn and the manner in which the subsequent "slot" is "filled" signal what we shall see to be Jackie's efforts to be allied with Lauren. Jackie demonstrates she is willing to take a supportive position by not only offering a sympathetic response to Lauren but also by mirroring Lauren's description of her boyfriend's tan (lines 3–4). Lauren is spurred on by Jackie's response; in her next turn, she recounts what is taken to be a plausible encounter in which she re-emphasizes how dark her boyfriend is (line 5).

Turning to my primary interest in the excerpt, what we see is this mutual support from Lauren and Jackie first slip but then reassert itself by way of the mobile phone. After some discussion of boyfriend troubles (lines 9–26), Lauren makes a concerted effort to change the topic of talk. Perhaps seeing her own vulnerability exposed, she seeks to maneuver the conversation, first by belittling the emotional commitment she has to her relationship with her boyfriend, Ali (lines 27–28), and then through an all out attempt to curtail the discussion by highlighting the inappropriateness of the topic (line 30). The summing up of her relationship with Ali is the first "hint" Lauren gives of preparing for a topic shift. Such *summary statements* "appear to be implicative of closure of topic, and are recurrently deployed prior to various forms of topic shift" (Jefferson, 1994, p. 211). It is Lauren's subsequent turn, though, that explicitly reveals her desire to put an end to the topic-in-progress. Her use of "But anyway" serves as a *positioning marker* to bring closure to the topic (Schegloff and Sacks, 1973), and her utterance "this isn"t about …" invokes the *known-in-advance* status of the interview and can be seen as an attempt to draw us back to what has been previously agreed as *on-topic talk* (Button, 1991) – talk about mobile phones.

Having provoked Lauren with a teasing "Really?", my response is to accept Lauren's bid to shift topic. However, this effort is thwarted. Possibly in an attempt to console her, Jackie raises Lauren's friendship with another boy, Imie (line 32). Lauren publicly recognizes her strong bond with Imie, but expresses discomfort with this topic – maybe because it raises the possibility of disloyalty to Ali. In talking, she reaches across the table to pick up Jackie's phone. Having initially failed to support Lauren and possibly feeling discomfort herself, Jackie attempts to come to Lauren's rescue again. This time, rather than seeking to console Lauren, she bids for neutral ground. Taking up what is possibly a topic marker offered by Lauren, she attends to the phone and a feature that she first describes as "cool". She then retracts the expressive descriptor – possibly believing her attempt to change tacks as over zealous. Whatever the case, Jackie's efforts serve to draw attention away from the topic at hand. They also provide further evidence of the bond that the two have established. The attention given to Jackie's screen-saver, their mutual laughter, close physical proximity and glances towards one another all work to reinforce their solidarity.

It is this sequence that brings us to the main point of interest – the marked change of topic in conversation. Having made one explicit attempt to put an end to the topic in progress, Lauren produces another summary statement (this time about her relationship with Imie) and concurrently takes the phone placed beside her on the table. This coordinated act that gives rise to the diversion, or *topical disjunct* (Jefferson, 1994), she has been seeking. There is no strong evidence to claim that Lauren intentionally uses her turn or the phone to furnish a departure from the conversation's topic but, regardless of intentionality, Jackie and Lauren are jointly able to accomplish the maneuver. Lauren's move for the phone, her visible interaction with it and the tailing off of her turn with the quietly spoken "stuff" (lines 33–35), all operate to construct a *transition-relevance place* where Jackie is able to self-select her role as next speaker (Sacks et al., 1974) and bring about a change in topic. Notable in the above instance is that in this moment, and in a matter-of-fact manner, the *phone* is occasioned as a practical means to manage such an achievement.

This systematic, yet implicit, transition to phone talk is significant because it reveals the phone's taken for granted character in talk. With relative ease, phone talk enables uncomfortable gaps to be filled and, relevant here, uncomfortable topics to be supplanted. By invoking the topic of phones, Jackie relies on the conventional wisdom that phone talk is an acceptable topic in its own right and holds a legitimate place in conversational work. If it did not, her bid to "rescue" Lauren would have at best seemed out of place and at the very worse been publicly challenged. The mobile phone's very presence in everyday talk is what makes it such a valuable resource in these terms. Specifically, it is its presence as a tangible object that contributes to the conversation's reorganization and places the conversation in a position ready for change – nothing so coarse as an explicit statement of intent is needed.

Later, I will address what can be accomplished in using the mobile phone to manage topic and explain how this is made available through specific features of the phone. Before moving on, however, I want to unpack this notion of the taken-for-granted character of the mobile phone a little and examine why it might be that the phone, in particular, is enlisted as a wholly observable yet taken-for-granted device through which topic closure is routinely managed – why its very "taken-for-grantedness" in talk makes it something that is so suitable for such a purpose.

As I have already made clear, the phone's physical presence – its *materiality* – in everyday talk enables it to provide a means to manage and organize topic. Its ubiquitous presence, amongst teenagers in particular, has allowed it to become not only a commonplace subject of talk, but also a routine and legitimate means to mediate conversational work. The phone, in some situations, becomes an extension of those non-verbal, conversational resources we use, such as gaze, gestures and posture. The question to be asked, then, is what are those properties or features of the phone that have enabled it to become commonplace, or taken for granted, in conversation? To some extent, this question can be answered simply. It is the phone's size, shape, aesthetics, portability and all those features that make it perceptively mobile to begin with.

Although seemingly simple, I would argue that this response provokes a subtler understanding of the phone's physical presence as an acceptable resource in talk. It suggests that not only does the phone's ubiquity contribute to its use in conversation, but so too does its design as a device that is specifically meant to be handled. Watching teenagers engaged in talk provides evidence for this assertion. In my observations, I saw teenagers routinely holding their phones during their interactions, not using them for anything in particular, but rather stroking, caressing, moving and positioning them whilst talking. It would seem that the phone's material presence penetrates talk because it has been designed in weight, size, shape, etc., to be held. Lauren and Jackie help to explain the reasoning behind this "use" of the phone:

(2. Interview – simplified: local shopping mall)

1	Alex:		So what is that about?
2	Jackie:		It kind of takes your mind=
3	Lauren:	→	=It's like smoking. You know when you get a nervous twitch=
4	Jackie:		=Ya doin something with your hands
5	Lauren:		=and ye like need ta smoke or woteva and you want to
6			give up-like not that I'd know but hhhhum=
7	Jackie:	→	=or if you're tapping your fingers like this- it's like doing
8			somethink with your hands
9	Lauren:		It's a whole psychological nervous thing ((both laugh))
10	Alex:		But also are you detracting us from- to some other topic?
11	Lauren:		I don't know.
12	Alex:		Well! What d'ya think?
13	Lauren:		Umm. Yeah- Well- I'd rather not talk about it=
14	Alex:		=that's wot I mean. So you go
15	Lauren:		yeah- So I like change the subject by doing something else=
16	Alex:		=and why the phone? Wha'd you think it is about the phone=
17	Lauren:	→	=cos it was there.
18	Alex:		Right. So having something out like that allows you to refocus
19	Lauren:		Yeah! And I'm not goina stick out my wallet or something-
20			that sounded awful.

Although I do much, as the interviewer, to direct the subject matter of this sequence, Jackie and Lauren raise some interesting points about the phone's material presence. By drawing on smoking as an analogy (lines 3–6), Lauren evokes what might be thought of as the habitual nature of phone use. Jackie's own analogy furthers the sense of the phone having a taken-for-granted status (lines 7–8). The phone's material presence and its apparently seamless fit with the body make it as commonplace and acceptable as tapping your fingers or doing something with your hands.

Unlike a wallet, the phone is ripe to use because it is present and it does not have to be conspicuously "stuck out". Put simply, the phone can be thought of as a device that is "at-hand" in talk. By talking of the taken-for-grantedness of the phone in everyday talk then, my intention is not to suggest that it is rendered invisible; on the contrary, my argument is that the phone – through its routine presence and at handedness – has become a very real and yet unremarkable feature of teenagers' talk. It *is* something to be used in talk as a matter of routine.

10.3 Participation Status

This routine, taken-for-granted character of the phone in talk is further illustrated when we consider its part in managing and organizing *participation status*. That is, through its routine presence in talk, the mobile phone can be used to determine who is engaged in a conversation and the conversational roles that are taken. This section will demonstrate how the phone is used in recognizable ways to achieve such conversational work, and set the scene for later contemplating how the phone becomes a resource for subverting the known-about order of talk.

The following extract is taken from field notes that were made observing students using their phones in their school canteen. In the extract, there are several instances in which the phone provides a means to detract from the larger group engagement. What I wish to highlight are the possibilities that are provided for through the phone's material presence. I aim to show that participation status – demonstrated by what are referred to *as engagement displays* – is managed using the phone. By engagement displays, I make reference to those observable actions that are performed to demonstrate one's status or role in talk (i.e. speaker, listener, bystander, etc.).

(3. Observations: school canteen)

1 Five girls are at a table talking. One girl, G1, is navigating the menus on her Nokia 3210.

2 She passes the mobile to another girl, G2, who takes the mobile and starts to interact with it

3 with both hands. She holds the mobile between her index fingers and uses her thumbs to

4 press the buttons. She leans forward in her seat and rests her elbows on the table. The others

5 around the table are looking about aimlessly and occasionally talking amongst each other.

6 G2 looks as though she is playing a game. Although she seems to be able to contribute to

7 the group's ongoing conversation, she doesn't seem to be too engaged. Another girl, G3,

8 has her mobile (in a black cover) out. She plays with the cover's strap and holds the mobile

9   with the other hand. She talks to another girl at the table and fiddles with her phone as she

10   talks. As the talk at the table continues, she opens the mobiles case and pulls out a scrap of

11   folded paper, she looks at it and then shows it to the girl she is talking to – they talk and

12   smile at each other. G3 puts the paper back into the phone's case and the two return to

13   attending to the group's conversation. G4, another girl at the table, has her Nokia 3310 in

14   her hand. She looks at it with G3. G3 holds the phone with G4 and moves her head towards

15   it as though she is listening to something on the phone – perhaps a ring tone. This is

16   different to the way she would listen to something through the earpiece – she leans down

17   towards the phone and holds her ear against the side of the mobile. They look at the mobile

18   for a few moments longer and then put it away.

Although it was not possible to record the content of the conversations that occurred between the girls above, there is some readily observable, "performative" work being accomplished. What is apparent is that phones provide for small, localized forms of interaction that can be taken as a sign of exclusive engagement. The phones do not only mediate these forms of interaction, but can also be the focus of them.

In the above extract, G2's possible game playing is the first observable example of how engagement displays are managed via the phone (lines 1–7). Her use of the phone demonstrates how a barrier of sorts can be erected between the user of a phone and the situation at large. In contrast to the one-handed use of the phone found to be commonly performed by teenagers, G2 leans forward, holding the phone with two hands, thereby establishing a limited field of view.[3] This use of posture and gaze is what signifies her own separate status in the group and her willingness for her status to be seen as such. The phone is therefore not only used to display her reduced ability to attend to the ongoing talk at the table, but also serves to separate herself spatially from those around her.

G2's apparent withdrawal from the conversation would, at first sight, seem to be indicative of an altogether anti-social and unwarranted departure from the conversational cluster. However, in attending to her displays

---

[3] Throughout the fieldwork, it was common to find teenagers using their phones single-handedly whilst adopting particular postures. I began to see particular postures and gestures in performative terms, serving to demonstrate the degree of engagement or disengagement with the phone. For example, I read single-handed use with a leant-back posture as a sign of openness and willingness to be disrupted. At the other extreme, two-handed use, with the person hunched over the phone, signalled a high-level of engagement with the phone and a visible disengagement with the situation at large.

of engagement and their place in the group talk, I suggest that G2 is participating in some special work that makes her withdrawal unremarkable to the other conversationalists. By operating the phone in the way that she does, G2 is displaying her participation in the legitimate and warranted business of mobile phone use – whether that be game-play or any other form of one-person interaction. G2's display of engagement with the phone, and disengagement from those at the table, is thus the method she employs to demonstrate her participation status; it operates as a recognizable marker, projecting her availability (or unavailability) for co-participation.

In the subsequent interaction (lines 7–13), G3 and another girl at the table reveal how the phone, along with a folded piece of paper, precipitates a *side play* (Goffman, 1959) or *side sequence* (Jefferson, 1972). By drawing attention to the two artefacts – specifically the paper she has stored in the phone's cover – G3 initiates a focused and coordinated interaction. Notably, the material artefacts manage the subordinate conversation between the two girls. The artefacts come to provide a focus of attention for the conversants – turning their minds to the same exclusive subject matter – and in doing so bring about what Sacks et al. (1974) refer to as a *schism* from the original group-wide conversation. The mutual attention paid to the objects confers an agreed consensus of relevancies and irrelevancies, and operates to circumscribe an observably exclusive interchange. By attending to both the phone and paper, the two girls thus alter their participation status; they move from being members of the larger conversational group to select participants in a two-way exchange.

Turning to G4's interaction with G3 (lines 13–18), we again see that the phone is the focus of a move away from the ongoing conversation at the table. In this case, it is the phone itself that appears to be the subject of the conversation. The two girls seem to be listening to a phone's ring tone. By engaging in this interchange – leaning over the phone and consequently towards each other – the two disengage from the others at the table. Amongst the occasioned glances, gestures and posturing, the phone is bound up in the expressed signs of intimacy so as to become a basic element of the interchange. The phone comes to be included in that class of signs, or *back channel cues* (Goffman, 1981), that form part of the non-verbal communication that occurs during conversations.

All three of the above interactions hinge on the material artefacts in question being taken-for-granted resources to negotiate participation status. Specifically, by permitting mutually agreed upon topics of talk to be established and distinct foci of attention to be formed, the folded piece of paper and the mobile phone are both seen to warrant the withdrawal from one engagement and the participation in another. The successful transitions in participation status, systematically accomplished in each interaction, suggest that some common feature (or features) possessed by both artefacts is put to good use.

A common feature of both the piece of paper and mobile phone is the capacity that each has to establish a spatially limited focus of interaction that demarks the boundaries of participation. This is associated, at least in part, with the artefact's size and form: because a piece of paper and a phone are both small, only a limited number of people can attend to them at any one time. Leaving the paper to one side, the phone thus provides a legitimate reason to manage participation status, not only through its presence in occasioned talk, but also because of its particular physical characteristics.

Over and above its physical form, however, there is also the nature of the device and what it is used for. For teenagers in particular, the phone is an inherently social artefact used to communicate with friends (Weilenmann, 2001; Grinter and Eldridge, 2003). The phone is all about intimacy and sharing and it is this emotional character that further warrants its use as something to manage who is included or excluded from talk (Taylor and Harper, 2003). As we shall see, the legitimate use of both the physical and emotional characteristics of the phone has important implications for how context-dependent, multi-participant talk is practically accomplished. Specifically, it provides a means for phone users to manage the statuses that are available to those present in a situation.

10.4 Covert Phone Talk

Thus far, I have sought to show how the mobile phone serves to manage ordinary talk and how, specifically, it has become a taken-for-granted and practical means to accomplish topic management and organize participation status. This section will go onto reveal how these two practices, mediated by the phone, can be combined to engage in talk that is both subordinate to and concealed from a primary interchange: covert phone talk. In the excerpt below, taken from the interview transcripts, Lauren and Alice describe how the phone can be used to undermine a member's participation status in a conversation and engage in a topic-in-parallel.

(4. Interview – simplified: "The Mound")

1	Alex:	Is there something about the way a phone is designed
2		that allows you to share it between smaller groups as
3		opposed to larger groups? What is it about the phone that
4		sort of what I call affords certain ways of using it?
5	Lauren:	→ Well you can shield it can't you? It's not like it's so big that
6		you can't cover it-up and (.)
7	Alex:	What sort of situation would you be in where you want to
8		shield it?
9	Lauren:	Well if you were sittin next to two people and you only
10		wanted to talk to one of them about something (.) then
11		you can like cover it with your hands. But also it's quite

12		open so that you can show other people.
13	**Alex:**	So you have options for sort of shielding it or sharing it?
14	**Lauren:**	hmm
15	**Alex:**	Susie mentioned last week– I didn"t really pick up on it
16		until I listened to it the other day– that sometimes you'll
17		talk to someone and instead of speaking they'll write a
18		message and just show the message. Does that
19		happen a lot?
20	**Alice:** →	Yeah. You might be sitting next to each other and you
21		don't really want to say- oh, the most recent time this
22		happened I waz sittin at a table and three-people-and-
23		chairs Ther-s a girl sitting here and-a-girl-sittin here
24		((points to either side of her))- we don't talk to the girl
25		sittin here ((points to her right))- So she [points to left]-
26		instead of sending me a text message cos she- I dnt know-
27		she had credit, but there was just no point-
28		although sometimes we do send messages to each other
29		while we're sitting next to each other. She just typed
30		whatever she had to and she just gave it to me and
31		I look'd a'it and thor hmm. And I erased it and sent it
32		back to her. And withaut actually sending the
33	→	message you just- So i's like passing paper in a class
34		basically except you're not actually using paper, you're
35		using the phone.
36	**Alex:**	So what is it about a phone that allows you to do that-
37		that makes it something that you would do=
38	**Alice:**	=it's more- what's the word- conspicuous or something.
39		Cos with the paper it's more obvious that you're writing
40		something about somebody or you're sharing some kind
41		of secret or something. But with a phone whoever's
42		around you will just assume that she's looking at a text
43		message that you have. And she might think okay she
44		wouldn't even think twice about whatever you're doing,
45		but you've got some paper and you're reading it and
46		shielding an everything an she'll be like oh yeah that must
47		be something about me So you can afford to be nasty
48		about someone without them actually knowing and that's
49		nice.
50	**Alex:**	So it's not so obvious that it's something that's been
51		written in that situation?
52	Alice:	Yeah. Yeah. Coz for all you know it could just be some rude
53		text that you're looking at that you're going back to.

In this excerpt, Lauren reiterates how the phone's size provides a means to include or exclude people from an exchange. Importantly, she introduces a further point of interest. By describing the way a phone can be "shielded" (lines 5–6), Lauren shows that it can be used to produce a subordinate interchange where not only participation status but also conversational topic is managed (lines 9–12). Using "shield it" and "cover it with your hands", Lauren works up a sense of covertly managed information sharing that points to the phone's role in concealing information in subordinate talk. The real work of invoking the talk of concealment is accomplished by Alice, however. Alice chooses to describe how the phone can be used to conceal the topic of a subordinate interaction and consequently position particular members of a conversation as "outsiders" (lines 19–31). Like a piece of paper passed in class (line 30), the phone is seen an acceptable means to share private information because it is necessarily exclusive; its size and form simply do not allow numerous people to view it at once.

Alice's second turn suggests that there is more to this concealed form of information sharing (lines 34–43). Alice makes a crucial distinction between paper and mobile phones that has important implications for how they contribute to the accomplishment of subordinate talk. She reveals the "assumed", morally implicative character of the two artefacts is constituted differently; whereas the written-on piece of paper is seen to infer some secretive and possibly underhand exchange, the sharing of the phone is understood to be far more benign. This reveals the taken-for-granted status of each artefact. The exchange of a paper note in class, for whatever reason, has come to infer some activity with "sinister" intentions, while the use of the phone, as Alice explains, is potentially seen as innocuous because the underlying motivations of the interaction's participants are ambiguous. It seems that because the phone is commonly used to share messages with co-proximate others, there is no good reason to assume it is being used to conceal anything.

These taken-for-granted assumptions are what make the phone such a good tool for concealment. Unlike the piece of paper that must be concealed in a conspicuous manner, it is the phone's design in addition to its social character that is seen to encourage exclusivity. Broadly speaking, the phone, as an artefact that is routinely available in everyday talk, comes to afford particular forms of interaction that are plainly seen and understood by anyone. The interactions go unquestioned because they are thought to be a "natural" result of the object – to question them would be to question that commonsense that is available to anyone.

10.5 Classroom Talk

In support of the points made thus far, I now wish to turn to one further example from the interview transcripts. Although the participants in this excerpt refer to concealed interactions in the classroom, as opposed to ordinary conversational talk, the described use of the phone illustrates the role the phone can take in a subordinate interchange that subverts the ongoing accomplishment of the primary topic-in-progress. Indeed, because classroom teaching is in the business of imposing an explicit order and structure to talk (Freebody and Herschell, 2000; Macbeth,

2001), the concealed-subordinate interchanges are made all the more salient.

(5. Interview – simplified: In empty classroom)

1	Susie:	If they're doing it in lesson they tend to do it with one
2		hand so looks like you're just goin like that- you know-
3		chick chick chick ((mimics using phone under table))
4		And if you can do it without looking at your phone=
5	Alex:	=but some people can do it?
6	Paul:	→ Susie's a pro at this cos she sits in law like this ((mimics
7		texting with legs up on desk))
8	Susie:	just cos you do it huh-huh-huh
9	Alex:	So you can just nod your head and everything and still-
10		((mimics using phone))
11	Susie:	Yeah!
12	Alex:	Are there people who can text message without looking?
13	Susie:	mm ((in agreement))
14	Paul:	Not as much, cos my spelling's awful and I need predictive
15		text messaging. That's pretty bad.
16	Susie:	That was very funny. We spent the end of a law lesson
17		texting each other.
18	Paul:	→ cos we were bored and there was time to go
19	Susie:	I sit here and he sits like right there.
20	Alex:	So you sit right next to each other and sending each other
21		text messages
22	Susie:	→ I was like Paul I'm BORED! huh-huh!
23	Paul:	And Susie sits right in front of the teacher as well and he
24		never notices
25	Lauren:	Is this (teacher's name)?
26	Susie:	I might as well just- I might as well just turn around and
27		have a chat with you and he wouldn't-
28	Paul:	Yeah
29	Alex:	So how are you- how are you doing it? Are you still under
30		the table or something-
31	Susie:	No I tend to sit with my sort of feet up anyway and I will sit
32		and write cos- ((puts legs against table)) Dun know why I
33		do. So I can just hide my phone there anyway and jus-
34		hide it behind my legs.

Beginning the sequence, Susie demonstrates that the phone can be easily concealed and attention to the class can be feigned through displays of engagement, managed by using posture and gaze. Paul's turn, "Susie's a pro at this" (during which he mimics one-handed messaging), implies there is some status associated with "faking" classroom participation while texting (lines 6–7). He goes so far as to emphasize Susie's skilled use of posture to achieve the display of "civil" attention. This compliment, of sorts, is reciprocated in Susie's following turn (line 8). She replies, "Just cos you do it", inferring that Paul is equally accomplished at such acts of engagement. This two-part fragment, in which the two bestow praise upon each other, provides what might be seen as an indication of the value that is placed upon covert phone use in the classroom.

Through the remaining turns in this first sequence (lines 9–15), I seek to clarify these points. The sequence reveals that concealed interactions with the phone are in some way revered. In essence, the covert interchange is valued because it is seen as just that – concealed and subordinate to a formally ordered state of affairs. The phone, paradoxically, provides an orderly way to demonstrate disorder or to subvert what Paul and Susie know, perfectly well, to be routine classroom conduct. Through their talk, they explain that this subversion is achieved through the skilled management of engagement displays – and thus participation status – and concealed parallel-topic talk. The phone-mediated interaction is thus positioned as a valued symbol of opposition to the progress of the setting at hand.

In the second sequence of turns in this excerpt (lines 16–33), Susie first begins to produce a narrative account of a recent incident in a law lesson where she and Paul have exchanged text messages. Paul chooses to collaborate with Lauren, but elects to present a more general description of texting in law class. Both describe the boredom that is incurred by sitting in a law lesson. Their issuing of the term "bored" serves two purposes (lines 18 and 22). First, it provides a legitimate reason for their subordinate exchange of messages. Second, it can be heard as the document of a particular orderly state of affairs in the classroom that is commonly known about and adequately understood for all practical purposes (Garfinkel, 1967). It is this orderly situation – one that is boring – that is shown to warrant the concealed exchange of messages.

Paul goes on to describe how Susie is able to text in front of the teacher without being noticed (lines 23–24) and, in doing so, presents her achievement as worthy of recognition. Susie's response is to exaggerate the possibility for concealed parallel-topic talk in class by suggesting she could very well have turned and spoken to Paul without being noticed. This statement might be seen to achieve two things. First, it accepts Paul's "compliment" but positions message exchange as something that can be easily accomplished in class. Second, it dismisses the law teacher as inattentive and possibly deserving of such antics. Again, the subordinate interchange is set against the orderly conduct expected in class and performed in stark opposition to it.

We must, of course, accept Susie's word if we are to take her description as a reasonable indication of actual practice. To my mind, what is more important than determining whether Susie's depiction is accurate, however, is the recognition of the role her account plays in the interview. By presenting her behavior in such a way, Susie is engaging, at least in part, in the business of establishing her credentials as a skilled phone user (one of a particular type); she participates in the work of rendering her phone-mediated actions the actions of one "who knows". As one in the know, Susie, with help from Paul, demonstrates that the phone is a legitimate resource for organizing and managing participation status and parallel-topic talk. She also reveals that the phone provides a means to achieve talk that subverts classroom order.

10.6 Subversion

Moving on, I would like to attend, more closely, to what I term *localized acts of subversion*. In this section, I will briefly contrast local acts of sub-version with the broader discourses on sub- and counter-cultures, and then summarize how the material I have presented relates to the former. Although I cannot claim that acts of subversion are explicitly referred to in the source material, I believe I have provided sufficient grounds to con-template how such a term might be used to interpret what has been dis-cussed thus far.

The literature that attends to teenagers' subversive activities tends to be concerned with large-scale social movements. Subversion has been understood in terms of teenagers' expressions of resistance to "popular" culture – invoking and aligning themselves with counter- or sub-cultures through cultural objects such as music (e.g. Thornton, 1995; Redhead et al., 1997) and dress (e.g. Hebdige, 1998) or drugs (e.g. Willis, 1976). Such theorizing speaks of "themes of resistance" or "alternative narra-tives", politicizing the daily lives of teenagers and providing the backdrop for critical investigations into class, wage, locality, gender, age, etc. (Cohen, 1987; Widdicombe and Wooffitt, 1995).

The products of these works are, without doubt, important commen-taries. However, their end result is to position teenagers' locally accom-plished, routine activities as political events contributing to and arising from rarefied, anti-egalitarian visions of modernity: it is to understand everyday activity in terms of such totalizing ideas as cultural hegemony (Hebdige, 1998) and the altogether disparaging spectre of the *New World Order Inc.* (Haraway, 1997) – subjecting them to and incorporating them into the grand narratives of contemporary society.

The data presented here might be seen to reveal subversive behavior in a different light. I am inclined to interpret the transcript and field note excerpts as examples of the methods that teenagers can employ to coun-termine the recognized order of an occasion. One way in which teenagers make their actions accountable is to situate them in opposition to partic-ular practices and as concealed from specific members who participate in the orderly accomplishment of these practices. To my mind, this means of making sense of patterns of behavior can be seen to invoke a form of sub-version. This view depicts the subversive act as a concealed, locally assem-bled resistance against an established set of social structures or "rules".

The excerpts presented indicate that there are two discursive devices used to manage these local acts of subversion. One, that the management of topic – topic-termination, -change, -in-parallel and such like – offers a means to engage in talk that is subordinate to a group's talk in progress. The second, managing participation status, reveals subordinate talk can be organized to exclude particular members of an ongoing conversation. These discursive devices, used together, make available a means to subvert group talk by allowing particular members of a conversation to engage in sub-topical talk that is plainly exclusive.

The excerpts further indicate that the *mobile phone* and its material features provide teenagers with a means to participate in locally subversive forms of talk. The phone, because of what it is, how it is used and how it has come to be understood, makes available a means by which teenagers can change the topic of talk, conceal subordinate talk from particular members of a conversation and use talk to challenge the ordered progress of an ongoing conversation. Indeed, I would go so far as to suggest that such acts of subversion are routinely provided for by way of the phone's taken-for-granted presence in talk between teenagers.

10.7 Conclusions

In summary, the presented research has aimed to show that teenagers use their mobile phones to assemble and organize the topic of talk and participation status in local group settings. It has further revealed that, in accomplishing these conversational achievements, the phone provides teenagers with a means to subvert the ongoing progress of group-wide conversations. Such local forms of subversion are accomplished through phone-mediated interactions that exclude particular members of the conversational group and allow the participation in topic-in-parallel talk. This interpretation is seen to be interesting because it suggests the mobile phone has particular properties that allow it to be enlisted as a resource in teenagers' ordinary talk.

In conclusion, I must emphasize that the above analysis is presented as one means of understanding the situated use of technology and specifically the use of mobile phones amongst teenagers. The phone, I have shown, has particular features that enable it to be used, by teenagers, as a taken-for-granted resource for "doing" ordinary talk. Without doubt, there are a multitude of methods that might be applied to accomplish such analytical work. Also, the roles played by many other objects might be examined to see how they mediate teenage talk.

10.8 Acknowledgments

I am indebted to the students who participated in this study for their time, enthusiasm and invaluable insights. I would also like to thank the sponsors of this study: BT Cellnet, Granada Media Services, Mercury One2one, Orange and Vodafone. Funding for this project was arranged by the DTI (Department of Trade and Industry) under the Foresight Link program. My thanks also go to Nokia for supplying new phones for the students who participated in the interviews. I am also very grateful to Richard Harper for his supervision and support on this project, and Anne Cohen Kiel for her time and patience in helping me to present this work.

10.9 References

Button, G. (1991) Conversation-in-a-series. In Boden, D. and Zimmerman, D.H. (eds), *Talk and Social Structure: Studies in Ethnomethodology and Conversation Analysis*. Polity Press, Cambridge, pp. 251–277.

Cohen, S. (1987) *Folk Devils and Moral Panics: the Creation of the Mods and Rockers*. Blackwell, Oxford.

Freebody, P. and Herschell, P. (2000) The interactive assembly of social identity: the case of latitude in classroom talk. *International Journal of Inclusive Education*. 4, 43–61.

Garfinkel, H. (1967) *Studies in Ethnomethodology*. Prentice-Hall, Englewood Cliffs, NJ.

Goffman, E. (1959) *The Presentation of Self in Everyday Life*. Doubleday, Garden City, NY.

Goffman, E. (1981) *Forms of Talk*. University of Pennsylvania Press, Philadelphia, PA.

Grinter, R.E. and Eldridge, M. (2003) Wan2tlk? Everyday text messaging. In *Conference on Human Factors in Computing Systems, CHI 2003*, Fort Lauderdale, FL. ACM Press, pp. 441–448.

Haraway, D.J. (1997) *Modest_Witness@Second_Millennium.FemaleMan©_Meets_OncoMouse™: Feminism and Technoscience*. Routledge, New York.

Hebdige, D. (1998) *Subculture: the Meaning of Style*. Routledge, London.

Jefferson, G. (1972) Side sequences. In Sudnow, D. (ed.), *Studies in Social Interaction*. Collier-Macmillan, London, pp. 294–338.

Jefferson, G. (1994) On stepwise transition from talk about a trouble to inappropriately next-positioned matters. In Atkinson, J.M. and Heritage, J. (eds), *Structures of Social Action: Studies in Conversation Analysis*. Cambridge University Press, Cambridge, pp. 191–222.

Macbeth, D. (2001) On "reflexivity" in qualitative research: two readings, and a third. *Qualitative Inquiry*. 7, 35–68.

Redhead, S., Wynne, D. and O'Connor, J. (1997) *The Clubcultures Reader: Readings in Popular Cultural Studies*. Blackwell, Oxford.

Sacks, H., Schegloff, E.A. and Jefferson, G. (1974) A simplest systematics for the organization of turn-taking for conversation. *Language*, 50, 696–735.

Schegloff, E.A. and Sacks, H. (1973) Opening up closings. *Semiotica*, 7, 289–327.

Taylor, A.S. and Harper, R. (2003) The gift of the gab: a design oriented sociology of young people's use of mobiles. *Journal of Computer Supported Cooperative Work (CSCW)*, 12, 267–296.

Thornton, S. (1995) *Club Cultures: Music, Media and Subcultural Capital*. Polity Press, Cambridge.

Weilenmann, A. (2001) Negotiating use: making sense of mobile technology. *Personal and Ubiquitous Computing* 5, 137–145.

Widdicombe, S. and Wooffitt, R. (1995) *The Language of Youth Subcultures: Social Identity in Action*. Harvester Wheatsheaf, New York.

Willis, P.E. (1976) The cultural meaning of drug use. In Hall, S. and Jefferson, T. (eds), *Resistance Through Rituals: Youth Subcultures in Post-war Britain*. Hutchinson, Birmingham, pp. 106–125.

Mobile Camera Phones: A New Form of "Being Together" in Daily Interpersonal Communication

Carole Rivière

11.1 Introduction

Roland Barthes concludes his essay "La chambre claire" by stating that "what characterizes so-called advanced societies is the fact that Society consumes images now instead of beliefs." He writes, "Pleasure is experienced via the image: this is the great difference" (Barthes, 1980). In analyses of the transition from modernism to post-modernism, society's saturation by images and their increasing impact on relationships between people, things, knowledge, imagination, events and information is essential to understanding cultural and social changes. The "image" industries have been the subject of close examination and endless interpretation. Attention has focused on industries, such as the television industry, advertising and the cinema, and their mass effect. The more intimate role of the image in interpersonal relationships has remained marginal. Put simply, we could say that two key issues in the social sciences are involved here: first, the analysis and study of specular (mirror) images and their imaginary role in the development of the self and the identification process; and second, sociology and ethnography with studies on the social and family-related function of photography following Bourdieu's definitive study on "the average art" (Garrigues, 1996). However, even in the second domain, there are still very few sociological studies on photography as a social, domestic or everyday practice, possibly because of its status as an artistic practice and as a stylistic and singular representation of the world.

With technological progress, the role of the photographic image has now become central to interpersonal communications, which apparently involves a radical transformation of its everyday social function. In fact, by the mid-1960s, Bourdieu described the photograph's social function as a way of celebrating and making the important moments of family life last, strengthening integration within the family group by reasserting the feelings that it has about itself and its unity. Considered as joint family property, the camera, widely used at family celebrations, thus served above all to fix the image of the family unit's major events (Bourdieu, 1965).

Considered to be the most faithful means of reproducing reality, photography in the context of personal uses was called upon to represent the family group at the times and in the places and poses defeating the passing of time, which best symbolized the memory of good times spent together. Although this family-centered function of photography is far from having died out, technological progress has extended the use of photography to more and more diverse situations and rendered its use commonplace where it was once reserved for special occasions. According to a study carried out by the Research Centre for the Study and Observation of Living Conditions (the CREDOC, cited in *Le Monde*, 2003), "in 1965, 40% of individuals stated that they took photos; today, 81% of the French population practices, at least occasionally, amateur photography." According to an earlier survey carried out by the INSEE (France's National Institute of Statistics and Economic Studies) in 1989, the rate of photographic practice was 62% at the time, which leads us to conclude that disposable cameras have had a major impact on the increasing popularity of photography since 1989. Furthermore, and still according to this same survey, "different social groups and age groups now practice photography in a similar fashion. The youngest and oldest members of the population have made up for their relative delay compared with the age groups in between (30–50 years old)" (*Le Monde*, 2003). Finally, since 2001, sales of digital cameras have exceeded those of conventional cameras at an adoption rate that is "faster than that for the refrigerator and the television and similar to that of the mobile phone" (CREDOC, 2002).

The development and availability of digital cameras and the current fusion of digital photography and the mobile phone – the technological innovation that enables you to take photographs with your mobile phone and send them to friends and family – brings photography into interpersonal relationships as a new form of "scripto-visual" communication. Continuing on from the success of SMS (short message service), telecoms operators now talk of MMS (multimedia messaging service) in Europe to describe these new services, which, furthermore, lie at the heart of the scenarios being developed by the manufacturers of tomorrow's mobile society (Koskinen and Kurvinen, 2002).

From our point of view, this enriched form of communication is of interest in examining the image's function as a medium for social relationships insofar as concerns its visual and imaginary aspects. Based on analysis of the representation of the technical object that produces it, we will first show how the mobile phone creates the conditions for a new photographic practice. We will then give examples to illustrate changes in the use of and the social significance of photography. Nervertheless, there is an opposite issue of bullying or victimizing people through the use of mobile photography. While the total effect of mobile photography may be to bring people together, it may also estrange people or be used to enhance power differences. This point of view is not discussed in this chapter for several reasons. First, it did not appear strongly in the respondents's speech, and second, because we chose to develop an analysis around the new function of photography in a context of friendly relationships and then because our study took place in Japan, where a large number of persons have a mobile-camera (sha-mail), so that it seems to be very usual and considered like a sign of integration and not like a power sign of distinction.

The empirical material that we have consists of a qualitative survey, carried out in Japan in July–August 2002, on the reasons for using mobile phones and the situations in which people used them within the framework of a study carried out by the UCE laboratory (Usage Créativité Ergonomie – ergonomic design and use) at France Telecom Research and Development's (FTR&D). A series of semi-structured interviews lasting approximately 2 hours were held with 41 people, in the capital, Tokyo (32), and in Osaka (9) on the reasons for using mobile phones and in what situations they are used, and also on the sociability of the participants surveyed (personal relationship networks, going-out "rituals", etc.). More specifically, 25 interviewers owned a phone with a camera called "sha-mail" in Japan that we discuss in this chapter. The interviews were held in Japanese in the presence of a professional simultaneous interpreter. A recruitment company selected the people who took part in the survey based on socio-demographic criteria and the equipment used.

The fact that the study was carried out in Japan and performed by a western sociologist in the presence of an interpreter obviously had some effect on the nature of the results obtained. How far does analysis of the discourse produced remain relevant as an aid in examining the function of the photographic image in social relationships? Insofar as Japan has a very specific social structure, where social relationships are organized according to a traditional system of relating to others which is very different to that in, for example, France, we shall not refer in this chapter to the analysis of this social practice from the point of view of the specificity of Japanese culture and intercultural differences. Nonetheless, we will consider the image's new function in interpersonal communication as a common trend in advanced industrial societies. We will analyse the uses of photography described by the people questioned as part of the Japanese survey, to examine this medium from the point of view of the changes it brings about in everyday exchanges. Two other surveys showed similar results on the same subject: the first was an "experimental" Finnish study, carried out between July 1999 and February 2001, that used an eth-nomethodologically oriented conversation analysis to analyze the mobile visual messages between five groups of individuals who were lent camera phones for a period of several weeks (Koskinen and Kurvinen, 2002). The second was a more general French study on new interpersonal communication practices, in which a subsample group owned a mobile camera phone (Rivière, 2003). The common points in the results of all three studies concerning photography's uses as described and situated in context in daily mobile interaction seem similar enough to us to validate the value of the Japanese survey as empirical material for our study of the mobile image in daily communication. Given the experimental nature of the Finnish survey and the limited uptake of mobile photography which is still in the process of developing in France, we decided to retain an element of consistency in the observation and analysis of real and well-established uses in Japan, in order to give concrete examples in our analysis.

11.2 The Effects of Integrating Photography into Mobile Phones: A New Space for the Social Representation and Function of the Photographic Object

With the photograph, everyone recognizes that it is able to represent reality, which distinguishes it from other forms of pictorial representation. In their analysis of press photography, Castel and Schnapper (1965) emphasize "that everything contributes to bring photography, which is, above all else, an unbiased way of recording reality, closer to the Press, whose role is to communicate real human action." This assumption that there is a "natural" connection between using an image for bearing witness to an event and using a communication media to broadcast it may be extended to the combination of the mobile phone and the photographic image. Once we have got over our initial wariness of the idea of such sophisticated and innovative technology and the possible perception that it is a worthless gadget, using images to communicate with close acquaintances by means of a mobile phone leads to a wide range of daily uses, maintaining the mobile's social function of "reliance".

The technological development and marketing of the latest generation of mobiles has given us handsets with colour screens of a decent size and quality, making digital photographs both legible and acceptable on them. Moreover, such handsets enable us to communicate by e-mail and/or SMS, connect to an increasing number of mobile Internet sites, connect to our computer's e-mail inbox and provide more memory space for saving and keeping texts, music and pictures. Integrating a camera with a mobile phone means that we can take pictures, framing them with the help of the phone's screen, view them on the screen, select one as our screen background image, store and archive them in your phone's memory and, finally, send them to our friends, to their mobiles or PCs.

In addition to such hi-tech performance, it is their effects on photography's social function and use that are of interest to us. By integrating with a mobile phone whose characteristics are compatible with the tool's portability and the correspondent's permanent availability in addition to the instantaneous nature of the exchange, the photographic image will gain an everyday usage value very similar to photography's function of representing events in the daily press., Furthermore, the mobile pushes the concepts of real time and the uninterrupted continuous link to the limit and photography, fixes and immortalizes the bond in time, thereby negating the passing of time and the distance that separates you from the Other. The meeting of these two imaginative modes creates the conditions for an emotive experience that is over-charged with connotations within a symbolizing function of being together. In their domestic use, the two practices are supported by representations that register the relationship with Time and the relationship with the Other within a positive imaginary of the interpersonal bond and, even more so, the intimate bond.

11.2.1 Making Photography Commonplace by Using It Daily Without an Aesthetic or Archiving Purpose

Combining a mobile telephone with a camera significantly transforms photography's social function by integrating the possibility of producing and exchanging photographs with an interpersonal communication medium. This does not mean that one becomes the other and, in common representations, the camera is still associated with a specific practice that is laden with meanings linked to its traditional social function. This is why we will talk here about making the photographic act, rather than photographic practice, commonplace.

When it becomes part of the daily experience of using a mobile phone, photography departs from the realm of the occasional, or even the exceptional, that gave it its traditional function. Moreover, the photographic act is disassociated from the possession of a unique, specific object, the camera, whose existence and representation consolidated the perception of photography as a specific practice reserved for certain occasions, for specific events. As Bourdieu reminds us, the camera, considered then as joint family property, was brought out at the family unit's special reunions, intermittently and relatively infrequently, and was limited to certain occasions (Bourdieu, 1965).

In contrast, the status of the "mobile" telephone is primarily that of a "prosthetic" object, it is part of its owner:

> We always take the mobile with us everywhere we go. (Mika, student, aged 21)

> It's less of a burden to carry around than a digital camera. (Mrs Takana, aged 28, working)

> It's the sort of camera that I can easily take out in comparison to a digital one. (Mr Aoki, aged 40, engineer)

This status of the mobile explains also the eruption of the backstage in public places analyzed by Fortunati (2003), for whom the "incoming or outgoing calls completely modify information about self-presentation". In the same respect the success of written communication by e-mail and SMS in France and in Japan has been observed (Rivière, 2002; Rivière and Licoppe, 2003). Furthermore, in new representations of the photographic act, it is not uncommon to find the need to make a formal distinction between the traditional practice of photography and the new sphere of meanings and uses with which it is endowed thanks to the mobile phone. The latter brings photography into the 21st century as an agreeable form of communication or language, one that can be used by anyone, anytime, anyhow. In this sense, it makes photography "commonplace", stripping it of every intention other than for one's own pleasure and the pleasure of expressing something in the immediate present.

> When I want to take serious photos, I use the digital camera. Sha-mail[1] is for photos that are for fun. When I want to keep a photo, I use my digital camera; it has one million pixels. For everyday use, when you have no specific reason for taking a photo, sha-mail is fine. (Sanae, aged 32, graphic designer, female)

> If I want to remember a scene, I take a photo with a camera. (Mi Isii, aged 20, student)

Here, the social perception of the technical object that produces it frees photography from its conventional uses associated with aesthetic or archiving purposes. Its everyday value is bound up in the chance encounter, the unexpected. It is in this sense that mobile photography gains a usage value similar to that of reportage photography in the press.[2] It is the unpredictable that is preferential, regardless of the subject and however banal the subject that we photograph. On the other hand, anything that reminds us of the posed photograph is rejected and/or

---

[1] *Sha mail* (picture mail) is the Japanese name for a mobile with a camera.

[2] In his analysis of the photographers for *France-Soir*, R. Castel notes that "photography of everyday life does not entirely draw its intrinsic value from what is represented, but rather from the exceptional nature of the encounter between a chance event and the photographer ... As photographs of the dramatic, of the speed of the dramatic and of the unpredictable, *France Soir* photographs are necessarily instant."

perceived as relative to conventional photographic practice because it can be anticipated as the respondents tell us above. Any event in the private world can be used as the subject for communication and the account of one's self, the selection criteria for which, as we shall see later, carries a more intense yet spontaneous degree of emotion, strangeness and rarity compared with the expected norms of social behavior. The snapshot becomes an extension of the way in which one sees oneself and it gives value to communicating with other people.

> If I come across an event that surprises me, I want to be there to take some photos. In Shibuya, there are always protests and concerts that I come across by chance. If I know that I'll be there, I take my digital camera. (Mrs Seki, aged 30, employee, Foma)

> I photographed a cat that I saw by chance on the road. (Mrs Seki, aged 30, employee, Foma)

11.2.2 A Visual Instant Communication Function and a Function for the Imaginary Symbolization of Being Together

Photography integrated into a mobile phone also entails a transformation of the photograph's relationship with the past and its emotional charge. In 1965, Bourdieu pointed out that "the need to take photographs is nothing more than a need for photographs that, thanks to their function of reproducing reality, bear witness to and express the truth of the memory" (Bourdieu, 1965).

Taken up by Barthes, this relation to time, described as "that has been", reflects the unique position of photography, which juxtaposes reality and the past, thus creating that emotional charge in which reality and life, reality and the present are blurred, imbuing in the authentic re-presentation of that which has irredeemably "been" with a particular intensity and brilliance (Barthes, 1980). We can use the juxtaposition described by Barthes to distinguish between two different aspects of the photograph: its power to authenticate and its power to represent. The former refers to its value in ratifying the world, its value as the indisputable truth, the certitude that what is represented in the photo really existed. The latter refers to the photograph's "punctum", that which is piercing, "which shoots through me" in the photograph and which feeds the emotional and often nostalgic (although not necessarily) perception that one may have of the object present in its absence.

By highlighting the photograph's power to record time, rather than the actual subject of the photo, Barthes emphasizes its power to ratify a past situation, its power to bear witness, the reason why we want to immortalize what is important to us by taking a photograph of it. Sontag (1983) expresses the same thing when she writes that, "in current practice, photographic activities are of little importance unless the photos are taken and kept with love."

With the mobile phone, the representational aspect is placed at the service of the real, not of the past but of the present. Almost as soon as it has been captured, the photographic image appears on the screen. It can then be shared with the people that we know, thus consolidating a way of being part of the world, portraying and symbolizing what is experienced together. The photo can be sent and received on a mobile phone, accompanied by text or not. Photography thus acquires the status of an instant communication medium that is nearly as sychronious as voice and text messaging (SMS and mobile e-mail).

> Most of the time, I send the photo straightaway. When I want to send the photo immediately, the fact that I can do so is great. Sometimes, when I'm traveling, for example, the sunset is a magical moment so I send it. At the same time, I don't know what the other person does with it, but for me, it is a way of showing what I can see. (Eita Kido, aged 25, working)

By laying the emphasis on a visual mode of communication, photography first acts to reinforce the viewer's capacity to perceive things in hyper-real mode. Integrated in actual messages sent, it then represents a relationship of substitution, since the criteria of resemblance and legibility that define photography work to enhance the capacity of the message one wishes to communicate to be objective and authentic.

> When I changed my hairdo, I took a photograph of myself. I sent it to my best friend. She would do the same. (Tie Satou, aged 31)

At the same time, photography operates at the level of emotional perception and increases our capacity for emotion and to feel "together" (as opposed to operating at the intellectual level). This implies that what is mobilized is less what we see objectively than what the photograph signifies symbolically and what it refers to, from the point of view of the imaginary. Whether it is in the immediacy of the exchange between mobile phones and/or in the instantaneousness of the photographic act that we see photography's power of representation, a new imaginary and visual language is introduced, one which upholds and reinforces the emotion inhering in an imagined "being together".

> I can imagine the atmosphere better with a photo. If they only send text saying "we're eating and drinking", it's not quite the same. (Kinebushi, sales rep., aged 39).

> It gives you a good feeling to receive a photo. If it was of a cat, I thought the cat was cute. If it was of friends in a restaurant, I can sense the lovely pleasant atmosphere. It's the same when you send photos, it's easier to convey what you are feeling, to get across your state of mind. (Mrs Seki, aged 30, employee, Foma)

If the photographic image here works as a symbol of "being together", it is because it refers to a familiar and shared world which each member of the group is able to decode. It is not so much the objective legibility of the photograph that produces the feeling of being together as what is referred to in each member's memory, the shared meanings that each gives to it. To use Castel's terms, we might say that the image functions here as a symbol in its passive aspect, i.e. by referring to pre-existing collective knowledge, nourished by a bank of memories and perceptions that is built up day in, day out. The image here makes a leap and expresses the existence of an imaginary with symbolic value (Castel, 1965).

In the same view, Harper and Taylor's analysis of text messages exchanges shows how mobile phones have specific meanings in young people's daily lives. The authors show how text messages can be thought of as forms of "gift-giving". The value of the "text-gift" is connected with the intention of cementing social relationships. By sharing mobile contents, mobile credit communication and the mobile phone itself, teenagers express their relationships with one another, "demonstrate and preserve their commitment to the relationships (...), offer each other an intangible show of trust and loyalty". So, say the authors "phones, in this sense, give young people something to talk about amongst themselves, providing them with yet one more mechanism for sharing their emotional experience and exchanging objects of personal significance (Harper and Taylor, 2003). A part of the value of the sha-mail (picture-mail) is created by the same model of reciprocity exchanges.

11.2.3 The Life of a Photograph: A Cross Between the Fleeting Temporality of the Memory and the Magical Presence of the Other (When Absent) Brought into Sight

The integration of the photographic image within a world of instant communication also comes with a lack of interest in the value of the photograph as a photograph-as-memory. This is another major change in the way we relate to photography. The digitization of the image that operates the switch from an object-image photograph to a screen-image photograph dematerializes the photograph as a memory-object that can be kept, handled and even displayed. However, there is nothing particularly new about this: indeed, there is nothing to stop owners of digital cameras from printing their photographs on paper and reserving a special place for them as the memory and record of a past event that is commemorated and maintained every time they look at them.

Mobile photography becomes a routine form of instant expression and communication because of the interactive model of the phone call, i.e., short oral communication that exists only for the amount of time that the interaction lasts. It is also conditional upon the fact that photographs taken and sent via mobiles are not of such good quality as those taken with digital cameras. Although we shall take this last factor into consideration, it is not, in our opinion, a determining factor. It is really photography's new social function, which consists in consolidating the feeling of an "image of being together" in the immediacy of the exchange and/or of the encounter, that goes to explain why the storage of the photograph is becoming less important.

Technically speaking, it is, of course, possible to conserve the photographs in the mobile phone's memory or as a background screen image. Depending on the phone's capacity, several dozen photographs can be kept or even sent to be stored on computer. Nonetheless, the purpose of taking a photo-as-memory remains marginal and the normal "life" of a photo is consistently deleted as more recent photos take its place.

> Generally, when a photo comes with an e-mail it is kept with the e-mail. Photos are deleted in time along with the mails. (Melle, Aya, aged 20, student)

> I don't save any photos because the image quality is not good. I delete as soon as the memory is full. (Isii, aged 20, student)

In the phone's memory, however, a special place is given to the photograph that may be used as a background image. This indicates a desire to maintain the presence of an absent loved one via their image, thereby ensuring the continuity of the symbolic bond.

> I have a photo of my baby as a background image. (Tie Satou, aged 31, housewife)

> I've got a photo of my girlfriend. (Igarassi, aged 20, student)

This form of transport (in all senses of the word) calls upon the memory not at the level of memories of the past but rather at a ubiquitous level. This relates back to the traditional practice of carrying a photo of a loved one with us wherever we go (in one's wallet, at the office, etc.), a practice that is more in line with a magical belief in the power of the image to represent, consubstantially as it were, the person or object represented.

To conclude this first part, we would say that the mobile phone captures the photograph and what it re-presents within a contracted and almost completely fused space–time continuum which transforms the role of the photograph without actually reinventing it. We will now go on to question the significance of the photographic act in more detail within various use situations that have been observed and see what they tell us of our relationships with the world and with other people and with regard to the values perceived to be important enough to trigger the desire to take a photograph.

11.3 Daily Use: Situations and Motivation – Meanings of the Photographic Act Become More and More Diverse

A photograph is always the result of a deliberate choice, a process of selecting what we perceive that results in a choice that is more or less conscious and through which we operate a foregrounding of one reality over another. What can be photographed is a reflection of social norms which are more or less obvious depending on the contexts (Castel, 1965). Bourdieu also described the social norms to which family photographs have to conform in order to be considered to be good photos. This might mean, for example, a posed attitude, wearing your Sunday best, not being taken by surprise in an ordinary outfit or task and, lastly, avoiding anything that would have gone against the rules of propriety and convention agreed by the group. More recent studies carried out in the USA show that:

> while photographs still play a central role in the imagery of US households, they show the family in quite a different light than that implied by the information analysed by Bourdieu. The photographs on display in the contemporary home do not show a romanticised version of the family and its value system. In fact, the people are shown doing things, rather than sitting and standing in symbolic poses as in the traditional family photograph. Family portraits have been replaced by photographs that show people acting and interacting in an informal manner. (Halle, 1993)

Hence we may well wonder whether the new liberty involved in the private use of photography, and its becoming a function of interpersonal communications, refers to a new set of social meanings. Does it define a set of specific expectations and norms shared by the group within which the photographs are exchanged? Does it, moreover, point to a way of representing oneself or the aspiration to be what one photographs?

In other words, what do you mean to yourself and to others when you take a photo? Psychoanalytical research such as that carried out by Tisseron (1996b) indicates that taking photos is an attempt to appropriate the world by means of each gesture involved. Thus, his definition of photographic "activity" includes the billions of hand movements required to press the shutter and not just the developed and printed photographs. The photographic act therefore backs up the act of thinking and participates in the process of making the world symbolize something, of psychically assimilating the world around us – in sensory-motor terms (the movements involved in framing the shot) and in emotional terms (the emotions felt) and in verbal terms (the photos discussed and described).

This finding seems all the more interesting for our purposes in that the ratio between the number of photos taken and number of photos printed has increased considerably since the emergence of digital photography.

People take twice as many photos with a digital camera (an average of 24 compared with 13 in France) but they only print between 5 and 10%.[3] The photographic act of using a mobile phone makes this phenomenon even more widespread since it contributes to the formatting and the communicating of the world as image without any intention to archive such images and with little desire to remember. The greater liberty and simplicity provided for the individual to take photographs and create images reflecting his/her perception of the world mean that there is a much broader variety of events to which he or she may wish to give such a special place.

So, what are the choices made by people who have mobile camera phones? By analyzing the situations in which this is used, we see that a special place is given to the desire to fix emotion insofar as it is unexpected and spontaneous, whether this is in the form of a festive atmosphere, an unusual situation or a sense of beauty when seeing a landscape. The photographic act itself becomes a fun way to explore the world that is shared together. In addition to this most popular configuration, there is a second, more practical, use category which consists in optimizing the efficacy of a call using the visual properties of the image. Lastly, there is an intermediate category in which the values assigned to the emotions and to efficacy merge in the practice of using your mobile background image to carry around a photo of someone you love. There is also the issue of economics. Although this may not be a priority, I would argue that all matters conflate to make the mobile picture right for that moment.

11.3.1 Fixing an Unexpected Emotion and the Pleasure of the Present Moment

Unlike the use of a camera, which is usually related to anticipating an event in daily life (a family celebration, holidays, an organized cultural event, etc.), the mobile camera phone is the tool for taking snapshots par excellence. The desire to fix a specific moment is triggered by the unexpected thrill of a situation, a mood or event, a feeling experienced spontaneously. Concluding their analyses of visual messages in an interactive situation, the authors of the Finnish survey mentioned earlier also point out that "humor and the expression of feelings appear to be salient features in the production of mobile images".[4] "In contrast", they continue, "a much lower number of 'useful' calls were reported" (Koskinen and Kurvinen, 2002).

---

[3] Source: report published in "20 minutes", March 11, 2003.

[4] In the examples quoted and analysed, the focus is mainly on photographs of landscapes, friends and funny situations intended to communicate feelings of well-being, to tease or defy, or for the purposes of sexual titillation or providing trivial information about oneself, one's friends and one's environment.

There are at least three interpretations that we could give to this way of capturing the present moment. Thanks to the simplicity and regularity with which it can be used, the camera phone helps to de-solemnize the exceptional as a group symbol. Based on the photographic choices made, focused on capturing inner emotions and/or shared pleasures, we shall also develop the idea that it makes possible a process of empathetically participating in the world and produces an effect of acculturating intimate experiences.

11.3.1.1 Festive Atmosphere

> When I go for a drink in a restaurant, I take photos of my friends. For example, one of my favourites is a photo taken in a restaurant with friends celebrating a birthday. They all look like they are having fun. There's another showing 5 or 6 friends that I went to the beach with. All 5 or 6 faces are in close-up. (Sanae, aged 32, graphics designer)

> I usually take photos of my friends and colleagues at the karaoke bar. (Kinnebushi, aged 39, working)

Parties have always played a central role in analyses of rituals that form cohesive bonds between group members and develop the feeling of belonging to the group. Durkheim analyzed the family party as a rite proper to the cult of domesticity whose function is to renew and recreate the group (Durkheim, 1897). Bourdieu associates photography with the party, demonstrating that it provides a way of rendering the high points of family life more solemn. In a certain sense, the decision to capture the party atmosphere as an image refers to a continuity in the role of photography through symbolizing each member's participation in the group. But difference with the camera phone is that this participation is sought outside any institutionalized form of ritual, and the experience is disconnected from its solemnization aspect which made the festive occasion coincide with a conventional special occasion. Where photography may once have been assimilated into a technique of repeating the celebration as a special moment in exceptional situations and outside of daily routine, it now becomes the tool used in the objectivation of the individual in all his or her uniqueness. The photographic and communicative act itself is imbued with value as a fun activity and a social sign of complicity.

> I usually take photos of my friends in restaurants and bars. For example, friends acting up in the bar. There was one time, some friends were really drunk. They were having fun in a karaoke bar, singing and dancing. Just for a laugh, two boys kissed each other (they were drunk). So I caught them on film. (Hosogai, aged 28)

One might almost say that the new way of using the photographic act is developing in opposition to traditional social norms, against which people adjust their position in order to conform to the social expectations shared by the group. At the same time, the photographic act is also freed from a specific awareness of time related to the event photographed. According to Castel, taking a photo implies a certain distance in relation to the present, intensified by the feeling that the instant you want to capture has already passed. It was from this that he observed a desire in "classic" photography to break with the everyday, with subjects related to leisure activities and holidays, for example. Such a break with everyday life introduced, he thought, a certain distance in time from the event. In contrast, with its desire to capture the moment, the everyday, and the lack of desire to keep a record of the event that quickly "turns into" the past, mobile photography blurs the awareness of time in which photography traditionally participates. Indeed, when we photograph ourselves in an inebriated state, when we have little control over our gestures and attitudes – this could even be regarded as anti-social – and what is revealed is a form of consciousness of the social constraint implied. It is as if one were acting on a desire to see oneself objectively as existing outside one's social role. The unexpected, spontaneous emotion then reveals a quest for self-authentication that the photographic image captures and authenticates.

11.3.1.2 Landscapes, Unusual Situations and Chance Encounters

We can look at photographs of landscapes from the same perspective. There is no aesthetic intention or intent to conserve that can explain the photographic act. It is a pleasure of the senses which is expressed through a landscape reflecting a more intensely happy moment that one wants to capture either for one's own pleasure or to share with one's correspondents. This may involve the beauty or the unexpected aspect of a landscape or situation that one comes across which triggers an emotion that is not part of the expected order of things.

> When I find something pretty, a field of sunflowers, a sunset or fireworks, I take a photo. (Mrs Tikada, aged 28, employee)

> I use it when I'm traveling. I get someone to photograph me with friends and in front of landscapes and I send them to other friends. (Isii, aged 20, student)

> Last spring I photographed the cherry trees in blossom. During a trip to Karuisawa, I took a photo of the Shiraito cascade. (Eita Kido, aged 25, working)

In general, as Tisseron recalls, when the individual is overcome by something he or she sees, he or she comes up against the problem of introjecting the resulting new experiences into his/her ego (i.e. accepting them, making them familiar, finding a place to put them in the psyche), regardless of whether the experience is one of strangeness, beauty or horror or simply an echo set up in the conscious or unconscious mind by the sight. This is the sense in which he talks of the photograph as an attempt to fix one's presence in the world. For Tisseron, photography is a way of empathizing with the world which is not so much an attempt to hold back the passing time (which would be its traditional family-oriented function – to fix and immortalize solemn scenes in a bid to halt the passing time) as a way of trying to touch the fleeting fragility of time as it passes (Tisseron, 1996 b).

Observing the use situations described above, we can say that to capture the emotion, the unexpected by means of the image is also to attempt the extraordinarily difficult operation of assimilating a sensitive relationship to the world and to other people. The fact that the photographic act is becoming more common and more commonplace may then make it another way for the individual to control the emotional events which are experienced but which cannot be fitted into a rational logic of understanding, which, in this sense, are "beyond him/her". Can we talk here of an attempt to "acculturate" inner emotion? If we base our opinion on a history of culture in the anthropological sense of the term, defining culture by means of its opposition to Nature, technical progress thus comes to constitute one of the primary vectors through which humans appropriate their environment and has developed ever more elaborate cultural systems in all areas of cultural activity (thought, language, the law and social institutions, etc.). Understood as a form of learning behavior, as defined by Tylor (1871),[5] capturing experiences photographically could be seen as a way of learning how to deal with our emotions and to recognize the existence of an inner life that usually escapes explanation. This hypothesis is based on the connected idea that the image and sight mediate more and more frequently in everyday life and photography's widely recognized role in the process of sensory, emotional and motor symbolization that is brought into play when the shutter is pressed.

11.3.2 Optimizing the Informative Efficacy of Communication: A Hyper-real Language

If visual enrichment exists in all cases, here it is optimized to the maximum and becomes the most effective means of expression for transmitting a piece of information that is difficult to put into words. Here, what is at stake is the power to represent reality, valued as a means of expressing reality in a way that is less ambiguous than writing. Paradoxically,

---

[5] "Culture is this complex whole that encompasses knowledge, beliefs, art, morals, the Law, customs and all the other abilities and habits acquired by Man as a member of society" (Tylor, 1871).

removing ambiguity does away with the need for "other" communication. As Baudrillard and Guillaume say, communication is nourished by mis-understandings such that "it relies on what is contrary to it and on the gap between people. If two people have everything in common, communication is swallowed up in too great an intimacy" (Baudrillard and Guillaume, 1994). To take this idea further, we could say that the less distance there is between the thing and the word, the less need there is for language to define and signify reality. From a certain point of view, photographic language belongs among the most primitive communication systems in which people show what they mean rather than say it, just like the first forms of writing that relied on the resemblance of reality and its representation.

Hence the situations that are photographed correspond to a concern for efficient and optimized communication in time and a concern for precision so that the correspondent can share a vision of the same situation without any possibility of error. Lastly, this new form of communication consistently brings the values of hedonism and play combined with the values of saving time, efficiency and profitability driven by a desire for performance in economic life into the private sphere of interpersonal relationships.

In practical, organizational situations:

> When I went shopping, a friend asked me to get something for him. I wasn't sure what to choose since there was a wide choice in the store. So I took a photo and told him to choose. (Melle Aya, aged 20, student)

To show people's faces:

> I was with a girlfriend in a café and we saw a guy who looked like a film star. Since we weren't sure, we wanted an objective opinion. We took a photo of the guy and sent it to a third person to see if he really looked like the film star. (Haya, aged 20, student)

To show clothes and accessories:

> I went to a wedding reception. I had made an outfit for my baby, which I photographed and sent. Also, for something a friend of mine bought. She sent a photo of it – it's more practical. (Tie Sato, aged 31, housewife)

11.3.3 To Carry Loved Ones With You and Keep Them Present Symbolically: The Intimate World

This third way of sharing an "image of being together" reflects intimate, personal boundaries and values that are recognized and authenticated beforehand as such. This has less to do with active involvement in the photographic act and more about the use of mobile phones as a container, enabled thanks to the background image as a permanent photograph holder. The choice of photographs here reveals a cultural dimension that reflects the extent to which typical Japanese society tolerates public displays of affection and feelings in general, given the fact that the background image is visible to the outside world, i.e. in public space. Hence the photos selected are bound to reveal a conventional aspect of this intimate world, secured and contained within given limits. We will therefore discuss them to illustrate how, thanks to the photographic image, the emotional force carried by the mobile object is intensified.

The choice of photographs is dominated by three main categories connected to intimacy in one's home life: family photos symbolized by pictures of the children, pets and, more rarely, of the couple.

> I took a photo of my son when he started at nursery. I sent it to my sister's mobile – she lives with my parents who don't have a mobile themselves. (Kinnebushi, aged 39, sales rep.)

> With my boyfriend. We each take photos of us together and then send them to each other. (Sanae, aged 32, graphics designer)

> I like taking pictures of my dog. Especially if he's lying in a bit of a funny position. (Mrs Takana, aged 28, working)

11.4 Conclusion

With mobile photography, the image as a form of interpersonal communication becomes, more than ever before, a part of everyday life. It comes in at least two forms: first, it is the visual image, iconic, a demonstration of hyper-reality via the screen of your mobile phone; second, it is the mental image, calling on the imaginary to structure exchanges between correspondents.

Photography intensifies the experience and perception of feelings within a relationship with another person, in other words, the capacity to share an experience at the same level (unlike thought, which is intelligible). By over-investing emotion as a form of relating to another, the instantaneous aspect of the interchanges encourages us to subscribe to an agreeable, shared and common imagination which will be sought after for itself and which creates the perception of "being together" founded on an affective reality that is shared at the same time and together. Every new photo and every new exchange are also incorporated into a brief temporality linked to a never-gratified quest for satisfaction, for the pleasure felt in the immediacy and intensity of the moment, inducing ephemeral, fluctuating modes of belonging, guided by a mode of empathy involving recognition of and identification with the Other.

Using the concept of the tribe to describe the individual's new mode of social existence, Maffesoli, for example, partly describes this scenario by highlighting the idea that an alternative mode of being, an alternative form of social existence is developing, one which places the emphasis on the emotional and affective, as opposed to the concept of the social contract with its emphasis on the rational and free will. Thus, he speaks of "Homo estheticus developing on the basis of shared emotions" functioning in the same way as the tribe, i.e. through belonging and identifying at many different levels to micro-communities (Maffesoli, 1988, 2003). In his analysis of post-modernity, Bauman (2000) uses the metaphor of liquidity to examine the present new phase in the history of modernity. In this liquid modernity, the author talks about "cloakrooms" community to describe the kind of identity of being together. "The community in question tend to be volatile, transient and single 'aspect' or 'single-purpose'. Their life span is short while full of sound and fury. They derive power not from their expected duration, but paradoxically, from their precariousness and uncertain future, from the vigilance and emotional investment which their brittle existence demands."

We believe that this finding carries on from one of the main trends discussed in post-modern theory, which relates ideological disillusion to the idea of progress, i.e. to an eventual end, a purpose that corresponds to the future achievement of great ideals, which has been superseded by a society for which the present, the present moment and immediacy are the end-purpose of the quest and the satisfaction of pleasure. This creates a new way of entering a collective communication space centered on exchanging content that is intimate, fun, and has no rational or informational purpose but is, rather, sensation-oriented.

11.5 References

Barthes, R. (1980) *La Chambre Claire. Note sur la Photographie, les Cahiers du Cinéma*. Gallimard, Paris.

Baudrillard, J. and Guillaume, M. (1994) *Figures de l'Altérité*. Descartes, Paris.

Bauman, Z. (2000) *Liquid Modernity*. Polity Press, Cambridge.

Bourdieu, P. (ed.) (1965) *Un Art Moyen, Essai sur les Usages Sociaux de la photo*. Les Éditions de Minuit, Paris.

Cadéac, B. and Lauru, D. (2002) *Génération Téléphone, les Adolescents et la Parole*. Albin Michel.

Castel, R. (1965) Images et fantasmes. In Bourdieu, P. (ed.), *Un Art Moyen, Essai sur les Usages Sociaux de la Photo*. Les Éditions de Minuit, Paris.

Castel, R. and Schnapper, D. (1965) La rhétorique de la figure. In Bourdieu, P. (ed.), *Un Art Moyen, Essai sur les Usages Sociaux de la Photo*. Les Éditions de Minuit, Paris.

CREDOC (2002) Cited in *Le Monde*, "Photo: le boom du Numérique." Supplement June 19, 2003. Paris.

Crook, S., Pakulski, J. and Waters, M. (1992) *Postmodernization, Change in advanced society*. Sage, London.

Durkheim, E. (1897) [1997 (9th edn)] *Le Suicide*. Quadridge Puf, Paris.

Fortunati, L. (2003), Mobile phone and the presentation of self. Presented at Front stage/back Stage: Mobile Communication and the Renegotiation of the Social Sphere, Grimstadt, June 22–24.

Garrigues, E. (1996) La photographie comme matrice de recherches. *l'Ethnographie*, Special issue on the family and photography, **92.2**, No. 120,

Halle, D. (1993) *Inside Culture. Art and Class in the American Home.* University of Chicago Press, Chicago. Cited by Koskinen I. and Kurvinen, E. (2002).

Harper, R., and Taylor, A.S. (2003). The gift of the gab?: a design oriented sociology of young people's use of mobiles. *Journal of Computer Supported Cooperative Work (CSCW)*, **12**, 267–296.

Koskinen, I. and Kurvinen, E. (2002) Messages mobiles visuels, nouvelle technologie et information. *Réseaux*, No.112–113.

Le Monde (2003) Photo: le boom du numérique. *Le Monde,* Supplement, 19 June, Paris.

Lipovetsky, G. (1983) *l'Ère du Vide.* Gallimard, Paris.

Lipovetsky, G. (1987) *l'Empire de l'Éphémère.* Gallimard, Paris.

Maffesoli, M. (1988) *Le Temps des Tribus.* Méridien Klincksieck, Paris.

Maffesoli, M. (1990) *Aux Creux des Apparences.* Biblio Essais, Paris.

Maffesoli, M. (2003), Perspectives tribales ou le changement de paradigme social. In Michaud, Y. (ed.), *La Société et les Relations Sociales.* Université de Tous les Savoirs, Vol. 12.

Rivière, C. (2002) Minimessaging in everyday interactions: a dual strategy for exteriorising and hiding privacy to maintain social contacts. *Revista de la Juventud*, June. English version at URL www.mtas.es/injuve/biblio/revistas/Pdfs/numero57ingles.pdf. French version in *Réseaux*, **20**, No. 112–113.

Rivière, C. (2003) L'influence des échanges numériques sur les pratiques de communication interpersonnelle. FTR&D internal document.

Rivière, C. and Licoppe, C. (2003) From the spoken to the written world: a comparative study of the use of mobile voice communications and text messaging in France and Japan. Presented at Front Stage/Back Stage Mobile Communication and the Renegotiation of the Social Sphere, Grimstadt, June 22–24.

Sontag, S. (1983) Sur la photographie. *Le Seuil*, Coll. 10–18, Paris.

Tisseron, S. (1996a) *Le Bonheur dans l'Image.* Les Empêcheurs de Tourner en Rond, Paris.

Tisseron, S. (1996b) *Le Mystère de la Chambre Claire.* Les Belles Lettres, Archimbaud, Paris.

Tisseron, S. (2000) *Petites Mythologies d'Aujourd'hui.* Aubier, Paris.

Tylor E. (1871) [ed (1994)] *Primitive Culture.* Routledge, London.

12

Tell Me About Your Mobile and I'll Tell You Who You Are: Israelis Talk About Themselves

Dafna Lemish and Akiba A. Cohen

12.1 Introduction

Much attention has been paid recently to the roles that media play in forming in addition to performing identity. This is particularly the case with studies of new technologies, such as the Internet (Turkle, 1995; Jones, 1999; Hine, 2000; Mann and Stewart, 2000), where users are presumably utilizing online reality as an opportunity to experiment with facets of identities suppressed in offline situations. Identity play, however, has also been studied previously in relation to other contexts and contents of media consumption in a variety of audience groups. Such, for example, is the focus on the role of popular music in the search by adolescents for individual as well as group identity and solidarity (e.g. Frith, 1983; Grossberg, 1983–84; Lull, 1992; Gilroy, 1993), or the feminist perspective on the place that romance and soap operas play in women's lives, inspired by the breakthrough studies of Modleski (1984) and Radway (1984).

The existing literature suggests that the notion of identity and identity politics is complex and non-uniformed. Emerging identities are presumed to be plural, fragmented and in a state of flux. Identity is presumably never fixed or fully resolved, and is therefore always open to transformation and change. McRobbie (1994) poses several challenging questions in this murky debate: "But what exactly is meant by identity? Is it a term which implies the psychic processes of acquiring identity as theorized by Lacan? Is it a term which somehow suggests the political shift away from class? Or does class identity constitute one among many identities of equal validity in the struggle for a pluralist radical democracy? Through what processes has identity in cultural studies come to replace the more psychoanalytical notion of the subject?" (p. 49). In her critique of theories of identity, McRobbie suggests that identity could be seen "as a kind of guide to how people see themselves, not as class subjects, not as psychoanalytical subjects, not as subjects of ideology, not as textual subjects, but as active agents whose sense of self is projected on to and expressed in an expansive range of cultural practices, including texts, images and commodities" (p. 58).

Such a discussion has moved cultural theory away from centering on the text to concern over identity formation in practices of everyday life. This is much in line with the growing literature on narrative identity, based largely on stories people tell about themselves and their life events (see, for example, the work of Gergen, 1991, 1994, 1999; McAdams, 1993; Randall, 1995; and Wortham, 2001). In the poetic words of Gubrium and Holstein (2000, p. 101), "we talk ourselves into being". Through the interaction with others, interviewers included, people telling stories about themselves are engaged in actively constructing their identity in a particular context and performing it in orientation to a particular listener. This quality – referred to by Bakhtin as "addressivity" (Bakhtin, 1981, p. 288) – emphasizes the dialogical nature of the narration. Goffman also echoes this notion when he discusses the dynamics of effective performance in terms of the listener as much as the teller (Goffman, 1974, 1981). The focus of the present study, therefore, centers specifically on the way users of a particular medium – the mobile phone – choose to use it to present themselves to others.

More generally, we also rely here on Goffman's (1959) dramaturgical approach to social life, according to which people as actors use encounters to sustain a "face" (Goffman, 1967). Ownership and display of commodities or objects – be they name-brand jeans (Fiske, 1989) or bedroom computers (Livingstone and Bovill, 2001) are part of the process of creating and maintaining a social face. Commodities or objects can be used by consumers to construct meanings of one's self, in addition to one's social identity and social relations. Throughout the life cycle, different goods acquire symbolic value for the self (Furby, 1978) and those may serve as a personal resource: "Valued material possessions, it is argued, act as signs of the self that are essential in their own right for its continued cultivation, and hence the world of meaning that we create for ourselves, and that creates our selves, extends literally into the objective surroundings" (Rochberg-Halton, 1984, p. 335). We suggest that the mobile phone is one such object serving as a facilitator for the presentation of a multi-faceted identity of Israeli users. As such, it offers us a site in which to explore how Israelis choose to see themselves and to present themselves to others.

In our on-going project on the place of the mobile phone in Israel, we have been studying the patterns and contexts of its use, habitual behaviors related to it and attitudes towards its role in the daily lives of Israelis (Cohen and Lemish, 2003; Lemish and Cohen, forthcoming). The findings suggest that the mobile in Israeli society has become an everyday, highly regarded, multi-purpose interpersonal communication device rather than a working tool. More than anything else, habits of Israelis tend to reflect a concern for gratifications and communicative needs of the "here and now" and emphasize preferences for individualistic and private behaviors. This is much in line with sociological analysis of the trends in Israeli society and the gradual breakdown of the collective ethos and the development of an individualistic culture similar to other capitalist societies (Horowitz and Lissak, 1992; Ram, 1993).

Israel has seen a tremendous growth of the mobile phone market in terms of both penetration rate and amount of airtime used. By 2002, there were more than 4.8 million mobile phones in use. In fact, the number of mobile phone subscribers and the expenditure for cellular telephony in Israel long ago exceeded that of fixed telephony services (Schejter and Cohen, 2002). Data for 2002 indicate that there were over 6.3 million mobile phones in Israel with 95.45 mobile subscribers for every 100 persons (International Telecommunication Union, 2004), 78.8% of Israeli households own at least one mobile and 44.0% own two or more (Central Bureau of Statistics, 2003).

The unique cultural characteristics of Israeli society may serve to explain this phenomenon. First, it is a society typified by close familiarity and cohesive social networks that are conducive to much interpersonal contact and communication (Herzog and Ben Rafael, 2001). Second, Israeli society has a history of intensive diffusion of and infatuation with other communication technologies (Caspi and Limor, 1999). Finally, Israel's political situation has created special security needs for both civilians and soldiers; therefore, many family members own a personal mobile just in order to be able to contact each other in the event of terrorist attacks or military activity. The mobile thus provides an illusion of "protection" for both parents and their children, and in worst-case scenarios each person knows where the other is, knowledge that creates a sense, albeit imagined, of control over an otherwise incomprehensible anxiety. The combination of all of the above creates a unique mobile phoning culture.

In this chapter, we focus our attention on the role that the mobile has for users' presentation of self. We ask the question: How does the ownership and use of the mobile facilitate the user's discourse about his/herself as an agency? What are the dimensions of "self" that become apparent through talk about the use of the mobile phone?

12.2 Method

The multi-method study from which the present findings are derived was conducted in Israel during 2001 in cooperation with Cellcom, Israel's largest mobile phone company. It consisted of three sequential stages: (a) survey research, (b) obtaining real-time interactive voice response (IVR) measures (i.e. responding to questions by pressing appropriate keys on the mobile phone following calls) and (c) in-depth interviews.[1] The sample consisted of 103 women and 137 men.

In the first stage, each participant completed a questionnaire (that was sent to his/her home), dealing with several topics. In the second stage, the participants responded to several questions following each incoming and outgoing call during a 5-day period using the IVR technology.

The third and final stage of the study consisted of an in-depth phone interview with each participant. Within two days of the termination of the IVR stage, Cellcom provided printed logs of all the incoming and outgoing calls made by the participants during the 5-day period. The research assistant then phoned the participant and conducted an interview using a semi-structured interview schedule focusing on the unique characteristics of the participant's calls based on the computer log. The interview also included other questions relating to the original purpose for purchasing the mobile and the development of user habits, and reflections over one's own and others' calling habits, and experiences, if any, with the mobile outside Israel.

The interviewers took detailed hand notes of the interview, including many verbatim quotes. Thematic analysis of the interviews originally followed the main open-ended questions, and later developed into a set of relevant categories as is commonly done in the analysis of such transcripts (Lindlof, 1995). Of the original sample, 203 participants completed the interview, including 88 women (85.5%) and 115 men (84%). The discussion below is based on this part of the study.

12.3 Mobile Phone Use and Presentation of Self

We came across the issue of the use of the mobile phone for the presentation of self accidentally as we were analyzing the data of the interview transcripts from the comprehensive research project. Three overriding and unexpected themes emerged from the data as we were engaged in making sense of the interviewees' responses, as is often the case following the principles of grounded theory. These three themes can be conceptualized as referring to mobile-phone use in (a) both constructing and performing gender differences, (b) distinguishing between adulthood and childhood and (c) defining a collective identity of "Israeliness".

---

[1] For a detailed methodological discussion, see Cohen and Lemish (2003).

12.3.1 The Mobile Phone and Construction of Gender

A social analysis of technology from a feminist perspective (Cockburn, 1992) suggests that technology is much more than hardware – it is also a process of production and consumption, a form of knowledge, a site of gender domination in addition to a power struggle. Gender relations in both the private and public spheres and their characteristics shape the way technologies – including communication and leisure – are adopted and used in everyday life. In our work we found that both men and women use the mobile phone first and foremost to talk to family members and friends (Lemish and Cohen, forthcoming). This development follows that of the mobile's predecessor – the fixed telephone. Indeed, women's social uses of the fixed phone are now credited as being responsible for the development of the "calling culture", that is, the use of the device not for goal-oriented activities but for social process-oriented functions (Fischer, 1992).

The concrete everyday use of the mobile, as studied in our case, suggests that it may be located within the no (wo)man zone. That is, the device is stereotypically masculine in the sense of being a mechanical gadget, yet it is stereotypically feminine as it is used mostly for networking.

The mobile phone, therefore, might be playing a role in the blurring of gender differences in the actual use of communication technologies, and not necessarily reinforcing existing social divisions. Supporting evidence comes from a recent cross-cultural study that suggests that the mobile is making men more chatty and communicative than they were without it (Plant, 2002).

With this analysis in mind, we were taken by surprise when closer analysis of the interview data unexpectedly revealed striking gender differences. Men differed from women on three dimensions, as follows.

12.3.1.1 The Mobile as an Extension of Man

For many, the mobile phone has become a true "extension of Man" – to use McLuhan's (1964) expression. The claim for a bodily appropriation of this technology was expressed regularly by men, and only men, claiming it has become a natural extension of themselves: "Once we used to get by without it, it was possible; but now I feel that the instrument is part of me, it is connected to me," argues one interviewee; "… the cellular has become part of my body," declares another. And a third admits, "When I walk without my phone I feel like I am missing a hand."

Many used the terms "addiction", "dependence", "restlessness" without it, as if the absence of their mobile leaves them with a physical as well as psychological void. Our male interviewees seemed to talk about themselves through a perception that defines them in relation to mobile phone

191

technology. It has become part of them – the subject, the "me", is being voluntarily blurred with the "mine" in the form of the mobile phone. This seems to be in line with an ideology of hegemonic masculinity, which is typically framed within several normative expectations regarding, among others, the use of technology for domination, subordination and/or protection of others – humans and nature alike – in addition to its role in occupational achievements and in daring activities outside the domestic sphere (Cockburn, 1992; Hanke, 1998). Historically, from the spear to the computer, from the plough to the rocket, technology has become part of the social construction of manhood. Artistic and media representations of men have traditionally reinforced this conception by associating men with weapons, gadgets, vehicles, communication and technological devices (Craig, 1992). It is interesting, therefore, to note how our male interviewees so easily conform to this facet of their manhood, and how natural it seems for them to describe their mobile phone as a body organ, an extension of their self.

12.3.1.2 The Mobile as a Symbol of Inclusiveness

Both men and women referred to the mobile phone's social status and its use as a form of inclusion within one's peer group. However, each group focused on very different aspects: whereas men mentioned the ownership of the instrument itself as a sign of connectedness, women were more concerned with the phone calls themselves.

This can be illustrated by the fact that only men reported adopting the mobile phone in the first place as a status symbol; of mere "excitement" over owning and displaying a new toy like everyone else; and as a result of peer pressure. Reasoning such as "because everybody had one," or "to be 'in'," were often mentioned by the male interviewees: "I am a materialistic type ... I didn't need it. I got excited, it's like a toy," says one. "I was among the first to buy one and I did it with show-off-excitement. The whole country got excited," admits another, apologetically. Still another interviewee said it openly, "Everybody bought it so I did too, this is what you may call 'social pressure'."

Women, on the other hand, did not mention social pressure or excitement over the novelty of the gadget as motivation for acquiring the mobile in the first place. However, once they owned it they became particularly sensitive to the social hierarchy it seems to convey: "The phone itself has become an issue in its on right – 'how many called you?' 'Why nobody calls me?' It states a social status. Someone who gets less calls is less popular." Similarly, the mobile, according to some women, takes part of the romantic sensation out of a relationship: "once upon a time when someone was coming on to me it was a big deal – 'has he called?' 'When did he call?' Today he simply says to me – 'call me on my mobile'. It loses all the interest." Both examples suggest a form of passivity on the part of the women, expecting to be called by others, and measuring their popularity by others calling them.

12.3.1.3 The Gendered Advantages of the Mobile Phone

Although both men and women discussed the advantages of the mobile phone for accessibility, efficiently, social and familial connection and sense of security, they differed in their framing of these qualities. Men tended to talk more about their accessibility to others, framing this quality of the mobile phone in active terms: "It is always possible to reach whoever you want whenever you want." They are pleased, for example, with their ability to call for assistance while on the road: "The cellular helps me when I'm riding my motorcycle and I have a flat tire and I can ask someone to come and help me." Similarly, men present themselves as conscientious and responsible citizens: " I live in a [remote] settlement and sometimes you get stuck here without a bus. It's always good if you need to report on a suspicious object [a potential bomb]."

Women, on the other hand, took the passive position as they were more concerned about others being able to reach them: "It's always possible to reach me;" "I am freer to leave the house, because even when I do errands I know that they can reach me and I can come home if necessary." The importance of the mobile for managing household roles from a distance was very apparent in women's discussion of the advantage of accessibility: "I need to be accessible all the time if something happens to the child and the caregiver needs to let me know urgently," explains one mother, and a second demonstrates this quality "in real time" when she says: "… even now when I am talking to you, the kids are bugging me with call-waiting. They're always trying to get hold of me. I'm supposed to order a pizza for them … today is pizza day … it's important for me to be accessible."

Clearly, although both men and women talked about the importance of being connected with other family members, only the women were concerned about the management of the household and used the mobile phone to facilitate these traditional roles. As Rakow and Navarro (1993) noted, "The cellular telephone, because it lies in that twilight area between public and private, seems to be an extension of the public world when used by men, an extension of the private world when used by women. That is, men use it to bring the public world into their personal lives. Women tend to use it to take their family lives with them wherever they go."

Another form of women's discourse of dependency is their arguing for using the mobile as an "authority substitute". Only women praise the device for allowing them to receive long-distance instructions for

193

handling a repair or driving directions: "Once when I got stuck and called my husband, he told me on the phone what to do," recalls one woman, and another one says, "It's convenient when I get instructions on how to travel somewhere, and they direct me all the way to the place itself."

In summary, regardless of users' actual calling habits that point to a pattern of domestication (Haddon, 2001) of the mobile, and some say even feminization of its uses and gratifications, both men and women discuss their perceptions of its role in their lives in a traditional gendered manner. The mobile phone is another site for them to perform their gendered identity – activity and technological appropriation for men and dependency and domesticity for women.

12.3.2 The Mobile Phone and Construction of Adulthood

The second theme that emerged unexpectedly from the data was the interviewees' evaluation of the role that the mobile plays among children and in youth culture. Our sample consisted only of adults and the interview schedule made no specific reference to younger users. However, in responding to an open-ended question about the use "others" make of the mobile phone, reference to non-adult users surfaced as a major concern.

Many of the users expressed strong criticism, even detestation, of numerous behaviors commonly associated with mobile phones. They complained, for example, about loud and idle conversations, being inconsiderate in public spaces, rudeness and invasion of privacy. Many used strong expressions of revulsion: "disgusting," "a disease", "unforgivable", etc. However, when they wanted to express their deepest and heartfelt shock over mobile phone culture, they enlisted the discussion of children's use of the mobile. Criticism of this phenomenon included mainly two intertwined themes: the prevalence of the mobile phone among children and youth and the monetary "waste" of children's talkativeness.

12.3.2.1 The Prevalence of Mobile Phones Among Young Users

When emphasizing the high rate of the adoption of the mobile phone in Israeli society, interviewees often chose to illustrate the extremity of the situation by enlisting children: "The cellular has become a disease in Israel. You can't find anyone without a mobile. Every child has one, even in first grade," said one man bluntly. "Every one owns a mobile phone. Every child holds one and talks on the bus and everywhere," complains another one in disgust. "This is worse than cigarettes, this mobile phone thing. My young brothers – they are all hooked. I have a 17-years-old sister who is hooked on the mobile phone and it's a plague."

A terminology of addiction and sickness was commonly used by these interviewees to connote a negative tone of moral denunciation of such a bad habit: "They take the mobile phone with them everywhere – to the bath, to the toilet, to bed … it rings and they immediately run to it … they are linked to it so they won't miss even one call … what a disaster if my older daughter misses a call!" complains one mother. The metaphor of addiction has often been used in discussion of children's other media habits, e.g. television viewing, computer game playing and Internet surfing. Similar to Smith's (1986) discussion of addiction to television, for example, our interviewees discussed children being seemingly "addicted" to calling, having a constant need to be on the phone; prioritizing calling over other activities and commitments, an inability to miss a call and withdrawal symptoms of restlessness, unhappiness and anger when that happens.

12.3.2.2 Monetary Waste Related to the Mobile Phone

Although the issue of expense involved in the use of the mobile phone was not mentioned as a major concern for their own use, adults seemed to be preoccupied with what seemed to them as plain waste of money among children. " Toddlers of three and four are now talking to their parents on the cellular phone. Every two-year-old now owns a phone. None of them are big investors in the stock market … I myself try to control myself and I bear the consequences of over-talking on the phone, but in the case of children – their parents have to pay the bill – this is bad education," preaches one man. "My nephew is only seven and he cried and cried and demanded a mobile phone and finally got one. Another nephew of mine is 14 and he calls about every tiny thing … it's really an exaggeration. I look at my daughter and I see that she talks a lot and the cellular bill is huge – unbelievable. It's superfluous and it's a waste," complains a mother.

This adult discussion of children's calling habits can be cast in the familiar discourse of "moral panic" regarding children's culture in general, and their communication and leisure activities in particular. Any new media that has been introduced elicited major public anxiety over its potential negative effects on children (Buckingham, 1993; Livingstone, 2002). Questions are raised both regarding each medium's moral effects through consumption of contents such as violence, pornography, racism and sexism, as well as regarding social and physical influences of living with mediated versus real-life communication culture. As Drotner (1992) suggests, each "moral panic" tends to move from "pessimistic elitism" to a more "optimistic pluralism", and with it from calls for technocratic and legalistic measures of control to advocacy for media literacy and "better" parenting.

In the case of the mobile phone, our interviewees immediately resorted to the parenting option, as many of them agreed, "there is really nothing else that can be done about it". Proper expectations regarding "good" parenting require control and supervision of children's leisure time, education for values of industriousness, and proper use of resources such as time and money. Interestingly, none of the interviewees suggested that parents should also be expected to serve as role models in their own use of the mobile phone, the assumption being that phoning norms of adults are appropriate and justified, whereas those of children are, by definition, not.

Following McRobbie's (1994) argument that "... difference is constitutive of identity ... identity is as much about exclusion as it is about inclusion" (p. 40), we argue that a "moral panic" such as the one regarding the mobile is an illustration of an adult's attempt to define adulthood as distinguishable from childhood and to legitimize the advocacy of adults' control and power over children. As one mother recalls, somewhat nostalgically, her own "phone-less" childhood: "it annoys me to see little kids with mobile phones – what is this retarded thing – we were kids without mobile phones and nothing happened to us!"

The two worlds are conceived as dichotomous – with children being either vulnerable and innocent and therefore in need of adult protection and supervision, or as natives in need of taming and controlling as their behavior is deviant and dangerous to the stability of society (James et al., 1998). Therefore, excessive use of the mobile by Israeli adults is perceived in positive terms such as important, helpful, revolutionary, whereas children's use is attacked for being wasteful and unacceptable. Whereas the use of the mobile by adults for networking, keeping in touch, and chatting is perceived as facilitating social connectedness, children's networking with their friends is criticized for being morally inappropriate and an indication of bad upbringing.

Children, once again, are being perceived as vulnerable to negative effects of the new medium, and/or "disturbants" of the "good order of things" and incapable of self-regulation. They are not recognized as autonomous agents. Their special needs and pleasures are passed unrecognized and disrespected. In many ways, the discourse surrounding parents' "neglect" of their out-of-control, mobile phone-addicted children echoes earlier anxieties over latch-key children, presumably left alone unsupervised to do their "own thing". The fact that they do "their own thing", i.e. constantly talk in public on their mobiles, transforms this parental neglect from spaces designated for children (homes, educational institutions) into the public space perceived as adult territory, and with it to the public spotlight. The expectation that children will "behave themselves" in public, that they will be appropriately contained and regulated, becomes particularly apparent when discussing other people's children. This form of dissociating the issue from oneself contributes to the association of the "moral panic" with "otherness" – be it class, age or ethnicity.

Looking down on children and their communicative desires and clearly stating the hierarchic boundaries between adults and children can best be illustrated in the following condescending monologue of a mother: "My son is 17 and he drives me nuts that he needs a mobile phone. So I tell him: 'excuse me, you are not that important, you are not the Chief of Staff or Clinton.' I think there is a big exaggeration with this mobile phone business – every punk has a mobile phone for blah blah. What do kids need a mobile phone for? Let them use a calling card – a kid basically just goes to school and back, what important things can happen to him?"

12.3.3 Mobile Phones and the Construction of Israeliness

Our final theme for analysis is the way mobile phone-related behaviors were used by our interviewees to construct their Israeli identity. Here, too, identity was mostly about creating oppositions between "us", Israeli mobile phone users, and "them", other non-Israeli mobile phone users. However, what is most striking about this discourse is that it was overwhelmingly cast in negative terms of "the ugly Israeli character". According to informants, the speedy diffusion and high rate of mobile use in Israel are reflective of the nature of Israeli culture explained, among other things, by the constant political conflict and the security situation. In order to illustrate this as a unique Israeli phenomenon, they offered direct comparisons with other cultures.

12.3.3.1 The Mobile Phone and the Nature of Israeli Culture

Our interviewees[2] were surprisingly non-compromising and harsh in describing Israelis as hysterical, pressured, vulgar, audacious, showing off, impolite, rude, talkative, extroverted, loud, egocentric, ... They tied the mobile phone-related behaviors to other "typical" Israeli inconsiderate behaviors in public places. Examples for this are plentiful: "People yell on their mobiles on the street; but Israelis yell in every situation," explains one man. Another one adds, "It's part of the Israeli mentality – for the same reason that Israelis always cut the line." "People here need to feel in control all the time, they don't respect anybody," argues another man. "It's part of the Israeli compulsiveness, the mentality, like other things – vulgarity, like opening a package of cottage cheese in the store to check if it's fresh," illustrates another woman.

---

[2] Although a special effort was made to recruit Israeli–Arabs in our sample, we ended up with only a handful. We are therefore unable to make any meaningful observations regarding this segment of society, and particularly concerning issues of national identity.

This mentality is rationalized, among other things, with the eagerness of Israelis to try new technologies: "Israelis shout all the time … we are curious people, we want to try everything new. It's a new invention, everyone wants one … ," explains one man. Another one says, "Israelis always want to be 'in' – to show off that they have the most advanced technologies – Internet, DVD, the smallest cellular phone … they want to show that they are advanced."

But most of all, mobile phone use was explained in the context of the unique life conditions of Israelis being constantly under pressure and existential anxiety, and the needs for involvement and caring for others. "Maybe it's because we're a small country and everyone knows everybody. Everything here is under pressure; everybody is under atomic pressure. One can't rest for a moment," argues one man. And a woman adds, Everybody here wants to be in touch with everyone all the time because of the security situation."

Our interviewees clearly framed the prevalence of the mobile phone in Israel and its related disturbing social uses as a phenomenon that is an integral part of the nature of life in Israel. For them it is one more aspect of what they perceive to be an abnormal way of life that justifies a variety of breaches of what they clearly know to be normative, civilized ways of behavior in public. Blaming it all on the security situation (rightfully or not) is an illustration of the overpowering nationalistic discourse in Israel at the present time.

12.3.3.2 The Mobile Phone in Israel in Comparison with the World

Most of our interviewees were very sure that the situation regarding mobile phone use outside Israel was fundamentally different from the case in Israel. According to them, people abroad talk much less on the phone and have a very different calling culture. Many have visited other countries and offered specific examples from their personal anecdotal experiences. Others cite something they heard or read in the media regarding mobile phone research around the world. However, interestingly, many have expressed very clear-cut opinions and attitudes without ever being abroad, claiming that they have heard from others or that they "just know" it to be a fact.

According to our interviewees, people abroad talk less and are significantly more polite and considerate of others in public places. Personal experience was provided as testimony from countries in Europe, the Americas and Asia. In addition, people made broad statements such as "abroad," or "in Europe". For example, one man explained, "I think we have more cellular phones here then in the US because of our mentality, as we are a lot more extroverted – we like to be heard, we like to yell, to speak, to express ourselves. The Americans are a lot more introverted and shy." A woman shared her travel experiences: "I just came back from Argentina – they have lockers there at the entrance to the store where you deposit your bag along with your cellular. In the theater and museums you deposit your bag and cellular along with your coat. I have American friends in important positions and they don't walk around with the cellular at all – only in the car."

The lively descriptions go on and on. Although the factuality of many of the testimonies can be challenged, it is the perceptions of the speakers that are of interest here. Clearly, the rest of the world, for most of them, is perceived as much more civilized and much less dominated by mobile phones. Furthermore, many assume that Israelis already have a bad reputation, as an anecdote told humorously by a man who recalled visiting a theater in Prague populated with tourists from all over the world demonstrates: as the lights went off, an announcement requesting people to turn off their mobile phones was made in English, as customary, and then specifically in Hebrew, for the benefit of Israelis in the audience.

Discussion of the mobile phone culture was clearly used by Israelis in this study as an opportunity to distinguish themselves from the rest of the world and define themselves as "unique", albeit in negative terms. The dominance of what is coined a "siege mentality" of Israelis (Bar-Tal & and Antebi, 1992), manifested among other things by the dominance of tourism and travel in the lives of Israelis (see, for example, Noy, 2002) was clearly visible in the discussion of the "we against the world" discourse. They were much aware of the norms of disrespect for rules of conduct prevailing in Israeli society, and an interpretation of individualism that borders on egotism (Bloch, 2003). The potential dissonance created by the need to present a socially desired "face" and the criticism they were voicing, was handled through rationalizing much of the negative behavior as circumstantial, a form of survival strategy in an extremely difficult existential situation. Related to this interpretation is the unique nature of interpersonal communication style in Israeli society – "Dugri" speech – characterized as being largely informal, direct, simple, assertive, sincere (Katriel, 1986). While it is often perceived, particularly by outsiders, as rude and inconsiderate, it is at the same time understood to reflect solidarity and care. This free flowing self-criticism in itself is a form of being "Dugri", therefore, constructing oneself as being part of the collective mainstream "Israeli".

In addition, it is important to emphasize that a close look at our transcripts suggests that the discussion of the Israeli "us" is removed from a discussion of a "me". Most interviewees did not criticize their own behavior, and did not suggest that they themselves are guilty of excessive calling and vulgarity. In a form of distancing themselves from the negativity, they talk about "Israelis" as if they themselves are other people. "You know how Israelis are," they would say with a sigh ... denouncing for a moment that they too are Israelis ....

12.4 A Concluding Note

The analysis of interviewees' discussion of various aspects of their mobile phone use reveals, once again, that the meanings of technologies are culturally constructed in the service of particular values and constantly negotiated through discourse. The three themes that we have highlighted here suggest that the mobile phone serves as a site in which Israeli users define their multiple identities as related to gender, adulthood and nationality. In two of the three sites they manage the tensions between the "me" and the "other": adult versus child, Israeli versus non-Israeli. This is done through a process of inclusion on the one hand and exclusion on the other. Their gender identity is constructed through their choice of gender appropriate talk in addition to the content of that talk.

Clearly, all three themes have in common a discrepancy between the data of actual use and attitudes, and the interviewees' discourse about them. The quantitative data reported elsewhere (Cohen and Lemish, 2003; Lemish and Cohen, forthcoming) suggest that the mobile phone is blurring gender differences, as it is becoming more of a networking medium rather than a goal oriented one. However, the interviewees still chose to manage their social "face" in a gendered way. The complaints about the calling cultures of Israeli children and adults raise a different kind of dilemma: if most interviewees are against children's use of the mobile phone, who then is responsible for purchasing phones for children and footing the bill? If most interviewees are disgusted with mobile phoning vulgarity around them, who then are the people who behave in that way?

As has often been the case with other unacceptable behaviors, people tend to recognize their existence and the social desirability to criticize them, but to distance themselves from negativity by associating them with "others". Such, for example, is often the case with parents' discussions of the behavior of "other" children, or of older children claiming that television indeed has negative effects – but only on their younger siblings. This is much in line with the "Third Person Effect" hypothesis (Davison, 1983), according to which people believe that the media's effects "will not be on 'me' or 'you', but on 'them' – the third persons" (p. 3). As has been demonstrated in other studies (Perloff, 2002), there is a perceptual gap between the beliefs that our interviewees hold about the role that the mobile has in their own behavioral patterns as compared with that of others.

Finally, on a more general level, we would like to suggest that our case study provides an illustration of the potential for studying everyday media-related practices for a better understanding of identity formation and management. It also highlights, once again, the benefits of methodological triangulation: what people *do* with media, what they *say* they do with media, and *how* they say what they do with media are not necessarily one and the same thing.

11.5 Acknowledgments

The authors thank Mr Oren Most, Vice President of Cellcom, for his invaluable help in supporting the project and Ms Michal Dvoretzky and Ms Neta Amorai at Cellcom, who coordinated the study. Thanks are also due to Mr Ziv Sharon for coordinating the study at Tel Aviv University and to our students who worked on the project with much diligence.

11.6 References

Bakhtin, M.M. (1981) *The Dialogic Imagination: Four Essays*. University of Texas Press, Austin, TX.

Bar-Tal, D. and Antebi, D. (1992) Siege mentality in Israel. *International Journal of Intercultural Relations*, 16, 251–275.

Bloch, L.R. (2003) When the frame takes over: Analyzing freier discourse in Israeli communication culture. *Communication Theory*, 13, 125–159.

Buckingham, D. (ed.) (1993) *Reading Audiences: Young People and the Media*. Manchester University Press, Manchester.

Caspi, D. and Limor, Y. (1999). *The In/outsiders: Mass Media in Israel*. Hampton Press, Cresskill, NJ.

Central Bureau of Statistics (2003) Table 5.35: Household expenditure survey-ownership of durable goods and housing data. Central Bureau of Statistics, Jerusalem.

Cockburn, C. (1992) The circuit of technology: gender, identity and power. In: Silverstone, R. and Hirsch, E. (eds), *Consuming Technologies: Media and Information in Domestic Spaces*. Routledge, London, pp. 32–37.

Cohen, A.A. and Lemish, D. (2003) Real time and recall measures of mobile phone use: some methodological concerns and empirical applications. *New Media and Society*, 5, 167–183.

Craig, S. (ed.) (1992) *Men, Masculinity, and the Media*. Sage, Newbury Park, CA.

Davison, W.P. (1983) The third person effect in communication. *Public Opinion Quarterly*, 47, 1–15.

Drotner, K. (1992) Modernity and moral panics. In Skovmand, M. and Schroeder, K.C. (eds), *Media Cultures: Reappraising Transnational Media*. Routledge, London.

Fischer, C.S. (1992) *America Calling: A Social History of the Telephone to 1940*. University of California Press, Berkeley, CA.

Fiske, J. (1989) *Understanding Popular Culture*. Unwin Hyman, Boston, MA.

Frith, S. (1983) *Sound Effects: Youth, Leisure, and the Politics of Rock'n Roll*. Constable, London.

Furby, L. (1978) Possessions: towards a theory of their meaning and function throughout the life cycle. In Bates, P.B. (ed.) *Life-span Development and Behavior*. Academic Press, New York.

Gergen, K.J. (1991) *He Saturated Self: Dilemmas of Identity in Contemporary Life*. Basic Books, New York.

Gergen, K.J. (1994) *Realities and Relationships: Soundings in Social Construction*. Harvard University Press, Cambridge, MA.

Gergen, K.J. (1999) *An invitation to Social Construction*. Sage, London.

Gilroy, P. (1993) *The Black Atlantic: Modernity and Double Consciousness*. Harvard University Press, Cambridge, MA.

Goffman, E. (1959) *The Presentation of Self in Everyday Life*. Anchor Books Doubleday, Garden City, NY.

Goffman, E. (1967) *Interaction Ritual*. Anchor Books Doubleday, Garden City, NY.

Goffman, E. (1974) *Frame Analysis*. Harper Colophon Books, New York.

Goffman, E. (1981) *Forms of Talk*. University of Philadelphia Press, Philadelphia, PA.

Grossberg, L. (1983–84) The politics of youth culture: Some observations on rock and roll in American culture. *Social Text*, 8, 114–126.

Gubrium, J.F. and Holstein, J.A. (2002) The self in a world of going concerns. *Symbolic Interaction*, 23, 95–115.

Haddon, L. (2001) Domestication and mobile telephony. In Katz, J.E. (ed.), *Machines that Become Us*. Transaction, New Brunswick, NJ, pp. 43– 56.

Hanke, R. (1998) Theorizing masculinity with/in the media. *Communication Theory*, **8**, 193–203.

Herzog, H. and Ben Rafael, E. (2001) The study of language and communication in Israeli social sciences. In Herzog, H. and Ben Rafael, E. (eds), *Language and Communication in Israel*. Transaction, New Brunswick, NJ, pp. 3–27.

Hine, C. (2000) *Virtual Ethnography*. Sage, London.

Horowitz, D. and Lissak, M. (eds) (1992) *Trouble in Utopia: the Overburdened Polity of Israel*. Am Oved, Tel Aviv (in Hebrew).

International Telecommunication Union (2004) http://www.itu.int/ITU-D/ict/statistics/at_glance/cellular03.pdf

James, A., Jenks, C. and Prout, A. (1998) *Theorizing Childhood*. Cambridge University Press, Cambridge.

Jones, S. (ed.) (1999) *Doing Internet Research: Critical Issues and Methods for Examining the Net*. Sage, Thousand Oaks, CA.

Katriel, T. (1986) *Talking Straight: Dugri Speech in Israeli Sabra Culture*. University of Cambridge Press, Cambridge.

Lemish, D. and Cohen, A.A. (forthcoming) On the gendered nature of mobile phone culture in Israel.

Lindlof, T. (1995) *Qualitative Communication Research Methods*. Sage, Thousand Oaks, CA, Ch. 7.

Livingstone, S. (2002) *Young People and New Media*. Sage, London.

Livingtsone, S. and Bovill, M. (eds) (2001). *Children and Their Changing Media Environment: A European Comparative Study*. Lawrence Erlbaum, Mahwah, NJ.

Lull, J. (1992) *Popular Music and Communication*. Sage, Newbury Park, CA.

Mann, C. and Stewart, F. (2000) *Internet Communication and Qualitative Research: a Handbook for Researching Online*. Sage, London.

McAdams, D.P. (1993) *The Stories We Live: Personal Myths and the Making of the Self*. Willam Morrow, New York.

McLuhan, M. (1964) *Understanding Media*. Mentor, New York

McRobbie, A. (1994) *Postmodernism and Popular Culture*. Routledge, London.

Modleski, T. (1984) *Loving with a Vengeance: Mass Produced Fantasies for Women*. Methuen, London.

Most, O. (2003) Oral presentation at a symposium for marketers and advertisers, Tel Aviv.

Noy, C. (2002) "You must go trek there": The persuasive genre of narration among Israeli backpackers. *Narrative Inquiry*, **12**, 261–290.

Perloff, R.M. (2002) The third person effect. In Bryant, J. and Zillmann, D. (eds), *Media Effects: Advances in Theory and Research*. Lawrence Erlbaum, Mahwah, NJ, pp. 489–506.

Plant, S. (2002) *On the mobile*. Motorola, USA.

Radway, J. (1984) *Reading the Romance: Women, Patriarchy, and Popular Literature*. University of North Carolina Press, Chapel Hill, NC.

Rakow, L. and Navarro, V. (1993) Remote mothering and the parallel shift: Women meet the cellular telephone. *Critical Studies in Mass Communication*, **20**, 144–157.

Ram, U. (ed.) (1993). *Israeli Society: Critical Perspectives*. Breriot, Tel Aviv (in Hebrew).

Randall, W.L. (1995) *The Stories We Are: An Essay on Self-creation*. University of Toronto Press, Toronto.

Rochberg-Halton, E. (1984) Object relations, role models, and the cultivation of the self. *Environment and Behavior*, **16**, 335–369.

Schejter, A. and Cohen, A.A. (2002) Israel: Chutzpah and chatter in the Holy Land. In Katz, J.E. and Aakhus, M. (eds), *Perpetual Contact: Mobile Communication, Private Talk, Public Performance*. Cambridge University Press, Cambridge, pp. 30–41.

Smith, R. (1986) Television addiction. In Bryant, J. and Zillman, D. (eds), *Perpsectives on Media Effects*. Lawrence Erlbaum, Hillsdale, NJ, pp. 109–208.

Turkle, S. (1995) *Life on the Screen: Identity in the Age of the Internet*. Simon and Schuster, New York.

Wortham, S.E.F. (2001) *Narratives in Action*. Teachers College Press, New York.

13

Mobile Telephone and the Presentation of Self

Leopoldina Fortunati

13.1 Introduction

As Simmel so acutely observes (Simmel, 1908, It. trans. 1998, p. 291), "all relations between human beings rest evidently on the fact that each one knows something about the other." However, it is equally true that for a large number of social relations – particularly those that are established in the course of our moving about – we actually need to know very little about the human beings we encounter. That is, we need to know that "in the other there only exist tendencies and qualities that are quite typical of human beings, which are usually noted by their absence". This is the reason why we cross social space behind a neutral and hazy form of our personality and we recognize and deliberately respect the un-said.

This is also true for the so-called "information society", where – whereas the media system generates a superabundance of information – individuals continue to limit the information they give about themselves in public spaces. Indeed, if anything, there is a process of gradual reduction. This tendency was already observed by Goffman, who considered it "a general sociological form, quite neutral in respect to the value meanings of its contents," historically elaborated to cope with metropolitan crowding. "Urban life," he underlines, "would become unbearable for many if every contact between two human beings meant having to share efforts, worries and secrets" (Goffman, 1959, It. trans. 1969, p. 61). On this plane, it is one of the greatest conquests of humanity that by limiting communication we reach what Simmel calls "an enormous broadening out of life". In fact, concealing information on one hand "offers as it were the possibility of a second world next to the one manifested, and this is influenced by it in the strongest measure"; on the other, it confers on the personality "an attractiveness determined in a purely social way" (Simmel, 1908, It. trans. 1998, pp. 308 and 311).

Today, we are increasingly becoming aware at a psychological level that human beings have many personalities. The individual of post-modernity, furthermore, as sociology teaches us, incarnates many and complex family and social roles. According to the various situations of daily life in which we find ourselves, one or more roles are incarnated and one or more personalities are revealed. In this context, the choice of reducing the amount of information to be transmitted in the social space (Simmel, 1908, It. trans. 1998) becomes one of the necessary strategies. We reduce information as an adaptation. It helps us through the relational networks, modes of sociability and the kinds of social space in which we find ourselves at a given moment, even if, as Goffman stresses, persons "tend to remain in the personage" (Goffman, 1959, It. trans. 1969, p. 195). This reduced self-presentation finds, furthermore, an adequate framework in the widespread appreciation of anonymity and social uniformity, but also equality (or a limited distinction) and freedom.

In self-presentation, we resort on the one hand to a series of "vehicles", to use Sorokin's expression (Sorokin, 1947). That is, we use devices that have a socially shared meaning, and also play into to the main systems of meaning, made up of the body, fashion, language and technology. In their unconscious phase, the grammar and syntax of these systems are very hard to control. In the conscious part, on the other hand, they are the product of individual decisions and choices, always within certain limits. We say certain limits, because there are elements of rigidity in the organization of daily life that set up resistance to any variation in the decisions and choices of individuals. This is why when we decide to put on a certain dress to go to work – perhaps to help us effect a serious image – the same dress goes on corresponding all day to the initial intention, even if it may have been better at a certain point to effect a more light-hearted image.

The system of meanings that says most, but which is most difficult to control, is language. Verbal language is an extremely revealing system as to what we are (gender, age), to what social strata we belong, from what region we come from, etc. Non-verbal language is also very revealing and is even less under control than verbal communication. This is because it involves automatisms that are beyond the control of the individual. A blush reveals – perhaps when one least wants it to – a moment of embarrassment or timidity. A tic betrays nervousness; perspiration under the armpits or a tremor in the voice can become a calamity. But our self-knowledge and experience that we acquire in post-modernity allow us to avoid these signs or to defuse their effect through various prevention mechanisms. Self-presentation is therefore a minefield inside which there are at work various strategies of production and control of the same information, inside a framework that one would suppose is somehow coherent.

On the whole, however, we can speak of the shrinking dimensions of the front stage in public spaces. We communicate our being men or women (even if this opposing dualism is less and less evident), while age has already been backstaged (think of the various techniques for representing ourselves as young). For the rest, anything can serve to say nothing, to cover over our background, status, flaws and tics – think for instance of casual clothes. A way of dressing that is based on a tee-shirt and a pair of jeans gives very little information about the particular individual (not sex, or age, or social or ethnic belonging, or walk of life or income bracket).

However, this lack of information is sustained not only by the awareness of being able to do the opposite, that is, reveal oneself, but also by the urge to do so. This urge to over-inform others can be the source of problems, some of which Simmel has already glimpsed (Simmel, 1908, It. trans. 1988, p. 307). He spoke of what happens, for example, to individuals in coupled relationships, when they squander information about themselves. The only ones that can afford to succumb to this temptation are those rare individuals that, in Simmel's words, "have an inexhaustibility of latent psychic processes," that is, those whose soul "consists in continual further developments which after each gift immediately bring forth new treasures." In conclusion, we might say that human co-existence needs a controlled amount of information.

There has been a critical point wherein this picture is modified. That is, there has been a sea change with reference to our control over "giving off" information about ourselves. With the intervention of the mobile phone into in the public space (Katz and Aakhus, 2002) we are experiencing a change in the tendency to downplay the front stage in favor of the back stage, if it is true that "opportunity makes the thief," as the proverb says. In this context, the convenience offered by the mobile becomes irresistible. It also is being played out in the parts of our private life that are being paraded across life's front stage.

13.2 The Oral Use of the Mobile Phone as Backstage Catalyzer

The mobile phone threatens to overturn the social construction of our intended self-image in various situations. One of the reasons is the still nascent domestication of this communication technology in daily practice. Thus, the calls that break into our lives, or which we make, often re-describe, correct or modify the broader façade we are trying to maintain. The dramatic effect of this is the eruption of the back stage, with all its capacity and strength of information, into the unassuming picture of the front stage. This is a "limelight" that Goffman appropriately describes as "the expressive standardized equipment that the individual intentionally or involuntarily uses during his own enactment" (Goffman, 1959, It. trans. 1969, p. 33). Obviously, certain coherence is expected between setting, appearance and manner of the individual. However, the discrepancies are actually increasingly frequent and so less and less preturbing.

The oral use of the mobile phone in a social context of deliberately limited information ends up producing an incredible short circuit. That is, it produces the public revelation of information that may be intimate and significant with respect to a role or person, expressed at a certain moment by an individual. This unplanned eruption is often involuntary and imposed on bystanders. It can generate great tension that can boil over into the social negotiation of self-presentation. In addition, it brings about upheaval in the social space (Sorokin, 1943, 1962), with its continuity and discontinuity (Bentley, 1931).

This intrusion of the backstage into public places was made possible via the mobile phone because people have been socialized into hearing, reading and seeing common people speaking of their private lives in the mass media. This is seen in talk shows, on the radio, in gossip columns in the news, etc. This eruption of a fictional backstage into the mass media is reflected in mobile telephone use. Indeed, this system of meanings constituted by the technological chain of ICTs can be seen in Meyrowitz (1985). Another explanation of this can be seen in the tension between anonymity and popularity. The mobile telephone allows the non-celebrity to stage their life for bystanders in public spaces. In this way "every person" can underscore their existence by playing out their backstage lives in front of a small audience.

With respect to the fictional back stage of the mass media discussed above, the mobile reveals a much more basic back stage. Although it is still artificially mediated by technology, it is less fictional because it does not imply a large, voluntary, paying audience, as does TV. With the use of the mobile in public spaces, the back stage is revealed in front of a limited number of people, only those few who are within earshot. The co-passenger, the bystander, the passer-by, is at best a "weak link". The revelation of the back stage is an imposition that is often irritating, but at times curious or interesting. These spaces are freely accessible and, to play on Granovetter, they are the realm of weak ties where you pass without stopping, or where you kill time. The title of this chapter has no quotation marks around the expression "presentation of self", because the intention was not to refer only to the Goffman tradition but also to an array of authors who have discussed the self and identity, such as James (1890, 1910), Mead (1934) and more recently Gergen (1999).

13.3 Aims and Methods of the Study

This chapter looks into the use of mobile telephones and, in particular, into the different situations in which the incoming or outgoing call re-describes, corrects or modifies the self-presentation of information. The material used in this analysis comes from a qualitative research project, based on:

- Twenty non-structured interviews based on a convenience sample of mobile phone users, half male and half female.
- 200 hours of observation in Italian trains travelling in various directions (north–south, east–west). Half of the trains were high-speed (Eurostar and Intercity) and half were regional or local trains. An observation grid was applied to these 200 hours, which documented the type of train, the kind of itinerary, and the situations in which oral use of the mobile phone re-described or corrected or modified the passenger's self-presentation.[1]

This approach, including the interviews and the observation grid, provide fresh and original subjects for discussion. At the same time, it is clear that the results are not necessarily generalizable.[2]

The aim of this research was to understand better:

- When the interference of mobile phone calls arrives.
- The mechanisms in the structure of the mobile call that allow this interference.
- Which are the most frail elements in self-presentation.
- Which strategies mobile phone users used to protect themselves from the perhaps inevitable revelation of the backstage.
- Which emotions, feelings and reactions were evoked when a mobile call caused the reformulation of self-presentation.
- The most common patterns of social narration that pass through the mobile phone.
- The social consequences of mobile phone use on the performance of our multiple personalities and social roles.

The results of this research might help us not only to understand better the social use of the mobile phone, but also to understand some other important elements, such as how self-presentation is negotiated today in the public space.

---

[1] It must be emphasized that the observation grid had only one aim, that being to annotate only those cases in which the use of mobiles in trains revealed a back stage that in some way contradicted the user's front stage.

[2] This chapter does not reflect on the subject's possible national differences, as both the area of residence of the sample and the train routes in which the observation grid was applied did not cover the whole of Italy.

13.4 Results from the Analysis

13.4.1 Interview Material

The rich discourse material that emerged from the interviews illustrates the great complexity of mobile telephone use. This material makes necessary the presentation of some preliminary considerations. To say "public space" means to speak of spatial contexts that are extremely flexible. They are governed by very different rules, norms, habits, and kinds of behavior. As Chiara F. underlines (19 years, a secondary school student):

> It's one thing [to use the mobile telephone] if you're in a bus or on the train and another if you are in a shop, in church, in a bar, a restaurant, in the gym, in the beauty parlor, at the hairdresser's, etc. The self-control required in the various situations is different. Then it depends a lot if they are public places that you know and where you are known or, vice versa, if they are new places and you are among people you don't know. It is also fundamental if you are by yourself or in the company of friends, that makes a big difference.

These comments illustrate that it is necessary to map the various kinds of public space and to apply this kind of survey systematically to the various spatial contexts. There are, for example different dynamics if a phone call is made or if it is received. In the former case there is much greater area in which one can maneuver. In the case of the telephone call being received, one's ability to maneuver is more limited. Bystanders are also subject to different dynamics, according to the more active or passive role played by who it is that makes or receives the call.

These considerations make it difficult to fulfil the aims of this chapter. We shall try, however, to set out, point by point, the most relevant data that emerged from our interviews.

13.4.1.1 Interference

When does a mobile call – and the back stage that it reveals – interfere with self-presentation? The material indicates that a call on a mobile can cause interference to the image of the person calling in two cases. These are when there is a "passive" self-presentation that is discordant with the call and when there is a self-presentation that is "active", that is, when the person acts to create a situation that is completely overturned by the intervention of the call itself. In the latter case, the call belies the person, and reduces his or her credibility. According to our respondents, the mobile call is likely to be troublesome because of what is said, the way in which it is said, the language used, the kind of code and linguistic register used. These are all elements that fill in, specify further or modify the image of that given person. In addition, the kind of mobile handset that one uses reveals their "back stage". We can see this in the comments made by Egizia B. (45 years, financial consultant),

> It was the early days of mobile phones and I was at Cortina. One day I went on an outing with my son, a companion of his and this companion's father. To look at him, he could have been a schoolteacher, a clerk, a shopkeeper or something like that. At a certain point this man was called up on his mobile (which was a satellite phone) from Malaysia and he then called Argentina. In a short time he had spoken with half the globe, because there was some sort of business deal going on. You could make out that the subject of the conversations was the building of a gas pipeline. I would never have imagined that he was such an important person.

Other dimensions also provide a glimpse into a person's back stage. These include the meta-messages, such as use of intonation, the properties of the language and the use of certain expressions or words. Dino D. (49 years, Health District Director) recounts:

> One day on the train I happened to be following a conversation of a man who to all appearances was absolutely normal. Well, this person was telling someone in a gentle voice to take some medicine. He was acting as a doctor but you could see that he wasn't one, because they were medicines that shouldn't have been prescribed in the way that he was doing it. He was alternating threats ("When you're like that I could just punch you") and tender endearments ("I would so much like to stroke you").

Ella D. (47 years, graphics artist) also remembers an episode:

> I heard a doctor friend change tone of voice completely during a work mobile phone conversation. Her voice suddenly became very reassuring and professional, she started giving advice very firmly.

In these cases, the mask of the front stage seems to slip at certain points in the conversation. As it slips one sees other, perhaps unexpected, dimensions of the individual. The number of calls that a person receives or makes is also a significant element of their backstage.

Says Andrea D. (50 years, rheumatologist):

> Normally, people have very few relationships and contacts. Often they don't know who to phone. However, a person that in an hour on the train makes twenty calls is a person that shows that they are full of social relations.

13.4.1.2 The Structure of the Mobile as Intrinsic Producer of Interference

When considering the mechanisms that structure the use of the mobile and that carry interference, our respondents were able to provide various instances. The material indicates that we are able to create a "bubble" of dual relations with the distant interlocutor. This seems to limit our concern for those who are physically nearby. Interestingly, this mechanism can also be observed with the landline telephone (Fortunati, 1995). However, this indifference to those present takes on specific characteristics with the mobile telephone since one intrudes into public space.

When we speak on our mobile phones, there is a sense in which the attention dedicated to the phone call negates the physical context. This might be expected since the mobile often puts us into contact with our network of intimate or tested relations. As Riccardo D. (17 years, secondary school student) declares, "the mobile projects the person into a world that is more familiar to him or her, where it is easy for intimacy to spring up." Another informant, Michele R. (16 years, a secondary school student), notes that the mobile call "projects users into another context, so into another role and/or personality, detaching them from that in which they find themselves at that moment."

The usefulness of the mobile rests on the presumption that you should always be reachable. Thus, if you are unavailable your network of relations may feel that they have the right to protest. At the same time, there is an implicit, and perhaps intensified, indifference towards bystanders, who are not considered very important, because they are unknown.

13.4.1.3 Self-presentation as a Porous Structure Towards Interferences

The mobile-induced exposure of separated roles and personalities is, according to the respondents, the most fragile element of self-presentation in public spaces. In addition, respondents noted the passion for prostheses as supporting elements in facing the public space, the recourse to a multiplicity of systems of meaning and the reduction of the information that individuals communicate through their appearance.

Socially, the tendency is to separate experiences and thus roles and personalities. Roles are not contiguous; rather, people traverse them. At work we are workers, at home husbands and wives, in our leisure time fans of an actor or archery champions, etc. The use of the mobile – and also the fixed phone – provides a new juxtaposition of the different social roles. The difference between the fixed phone and the mobile is that the latter allows this, not only at work and at home, but also in public spaces. In public spaces we are, in fact, taught to assume a neutral role as "a member of civil society". The use of the mobile phone, projecting as it does the sphere of work, family or friends into the public space, threatens this neutrality. It creates juxtaposition between the various roles that was not foreseen and so is not really accepted, at least in principle. As Ella D. says, "the mobile is a mixer of experiences, it carries the private and professional dimension, which are usually censured, into the public sphere."

A particular vulnerability of the back stage is constituted by social roles tolerated but not fully accepted, such as the role of the lover in all its facets. We can see this in the comments of Carlo F. (49 years, economist) who recounts:

> I was getting off the train, I was standing near the door. Next to me there were a boy and girl who were obviously together. At a certain point the boy received a call from a woman who said loudly over the phone "Hallo, darling". He tried to break off the conversation. His girlfriend asked, "But who was that?" and he answered nonchalantly, "It was Katia. She said to say hallo to you." At which point she said, "Liar! I heard perfectly well that she said hallo darling to you." At which point she slapped him.

Dealing with this public dimension is not easy. The 19th and 20th centuries were the period in history when urbanized humanity went through the difficult stage of building a public image, elaborating an etiquette that would regulate behavior and modes of living in public places in the midst of an anonymous crowd. To do this, people provided themselves with a series of prostheses that would help them to affect composure. There was the pressure to assume a self-assured and fluid behavior. Cigarettes were one of the most powerful prostheses to show how "cool" one was in the public space. Stewart (2003) acutely explains how cigarettes and the mobile have many things in common. The mobile helps one to acquire self-assured behavior, to fill in the gaps or the empty moments with a precise rituality. In other words, it helps to overcome timidity and reserve, traditionally part of experiences in contact with others in the public space. When we speak on the mobile we in fact become on the one hand lords of the surrounding space and on the other indifferent to it, because we cut ourselves from it (Fortunati, 1995). However, the mobile mediates not only dialogic communication, but also the performance of one's own image in the public space.

13.4.1.4 Defense Strategies

There are, of course defense strategies used to protect against revealing our back stage self via use of the mobile. The respondents provided various approaches here. These included the broad use of allusion when speaking via a mobile, deferring the call until later, speaking very low, standing apart from others and ignoring calls. This latter strategy is revealed in comments by Caterina G., who said, "One day I told someone that his mobile was ringing, thinking that I was doing him a favor. He thanked me, but neither answered the mobile nor switched it off." Further, the informants described sending text messages, filtering incoming calls by checking who it is, turning on the answering service and of course turning off the mobile.

It is interesting to note, however, that although practically all respondents describe these strategies, they rarely apply them. This seems to imply that we are reluctant to defend ourselves from revealing our back stage. It is seemingly a confirmation of the joy we take in communicating with interlocutors, showing off, of being on center stage of a performance. Caterina stresses this point: "It also becomes a kind of play-acting in front of others present. This is also a way of being at the centre of attention in any situation." This is not without its risks, however. Our push to be the focus can also be unmasked through an inopportune call. Elisa F. (24 years, opera singer) says, "I remember a gentleman who was holding forth in a drawing room. At a certain point he received a call from his boss. He became a worm."

13.4.1.5 The Emotions in Front of the Back stage Revealed

In addition to the other dimensions discussed here, the forced change in our self-presentation occasioned by a mobile phone call is of interest. Our respondents showed a great variety of emotions and reactions. These depend on place in which the call occurred and whom they were with – alone or with friends and acquaintances, etc.

A call can cause embarrassment, frustration or discomfort when it elicits information that we do not want revealed. At other times other co-located friends may have requested an explanation as to the content of the call. The call can become the motivation for a new line of conversation among those present. It is also the case that we must recompose and repair the effects of a call that has taken place in front of others. Alternatively one can make excuses in order to fit their image into a more articulated frame of reference. "In some rare cases," says Chiara F., "they apologize to those present." But there are also those who, like Giovanni F. (52 years, lawyer), maintain that these persons "feel nothing, because it is their real previously concealed aspect that emerges. If they realized the effects of their revealed backstage, they would try and hide it."

The rhetorical dimension involved here is the "reluctant revelation" of certain data, information that we may have decided to keep to ourselves. According to Simmel, revelation is on the one hand the solution, the release of the tension contained in the unsaid, and on the other hand it produces informational redundancy that ends up compressing the halo of mystery that pursues every urbanite (Simmel, 1908, It. trans. 1998, p. 311). According to our respondents, however, often those whose backstage has been revealed by a telephone call assume the pose that nothing has happened. "No one could care less," says Paula C. (51 years, GP), "there is total indifference." This comment is also echoed by Sandra P. (54 years, housewife): "they couldn't give a damn about those around them." Nicolò D. underlines that "speaking over your mobile has become so normal that no one takes any notice any more. And actually you find out all sorts of things, because you are relaxed over the mobile, as it is usually friends that are calling you." Adolfo D.'s opinion on the other hand is that "people don't realize, otherwise they would refrain from calling."

Emotions and reactions on the part of bystanders also vary greatly, depending on various factors, if they are with friends or alone, the place they find themselves in, etc. According to Teresa C. (46 years, teacher of informatics), "bystanders remain above all surprised." Ella D., "I have a feeling of strong irritation." For Filippo D. (13 years, a pupil in junior secondary school), on the other hand, "if you are with friends and someone receives a call, they are curious to know more about who it is." But there are also bystanders who are "indignant about how people dare to talk about their private matters like that in public," according to Diana G. (80 years, retiree), and finally Dino D. suggests that "bystanders couldn't care less, because they are in their turn waiting to be called on their mobiles."

The fact that the mobile is an extraordinarily mass technology creates a certain naturalness and casualness in its use at a social level. Individual practice is reflected and amplified in that of other people (Calefato, 2004). The high adoption rates illustrate the process of inversion (from the artificial to the natural) that, according to Barthes (1957, It. trans. 1974, p. 210), is at the basis of the contemporary myth surrounding this device. Even if in general we are little disposed to consider a machine as a natural means of communication, we often forget its artificiality. It is because it is a tool through which we practice our use of language that it is seen as being alive and we perceive its use as "natural".

13.4.1.6 Eavesdropping on the Narration

Those who were exposed to others' mobile conversations have described this as a type of forced eavesdropping (Ling, 2004). There are those who define this situation as hearing half a conversation, others speak of it as short and jumbled stories, and yet others say it is a dialogue that can be followed, though in leaps and bounds. For Riccardo D., "on the mobile they are always short things. Obviously, there are also anecdotes, generally information. At times you understand, at times you don't, it depends on whether you are close to the person speaking or listening." According to Adolfo D., "the narration is diffuse, so every now and then you realize that they are talking about something else." Paula C. suggests that "people talk a lot and in detail, describing all the details of what they are doing."

The kind of situation also has a great influence on narrative behaviour. For instance, continues Ella D.:

> The train is of necessity a moment of pause. You take advantage of this dead time to describe what you did in other moments. But you avoid describing the

situation in the presence of bystanders, as they constitute censure. But there is also the case of three-way telephone calls, between the telephone interlocutor and at the same time a bystander. In this case we have a multiple narration, which goes in various directions.

In general, mobile calls can be considered a mixture of dialogue with short narrations. However, at times a little cameo is produced, a short exchange that reveals a story. This story, although it is decontextualised from the framework of the relations and their history, is understood all the same. It springs from an oral story generally aimed not at a co-present audience, but only at the telephonic interlocutor. However, although the bystanders are not accredited members of the conversation circle, they are nonetheless an audience. This half-narration with bits missing is like one of those exercises used for learning a foreign language, in which the meaning of the text is recovered by filling in the missing words. It is a kind of game, a linguistic treasure hunt, which consists in being able to reconstruct the meaning starting from the few items of information provided. The cleverness of the person constructing the exercise consists in preparing a text in which the missing elements do not compromise the structure of meaning.

In its basic elements, mobile communication is characterized by a first interlocutor, who is in one place, and a second interlocutor, who is absent from that place, but who shares the contents of the dialogue, and by people present who are not the addressees, and who therefore play an ambiguous role as beings extraneous to the relational context; they are, that is, involuntary listeners, or veritable eavesdroppers. In the same way, the contents of mobile communication require further analysis. Too often linguists condemn mobile speaking as being "an evaporation of the prepositional contents in the mere announcement of their being there", as Minnini reports (Minnini, 2000, p. 57). This must be investigated thoroughly. Both from the perspective of our respondents' and from our observations on the train, the contents of calls on the mobile turn out to be very articulated and often full of meaning.

13.4.1.7 The Social Consequences of Revealing the Backstage

The social consequence of revealing our backstage dimensions is, according to the circumstances, the presentation of either an aperture or closure in respect to those present. At times the arrival of a call on our mobile may close us off to other co-present individuals. This is because it reinforces the maintenance of psychological contact with the person to whom the call is destined, and excludes the establishment of a psychological contact with those present.

At other times the telephone call causes aperture with respect to the bystanders "when the content of the call creates curiosity" (Adolfo D.).

According to Nicolò (16 years, secondary school pupil), "it depends on whether you know the person calling. In the first case the conversation can be fuelled, in the second you can feel excluded." According to Andrea, "it depends on where you are. The mobile on the train has withered sociality; in other situations it has broadened it [through virtual sociality]. In hospitals one tends to make short calls, which serve to broaden or maintain sociality." When, for example, the mobile conversation is included in a face-to-face conversation, a strange hybrid of mediated and non-mediated communication is created.

The public dimension has also become private and professional. It is as if a new code of behavior had been created, so that we accept that the audience will be increasingly invaded by the private and professional sphere. Ella D. recounts, "there is a woman for example who every morning makes a phone call to her son to get him to get up and go to school. Evidently she has to get up first to go to work and so can't share the experience of breakfast with her son. This ritual call is a way of showing him that although she is far away from him [she] is close by and is assisting him in all the operations of getting up." Hence Ella D. has inadvertently become a witness to the private inner workings of this woman's life as played out in the public sphere. There is, in this way, a growing tendency to bear other roles in work places and at home.

13.4.2 Modes of Encounter

Turning now to the observational portion of the analysis, the material indicates that two elements that are important: (1) that the modes of encounter between passengers are changing owing to the use of mobiles in public means of transport and (2) that the rules of social behavior between strangers are changing.

As regards the first point, the behavior that is produced in public transport can be analyzed, as Goffman recommends in Interaction ritual (Goffman, 1967, It. trans. 1988), according to three fundamental units of interaction: the social occasion, the reunion and the encounter. For the purposes of this Chapter the dimension that is most relevant in examining the inadvertent exposure of the backstage is Goffman's notion of the encounter. It is in fact the modes of encounter that are changing in all kinds of public transport. They are being transformed from "encounters with a specific centre to encounters without a specific center". Until recently, people tried to "collaborate to maintain a common center of visual or cognitive attention so as to show each their mutual willingness to talk or use a substitute for words" (Goffman, 1967, It. trans. 1988, p. 161). According to our analysis, the appearance of the mobile means that people are increasingly deciding against "being co-participants in a situation in which they can speak".

As for the second point, the change of rules involving the way not only of approaching and addressing strangers, but also the status of being with someone (Goffman, 1967, It. trans. 1988, p. 161) seems to have been changed. We say "seems," because unfortunately there is still little research (apart from the interesting works by Ling; see for example Ling, 1998) that have been produced in answer to Goffman's invitation to investigate "the normative requisites of pure and simple presence".

From the current research it emerges that some rules regarding how we approach and address strangers have been overturned. Early in the diffusion process of mobile telephones, one was reserved in their use. Now, however, this has become an arena for the display of the most intimate back stage. An exhibition that was in the early 1990s hesitant, then swaggering, has now become "natural". In this framework, the stranger becomes the depository of the most intimate secrets and the most delicate information. Anonymity is a guarantee of freedom, which individuals can exercise in talking about themselves (Fortunati, 2002). The obligation not to disturb others with noise has seemingly been transformed into the right to talk with a person who is absent, without previous assent from those present.

In addition, as with the traditional landline telephone, those who are not physically co-present have the right of interrupting those who are co-present (Fortunati, 2003). There is social resignation now at being interrupted by a telephone call, because we know that, in our turn, we will be able to interrupt the conversation. There is seemingly the inevitable fragmentation of relations and conversation while the notion of giving and keeping attention has increasingly been shredded away.

The observation grid applied during train journeys for 200 hours showed several different episodes where the use of the mobile telephone provided the observer with unexpected insight into the situation of the user. Examples include elegant women speaking of their children's bowel movements, intellectuals describing the steamy details of their partner's infidelities, parochial looking individuals who are suddenly revealed to master several languages and the intricacies of international business contracts, bullies who are taken down several pegs when their mothers call, individuals who shift dialect when accepting a call and mild-mannered people who suddenly become enraged when talking with others via their mobile telephone. In each case, the mask of the individual was partially removed and the observer was given insight into a different, and unexpected, dimension of the person using the mobile phone.

13.5 Conclusions

If the analytical categories of Simmel and Goffman, discussed at the beginning of this Chapter, are applied *tout court* to the present social situation co-constructed by users and the mobile, the picture that emerges is discomforting. First, by broadening out communication, life is being restricted. This follows from Simmel's assertion that the limitation of communication leads to a notable extension of life. Second, the personality is losing the attractiveness that derives from the concealment of information. Third, urbanized life is becoming more tiring, stressful, less anonymous and more constrained. Fourth, we have lost awareness of social space as a place of control.

Actually, if the categories used to understand the new lead us to condemn it without appeal, it means that the categories have to be changed. These categories, as they are now, are extremely useful for understanding the society of Simmel and Goffman, not ours. They must therefore be adjusted to help us to understand the changes that have in the meantime taken place. Society is today one great experimental laboratory, where the capacity of post-modernity to metabolize the dissonances of the image, and unexpectedly reveal certain stereotypes, is being re-confirmed. It is true that by broadening communication life is being restricted, but at the same time it is becoming richer, because we are discovering the intimacy of many people. Certainly, life is still a theater, but the difference between when we are acting and when we are being ourselves is on the whole less distinct, if only because the mobile gives us the possibility, when necessary, to "stage" ourselves. Finally, even when we are alone, we are not alone completely any longer.

People in the public space are less neutral, more complete in their emotions and sentiments, but also with friends and family, and their professional experiences. Thanks to the widespread use of the mobile phone, the public is becoming the place where experiences are available for circulation whereas those in private and the workplace are still kept artificially separate. Now it is becoming acceptable that people can be in civil society with all their relations, roles, identities and personalities. We have become snails in the sense that we carry our relational house on our backs. In the public space today we encounter people who are carrying whole worlds of relations, experiences and life. From this point of view we have also become freer to connect with the world that is our friend. In the end that neutrality, which is so well synthesized by Goffman in his famous definition (Goffman, 1967, It. trans. 1988, p. 162) "educated and deliberate disattention", and which was so abhorred by the Futurists as the legacy and synthesis of the Old World (Balla, 1914), has been put into serious crisis.

13.6 References

Balla, G. (1914) Il vestito antineutrale. *Manifesto Futurista*. Milan, 11 September.
Barthes, R. (1957) *Mythologies*. Seuil, Paris.
Bentley, A.F. (1931) Sociology and mathematics, in *Sociological Review*, 23 (2).

Bruner, J. (1990) *Acts of Meaning*. Harvard College, Cambridge, MA.

Calefato, P. (2004) Macchine e linguaggio: l'automazione segnica delle nuova tecnologie comunicative. In "Athanor", 7, n.s. "Lavoro immateriale", edited by S. Petrilli. Roma: Mettemi.

Eco, U. (1988) Fenomenologia di Mike Buongiorno. In *Diario Minimo*. Mondadori, Milan, pp. 30–35.

Fortunati, L. (ed.) (1995) *Gli Italiani al Telefono*. Angeli, Milan.

Fortunati, L. (2002) Italy. Stereotypes: true or false. In Katz, J. and Aakhus M. (eds), *Perpetual Contact: Mobile Communication, Private Talk, Public Performance*. Cambridge, Cambridge University Press, pp. 42–62.

Fortunati, L. (2003) The mobile phone: towards new categories and social relations. *iCS*, 6, 1.

Gergen, K. (1999) *An Invitation to Social Construction*. Sage, London.

Goffman, E. (1959) *The Presentation of Self in Everyday Life*. Doubleday, Garden City, NY [It. trans. (1969), *La Vita Quotidiana come Rappresentazione*. Il Mulino, Bologna).

Goffman, E. (1967) *Interaction Ritual*. Doubleday, Garden City, NY [It. trans. (1988), *Il Rituale dell'Interazione*. Il Mulino, Bologna].

James, W. (1890) *The Principles od Psychology*. Holt, New York.

James, W. (1910) *Psychology: the Briefer Course*. Holt, New York, pp. 177–188, 190–192, 195–197, 200–203, 205, 215–216.

Ling, R. (1998) "one can speak of common manners": the use of mobile telephones in inappropriate situations. *Telektronikk*, 98(2), 65–76.

Ling, R. (2002) The social juxtaposition of mobile telephone conversations and public spaces. Presented at The Social Consequences of Mobile Telephones, July 2002, Chunchon, Korea.

Ling, R. (2004) *The Mobile Connection: the Cell Phone's Impact on Society*. Morgan Kaufmann, San Francisco.

Katz, J. and Aakhus, M. (eds) (2002) *Perpetual Contact: Mobile Communication, Private Talk, Public Performance*. Cambridge, Cambridge University Press.

Mead, G.H. (1934) *Mind, Self and Society*. University of Chicago Press, Chicago [It. trans. (1967), *Mente, Sé e Società*. Giunti Barbera, Florence].

Meyrowitz, J. (1985) *No Sense of Place: the Impact of Electronic Media on Social Behavior*. Oxford University Press, New York [It. trans. (1995), *Oltre il Senso del Luogo*. Baskerville, Bologna].

Mininni, G. (2000) *Psicologia del Parlare Comune*. Grasso, Bologna.

Pennac, D. (1992) *Comme un Roman*. Gallimard, Paris.

Ricoeur, P. (1986) *Du Texte à l'Action. Essais d'Herménéutique II*. Seuil, Paris.

Simmel, G. (1908) *Soziologie. Untersuchungen über die Formen der Vergesellschaftung*. Duncker und Humblot, Leipzig [It. trans. (1998), *Sociologia*. Edizioni di Comunità, Turin].

Sorokin, P.A. (1927) *La Mobilità Sociale*. Milan.

Sorokin, P.A. (1928) *The Mechanicistic School*. New York.

Sorokin, P.A. (1943) *Sociocultural Causality, Space, Time*. Russell and Russell, New York.

Sorokin, P.A. (1947) *Culture and Personality: Their Structure and Dynamics – A System of General Sociology*.

Stewart, J. (2003) Mobile phones: cigarettes of the 21st century. http:/www.rcss.ed.ac.uk/people/james.html.

Von Wiese, L. (1924) *Sistema di Sociologia Generale*. [It. trans. (1968), Turin].

Part 3
The Psychological Dimensions of Mobile Communication

Introduction

Steve Love

In the last 15 years, two communication medias have had a dramatic impact on people's inter-personal behavior: the Internet and the mobile phone. For many people, the main function of the Internet is seen as being inter-personal communication with friends and family via e-mail, distribution lists and instant messaging. However, claims that the Internet is a very successful medium for maintaining and developing social relationships with friends and family are a controversial issue. For example, Nie and Lutz (2000) found that frequent users of the Internet spent less time with friends and family and Kraut et al. (1998) found that there were problems of psychological well-being for novice Internet users who were studied over a period of time.

In her book entitled *The Psychology of the Internet* (1999), Patricia Wallace highlights the fact that the Internet is an environment that allows people to act and interact with other people in ways that they not normally be able to. For example, in chat rooms people can adopt "masks" and project a self-image that is totally different to what they are really like in everyday life. This can include changing your age, gender or even projecting a different personality.

Taking Wallace's work as a starting point, one can see that this has relevance when one considers the behavior of people when they are using mobile phones. People can use the mobile phone to interact with other people in ways they were not able to before. For example, if you are running late for a meeting, you can call ahead and make your apologies. Alternatively, you may be a teenager and from the comfort your bedroom you can arrange and maintain your social life by making calls and sending text messages. People can also be seen to be projecting or adopting different "masks" when using their mobile phones in public places.

However, as with the Internet before, the question that has to be asked is, what kind of impact is the mobile phone having on our inter-personal behavior? This part of the book, therefore, attempts to provide some answers and directions for further research on the psychology of mobile phone use and its impact on social behavior.

The first chapter in this part, by Kathleen Cumisky, deals with an interesting phenomenon, known as social attribution theory, and how it can be applied to people's perception of mobile use in public places.

In social psychological research, studies of social attribution theory have shown that people tend to explain the behaviur of others in a different way to the way we explain our own behavior. For example, when asked to explain the behavior of other people, we have a tendency to say it's the person's "disposition" (e.g. stable personal characteristics such as personality) that drives the individual's behavior rather than considering the effects of situational factors on social behavior. However, not surprisingly, people tend to be more forgiving when it comes to their own behavior. We are very reluctant to attribute any socially inappropriate behavior to our own behavior in pubic places.

In Cumisky's paper, she applies the concept of social attribution to look at how people perceive the mobile phone behavior of others in comparison with their own use of mobile phones in public places. The results of the surveys carried out by Cumisky indicate that when asked to explain their own behavior, participants went to great lengths to either legitimize their public mobile phone use or to attribute it to situational factors. This was in contrast to their perception of the factors affecting the mobile phone use of other people in public locations. Cumisky suggests that this denial of personal responsibility may in fact work to protect the user from admitting to times when they acted "out of role" and risked being punished for that behavior. Cumisky concludes that we should make a concerted effort to remind ourselves of the physical presence of others, despite the demands of the non-present "others", and behave accordingly.

In the second chapter, Louise Misfud looks at the effects of introducing mobile and handheld technologies into the classroom and assesses the impact that this has on the existing learning and teaching culture.

The introduction of mobile technologies and wireless communications into the classroom presents teachers with challenges such as curriculum development and delivery, assessment and formal teaching. Misfud also highlights the fact that it has been suggested that using technology in this way challenges our existing concept of what is a "good" learning–teaching environment.

For her study, Misfud observed a group of pupils and teachers pupils engaged in a social science lesson followed by a language lesson. Informal conversations with teachers were also videotaped. Misfud's analysis suggests that it is the quality of control not the quantity of control that is important. For example, in this study, students were free to roam about the classroom and beam information to each other, the teacher and play games (an example of back-stage behavior), as long as they followed tacit rules laid down by the teacher at the outset (an example of front-stage behavior).

Misfud concludes that the personal, mobile and unobtrusive aspects of mobile technology in the classroom lead to the student and the individual being in the center and as such a variety in the teaching culture. In addition, contrary to what some people might think, the teachers did not appear to be threatened by the handheld technology. Instead, they made effective use of it by integrating it into the learning–teaching culture.

The third chapter, by Parker, introduces a concept that has been applied to other areas of technology use: addiction. Parker's study investigates (using college students) whether or not addictive phone use actually exists and what the reasons are for this, such as loneliness and ritualistic behavior (i.e. habit, passing the time). Parker's paper is also interesting because although there have been studies focusing on negative aspects of mobile phone use (such as using a handheld mobile phone while driving) there are relatively few studies available that look at the possible negative psychological effects of mobile phone use.

In order to explore this idea of mobile phone addiction, Parker employs a variety of methods. These include measures of mobile phone usage such as number of calls made and received, and also using psychometric tests such as the UCLA loneliness scale.

Parks results showed that there was a potential link between loneliness and ritualistic behavior, with participants using their phone as a matter of habit and also using the mobile phone as a means escape from situations they were unhappy with, such as a current relationship. Parker concludes that although the mobile phone is an important asset to everyday life in modern society, it is also important to remember that, depending on the individual and their circumstances, it has the capacity to interfere with the people's ability to go about their daily lives and could be identified as an addictive behavior.

The final chapter in this part is by Steve Love and Jo Kewley. In this small-scale study, Love and Kewley looked at the effects of personality on people's perception of mobile phone use in public places. In particular, the study investigated how people felt about being either a user of a mobile phone or a bystander to a mobile phone conversation in specific public locations (such as a café) and at specific distances from other people (i.e. less than 1 m).

The results obtained from this study suggest that although it is now commonplace to see people using mobile phones in public locations such as the train, there are a group of people (such as introverts) who may find this situation uncomfortable as they do not like the idea of going back stage with people they do not know in particular contexts and locations. Love and Kewley conclude by tentatively describing a process that may explain people's behavior in relation to mobile phone use in public places.

References

Kraut, R. Patterson, M., Lundmark, V., Kiesler, S., Mukophadhyay, T. and Scherlis, W. (1998) Internet paradox: a social technology that reduces social involvement and psychological well-being? *American Psychologist*, 53, 1017–1031.

Nie, N. and Lutz, E. (2000) Study offers early look at how internet is changing daily life. Report, February, Stanford, CA. Website: www.stanford.edu/group/siqss/Press_Release/press_release.html

Wallace, P. (1999) *The Psychology of the Internet*. Cambridge University Press, Cambridge.

15

"Surprisingly, Nobody Tried to Caution Her": Perceptions of Intentionality and the Role of Social Responsibility in the Public Use of Mobile Phones

Kathleen M. Cumiskey

15.1 Introduction

As public use of wireless technology becomes more widespread, the world is no longer divided into users or non-users. Most of us now maintain the dual role of mobile phone user and observer (of other people's use of this technology). This duality of roles creates a dual allegiance in the sense that, during time spent in public places, we may respond negatively to seemingly irresponsible mobile phone users, yet, as mobile phone users ourselves, we may promote protecting the unregulated use of wireless technology in public spaces. This chapter is based on a research study that examined the differences in how people respond to their own public mobile phone use compared with the public mobile phone use of others. The theoretical foundations for these differences will be explored and the data analyses will illustrate that the nature of people's emotional response to the mobile phone use of others indicates their belief in the intentionality of the mobile phone users' behavior. Differences in emotional response indicate that perhaps an ethical solution is needed to facilitate the integration of wireless technology into the public sphere.

15.2 Attributional Biases, Intentionality and the Shaping of Social Interaction in Public

In Erving Goffman's use of the dramaturgic analogy, our behavior and self concept are shaped by time spent on-stage in the presence of others. Goffman (1959) had sincere faith in an everyday presentation of a social self, a remote, idealized and public version of a private, truer self, rarely seen in public. Yet, even with this separation of selves, Goffman could not deny that an individual's social identity is constructed in conjunction with a real (in addition to imagined) Other.

Studies in attribution theory within the field of social psychology have shown that we explain the behavior of others differently than we explain our own behavior. When asked to explain the behavior of someone else, we have a tendency to overestimate the degree to which that person's disposition is at the root cause of his or her behavior and underestimate the role that situational factors play in influencing social behavior. The widespread prevalence of this phenomenon has been studied by social psychologists as "the fundamental attribution error", also known as correspondence bias (Jones and Nisbett, 1972; Nisbett et al., 1973; Storms, 1973; Monson and Snyder, 1977; Ross, 1977; Taylor and Fiske, 1978).

Conversely, we have a tendency to commit a "self-serving bias" (Aronson, 1999). When asked to explain our own behaviors, we attribute behavior that is consistent with our self concept to our disposition, and attribute behavior that is inconsistent with our notions of self to situational factors. The distancing of self from socially inappropriate behaviors may be a means through which we defend what Freud calls our ego-integrity. Denial of personal responsibility may be the most prevalent means of alleviating one's sense of guilt for "bad behavior" and therein may lie the social utility of the self-serving bias.

The prevalence of these attributional biases needs to be understood in conjunction with each other. As observers, we indicate our need for actors to take social responsibility for their actions. As actors, we acknowledge our understanding of social responsibility yet distance ourselves from our own socially inappropriate behavior. Could this stem from our fear of having our bad behavior perceived as intentional? Social behavior that is judged as intentional is seen as being motivated by internal attributes of the actor. Intentional acts are those performed with meaning, skill and awareness (Malle and Bennett, 1998). People ascribe positive or negative intentions (or even a lack of intentionality) to social behavior as a means of expressing an understanding of the phenomenological reality of norms that govern social interaction. To be "in face", according to Goffman (1982), is to be acting in accordance with what is socially expected. Goffman (1982) stated that an actor fears "losing face" because others may judge her or his behavior as intentional and therefore "as a sign that consideration for his feelings need not be shown in the future" (p 7) People who are judged as behaving outside of the norm of social expectations risk being excluded from society. Opotow (1990) states that moral exclusion "occurs when individuals or groups are perceived as outside the boundary in which moral values, rules, and considerations of fairness apply. Individuals, who are morally excluded, are perceived as nonentities, expendable, or undeserving" (p. 1). This denial of personhood is the ultimate form of moral exclusion. As a result, an actor's claim of personhood is reliant upon recognition by others.

Our desire to be outside the gaze of others underlies Goffman's theorizing around face-to-face interaction. While focused on "on-stage" behavior, Goffman acknowledged the individual's life spent "off-stage". The romance of being off-stage lies in an individual's ability to escape the demands of social interaction (Goffman, 1982). The ability to escape the pressures of social responsibility and to be free of any impediment to one's desire is a fantasy, and an extravagant self-indulgence. This realm beyond the reach of others characterizes our understanding of private space.

The use of wireless technology gives users and observers a sense of bringing the private sphere into public space (Cumiskey, 2003; Hsieh, 2004). This boundary crossing causes a type of code switching on the part of mobile phone users. Observers become vulnerable to the social transgressions of mobile phone users' off-stage behavior. As a result, the social scene within which public mobile phone use takes place becomes a potentially volatile one. Researchers have found that mobile phone use disrupts the use of shared space (Love and Perry, 2004). Public mobile phone users are perceived as rude and observers report their behavior to be outside the norm of social expectations (Ling, 2002). When wireless communication technology is present, the notion of shared social space goes beyond traditional conceptions. Users of this technology, located within a particular physical space, serve as a portal through which absent others make their presence known (Gergen, 2002). It places others who are physically present (when they are not using their own mobile phone) in a vast conundrum as to how they should behave. When on the mobile phone, users appear to be non-attentive to those physically present (Love and Perry, 2004). Consequently, the observer is then placed in a position outside the "normal" realm of face-to-face social interaction.

Whether or not one is using a mobile phone in public is at the center of this paradox. Since mobile phone users are the agents through which remote others enter the shared physical social space, the mobile phone user is still held accountable by others present for maintaining social norms. The demands of the technology itself may make it difficult for the user to maintain the social expectations of face-to-face interaction. All of the user's attention may in fact be focused on the person who has called them and the user may then indeed ignore (or fail to recognize) the physical presence of others. The results of a study that was conducted in 2003 will be utilized to explore the theoretical issues put forth thus far in this chapter. This study was done to investigate three main questions:

1. Do people evaluate the public mobile phone use of others differently than their own public mobile phone use?
2. Do people have different emotional responses to the public mobile phone use of others compared with their own public mobile phone use?
3. Do these differences indicate a belief in intentionality? And, if so, what does this say about how the presence of wireless technology is changing the expectations of public social behavior?

15.3 Method

A study was conducted in New York City, at the College of Staten Island.[1] The total number of participants was 171.[2] Two forms of a 30-item questionnaire were used in this study. The questions were essentially the same on both questionnaires. The only differences between the instruments were in the way in which the questions were framed. In the "self survey", the questions were worded so that they related to the participants' own public mobile phone use. In the "others survey", the same questions were reworded so that they would relate to other people's public mobile phone use.[3] Both questionnaires included items pertaining to the participant's own mobile phone use. Participants also rated 22 emotional indicators. In the self survey, participants were asked to rate the strength of various emotions as they may experience them while on their mobile phones in public places.[4] The other's survey asked participants to perform the identical task with regard to the emotions they may experience due to other people's public mobile phone use. In the self survey, participants were asked to relate a story about a time when they had a mobile phone conversation in a public place. In the others survey, participants were asked to relate a story about a time when they overheard someone else's mobile phone conversation in a public place.

[1] The College of Staten Island is part of the City University of New York. It is located in the borough of Staten Island.
[2] The mean age of the participants in this study was 23.7 years (range = 17–56 years); 112 participants were female (65.50%) and 59 were male (34.50%). Participants were diverse in terms of ethnicity and economic status. All of the participants lived in the five boroughs of New York City or Long Island, New York.
[3] Each participant was randomly assigned to take either the self survey or the others survey. As a result, 84 participants completed the self survey and 87 participants completed the others survey.
[4] Emotions were rated on a five-point scale with a score of 1 indicating "never felt that way" and a score of 5 indicating "always felt that way".

15.4 Results

15.4.1 Profile of Participants' Mobile Phone Use

All participants had some experience with either owning or using a mobile phone.[5] Participants reported that they liked using a mobile phone.[6] Participants reported that public mobile phone use is a common occurrence. Participants, on average, strongly disliked overhearing the public mobile phone conversations of others. They reacted to the public mobile phone use of others differently than how they reacted to their own use.

15.4.2 Emotional Responses to Public Mobile Phone Use[7]

15.4.2.1 "When Other People Use Their Mobile Phones in a Public Place, I Feel ..."

Participants reported that they felt annoyed,[8] disturbed,[9] angry,[10] disrespected[11] and ignored[12] by other people's mobile phone use. As a result, participants reported chastising (through verbal and non-verbal means) public mobile phone users for their seemingly rude behavior. The language used when recalling instances of the public mobile phone use of others characterized the average mobile phone user as worthy of rebuke, a characterization that was often shared by other observers around the mobile phone user:

> I was on the bus on my way home and it was relatively quiet on the bus then a woman's [mobile] phone rings she answers it and proceeds to carry a rather loud

---

[5] A total of 153 of the 171 participants (89.5%) currently owned and used a mobile phone. The length of time of mobile phone use varied from 2 weeks to 10 years, the average length of use being close to 3 years. Participants reported having between 0 and 30 mobile phone conversations per day, with over half of the participants having at least five conversations per a day.

[6] Participants rated how much they liked having a mobile phone on a five-point scale, with a higher score indicating a stronger liking. The average rating for this item was 4.21.

[7] A General Linear Model (GLM) was performed to reveal the significant differences between the 22 emotional ratings by survey type, gender, age category and length of time of mobile phone use. Of the 22 emotional factors rated, 14 factors emerged with significant differences in responses between the two surveys. The eight emotions that were found *not* to be significantly different by survey were "amused" [$F_{(1, 159)} = 0.000; p = 0.984$]; "included" [$F_{(1, 159)} = 2.49; p = 0.117$]; "indifferent" [$F_{(1, 159)} = 1.80; p = 0.182$]; "sad" [$F_{(1, 159)} = 0.586; p = 0.445$]; "embarrassed" [$F_{(1, 159)} = 0.043; p = 0.836$]; "sympathetic" [$F_{(1, 159)} = 2.39; p = 0.125$]; "shocked" [$F_{(1, 159)} = 0.190; p = 0.664$]; and "lonely" [$F_{(1, 159)} = 0.399; p = 0.529$].

[8] $F_{(1, 159)} = 11.07; p < 0.01$.

[9] $F_{(1, 159)} = 9.33; p < 0.01$.

[10] $F_{(1, 159)} = 12.11; p = 0.001$.

[11] $F_{(1, 159)} = 14.62; p < 0.001$.

[12] $F_{(1, 159)} = 4.03; p < 0.01$.

> conversation ... I did not understand a word she was saying because she was speaking in another language but she was so loud. The whole bus could hear her. *The woman across from me looked at me and we all shook our heads in agreement that she was being rude by holding such a loud conversation.* (Female, 20, Haitian, working class)

What is remarkable about this account is the fact that the observer, after being ignored by the mobile phone user, looked to others who shared the space with her for acknowledgment. This acknowledgment then included a joining together in the castigation of the offending character (seemingly beyond the user's awareness). In essence, the observer's face is saved by another observer who shakes her head in agreement, thereby signifying each other's place in the social space and excluding the mobile phone user as rude.

A sense of powerlessness was evidenced in the ways in which people felt "trapped" in the position of onlooker or even "voyeur" to other people's public mobile phone conversations:

> A student was discussing with a friend about the betrayal of another friend. [Presumably she snatched her boyfriend.] *The girl was visibly upset that she either did not care who was listening or not or maybe she just decided to ignore those around her.* She babbled for long on how she was going to deal with the so-called friend while at the same time was calling her some unprintable names. *At first I was amused and surprised that a girl would be so stupid as to engage in such a conversation in a public place (not caring who was listening or not). Surprise turned into embarrassment* because I felt that others will come to view or generalize other girls (including myself) as been [sic] capable of such act. *Finally, I took my books to another corner where I will be out of earshot because the more I hear, the more I become upset knowing fully well that not only was she disturbing others with her conversation, she is also a disgrace to women folk. Surprisingly, nobody tried to caution her.* [Female, 28, Igbo (African), working class]

This student's vivid description of her emotional response to overhearing another student's mobile phone conversation illustrates the psychological impact of unintentional "eavesdropping". What begins as amusement and surprise quickly turns into embarrassment as the observer, owing to the user's failure to respond to her, is rendered a "nobody". Her feelings and rights to share the public space with the user are infringed upon via the nature of the user's phone conversation. As a nobody, the observer tries to caution the user that she is acting in a disturbing and threatening way. By moving away from the face of the user, the observer is indicating that the user has transgressed the norms of appropriate social behavior. In the user's lack of response to the observer, the user is then rendered by the observer to be unworthy of consideration and "a disgrace to women folk". In the user's lack of response to those physically present, she not only voids the existence of others but her presence in the space is psychologically annihilated by those around her. The only Other that exists in this space is the person that is being responded to via the wireless device.

15.4.2.2 "When I Use My Mobile Phone in a Public Place, I Feel ..."

Participants reported that they felt self-conscious,[13] friendly,[14] scared,[15] happy,[16] comforted,[17] surprised,[18] safe,[19] nervous[20] and paranoid[21] by their own public mobile phone use. Based on this range of emotional responses, public mobile phone users may, at times, realize that they are violating expectations of social behavior. They may understand that their mobile phone behavior puts them at risk of being morally excluded by those around them. Not unlike observers, mobile phone users may also feel disrespected and disturbed by the person calling their mobile phone while they are in public. The mobile phone users themselves may be trapped by the situation in which expectations are coming from those physically present in addition to those who are remotely making contact with the person via their mobile phone. This was indicated in stories that participants told about their own public mobile phone use:

> Disagreement with my boyfriend, *I was very uncomfortable – having this conversation on the bus.* The matter (in my opinion) had to be resolved, right at that time. (I should wait to cool off – *if I didn't have this [mobile] phone with me, I would handle this situation better.*) *Other people saw me getting angry – I finished conversation before I got very angry – so I didn't have to be embarrassed ...* Stranger people saw me getting angry. (Female, 27, white, working class)

> *I got into a fight with my ex-girlfriend at work. The conversation got heated and so I started screaming on the phone while my co-workers and customers just start staring at me like I was crazy.*

> Q: What was it about the conversation that made it stick out in your mind?

> *The way everyone just looked at me while it all went on.* (Male, 21, Latino, working class)

> I was shopping and did not have a lot of time because I had to go somewhere and my phone rang while in the store. *I was annoyed because I was pressed for*

---

[13] $F(1, 159) = 20.05; p < 0.001.$
[14] $F(1, 159) = 9.04; p < 0.01.$
[15] $F(1, 159) = 5.02; p < 0.05.$
[16] $F(1, 156) = 26.08; p < 0.001.$
[17] $F(1, 155) = 11.98; p = 0.001.$
[18] $F(1, 155) = 7.73; p < 0.01.$
[19] $F(1, 155) = 9.93; p < 0.01.$
[20] $F(1, 159) = 4.24; p < 0.05.$
[21] $F(1, 159) = 5.84; p < 0.05.$

time and it was just someone who wanted to bullshit with me. At that moment I wished that I didn't have a [mobile] phone. When my phone rang it seemed loud. (Female, 27, African-American, working class)

15.4.3 The Intentions of Public Mobile Phone Use

Participants responding to the others survey were asked to an open-ended question that read, "Why do people have mobile phone conversations in public?" Participants responding to the self survey were asked an open-ended question that read, "Why do you have mobile phone conversations in public places?" All of the responses were coded into five major categories.[22] Participants made more dispositional attributions about other people's public mobile phone conversations and they made more situational attributions about their own public mobile phone conversations.[23] Table 15.1 depicts the differences in participants' evaluation of the intentions of other people's mobile phone use compared to their own use.

Table 15.1 Frequencies of "why do people (you) have mobile phone conversations in public places"

Why do people (you) have mobile phone conversations in public places?	Other people: frequency (%)	Self: frequency (%)
In case of emergency; to comfort self or others	1 (1.1)	26 (34.7)
Because it's mobile; they are on the go; it is a convenience; public use of phone is important	26 (29.9)	29 (38.7)
Neutral; just because; to save time	28 (32.2)	10 (13.3)
Out of boredom or loneliness	18 (20.7)	9 (12.0)
To be inconsiderate; to be rude; to indicate self-importance	14 (16.1)	1 (1.3)

---

[22] Explanations for public mobile phone conversations received a score of 1 if the response related to "being inconsiderate or rude", "don't care about others", "to make self feel important". Responses received a score of 2 if they related to "being bored" or "being lonely" or "nothing better to do". Responses received a 3 if they were a neutral response such as "because I can", "to save time", "to keep in touch", "same reasons as using any phone". Responses received a 4 if they related to "because it is mobile", "on the go", "convenience", "for work, school, social reasons". Responses received a 5 if they related to "important calls" and "in case of emergency". The range from 1 to 5 indicated a range in responses from dispositional attributions to more situational constraints.

[23] A GLM was created in order to compare the participants' score on the intentionality scale by survey type, gender, age category and length of time of mobile phone use. A significant main effect was found for survey type on the intentionality scale $F(1, 161) = 25.79; p < 0.001$. Participants in the others condition ($n = 87$) scored an average of 2.79 on the intentionality scale. Participants in the self condition ($n = 75$) scored an average of 3.93 on the intentionality scale. There were no significant main effect for gender [$F(1, 161) = 0.242; p = 0.624$], age category [$F(1, 161) = 0.000; p = 0.990$] or length of time of mobile phone use [$F(1, 161) = 0.719; p = 0.398$] for the intentionality scale.

Table 15.2   Frequencies of "how would you describe the average mobile phone user?"

How would you describe the average mobile phone user? ($n = 148$)	Frequency	Valid %
Annoying; loud; show-off; rude	44	29.7
Social; lots of friends; busy; carefree	32	21.6
Normal; fine; average; anyone	31	20.9
Talkative; any free time spent on the phone; bored; lonely; dependent on others	31	20.9
Trendy; rich; popular; uses it for emergencies; has important reason to use phone	10	6.8

15.4.4 Analysis of Participants' Perception of the Average Mobile Phone User

All the surveys contained an open-ended question that asked participants to describe the average mobile phone user. A majority of responses negatively characterized the average mobile phone user.[24] Table 15.2 reports the frequency of responses.

15.5 Discussion

Participants' negative response to the public mobile phone use of others may be due to their feelings of being morally excluded by the mobile phone user. If, while in the physical presence of another, one feels ignored or excluded, and not worthy of consideration, negative feelings are sure to be generated between individuals. Observers labelled an actor's behavior as rude when the observer felt that they were being ignored by the actor. Observers' sense of well-being appears threatened by others' failure to respond to them.

---

[24] A total of 148 responses were coded on a scale from 1 to 5 in order to categorize the reported attributes from negative to positive. Responses received a score of 1 if they described the average mobile phone user as "annoying", "loud", "a show-off", "rude" or "inconsiderate". Responses scored a 2 if they described the average mobile phone user as "talkative", "talks for hours/any free time spent on the phone", "bored", "lonely" or "dependent on being connected to others". Responses scored a 3 if they described the average mobile phone user as "normal", "fine", "average", "anyone" or "has a reason to use the phone". Responses scored a 4 if they described the average mobile phone user as "social", "lots of friends", "on the go", "busy" or "carefree". Responses scored a 5 if they described the average mobile phone user as "good looking", "trendy", "rich", "popular", "uses it for emergencies" or "has important reason to use phone, not just to socialize".

To respond negatively to the public mobile phone use of others may be a means through which people defend themselves against the harm of being placed in a face-threatening situation. In observing the mobile phone use of others, the observer is not only ignored, but is also often subjected to the intimate details of a stranger's life.

As a mobile phone user within a social setting, one may be thrust into a crisis of personhood. As was indicated in the various emotional responses to "When I am on my mobile phone in a public place I feel ...", it is difficult to interpret exactly what users are responding to when on their phones in public. Their "selves" may be caught between the social demands of the physical context and the demands of the remote other who is contacting them via the mobile phone. Despite the reported acknowledgments of mobile phone users bridging the space between off-stage and on-stage via their use of phones in public, participants judged the behavior of other mobile phone users to be more socially inappropriate than their own. The results of the surveys indicated that when asked to explain their own behavior, participants went to great lengths either to legitimize their public mobile phone use or to attribute it to situational factors. This denial of personal responsibility may in fact work to protect the user from admitting to times when they acted "out of role" and risked being punished for that behavior.

In order to protect ourselves from the scrutiny of others, we may choose to ignore those around us. As a result, we psychologically eliminate them. Being treated as though one is invisible (when one is not using a mobile phone), while in the presence of remote others (those made present via the mobile phone use of others), forces us to create some psychological distance between ourselves and the mobile phone user. We (as mobile phone users) may become surprised when those around us try to caution us about our non-conforming behavior, via gestures such as leaving our most immediately shared space or shaking their heads. When one uses a mobile phone in our presence, we may then shift our attention away from the user and seek the responsiveness of another person (who in return will respond to us). This social response may come via the presence of other non-users in our shared space. If not, we may then turn to our own mobile phone to avoid the awkward experience of feeling ignored (or being judged as a "voyeur"). The proliferation of public mobile phone use then perpetuates a negative space for others who may physically enter the space. As a result, our ability to be responsive to others, to act in accordance with the ethical demands of shared public space, is gravely diminished.

As citizens, we do share in the social responsibility of maintaining the well-being of others. Although this chapter may have cast the public use of mobile phones in a negative light, the move towards the integration of wireless technology into public life may serve to remind us of the psychological importance of shared public space. Adjusting the social mores of the use of public space may then require us to recognize those with whom we are face-to-face despite the demands of those remote others. As a result, we may also need to caution those who contact us via our wireless devices that the use of such immediate communication technology does not preclude us from the social responsibilities of our immediate surroundings.[25] Concerted efforts need to be made to remind ourselves of the physical presence of others, despite the demands of our remote others, and to understand that in reciprocity our own humanity is affirmed.

15.6 References

Aronson, E. (1999) *The Social Animal*, 8th edn. Worth, New York.

Cumiskey K.M. (2003) "Can you hear me now?": The paradoxes of techno-intimacy via the use of personalized communication technology in public. *Dissertations Abstracts International* 64 (8-B) (UMI No. AAT 3103097).

Gergen, K.J. (2002) The challenge of absent presence. In Katz, J. and Aakhus, M. (eds), *Perpetual Contact: Mobile Communication, Private Talk, Public Performance*. Cambridge University Press, Cambridge, pp. 227–241.

Goffman, E. (1959) *The Presentation of Self in Everyday Life*. Doubleday, Garden City, NY.

Goffman, E. (1982) *Interaction Ritual: Essays on Face-to-Face Interaction*. Anchor Books, Garden City, NY.

Hsieh, A. (2004) Blurring boundaries: home, technology, and the public/private dichotomy. Unpublished manuscript.

Jones, E.E. and Nisbett, R.E. (1972) The actor and the observer: divergent perceptions of the causes of behaviour. In Jones, E.E., Kanouse, D.E., Kelley, H.H., Nisbett, R.E., Valins, S. and Weiner, B. (eds), *Attribution: Perceiving the Causes of Behavior*. General Learning Press, Morristown, NJ.

Ling, R. (2002) The social juxtaposition of mobile telephone conversations and public spaces. Presented at a conference on the social consequences of mobile telephones, Chunchon, Korea, July 2002.

Love, S. and Perry, M. (2004) Dealing with mobile conversations in public places: some implications for the design of socially intrusive technologies. Paper presented at CHI 2004, 24–29 April 2004, Vienna, Austria.

Malle, B.F. and Bennett, R.E. (1998) People's praise and blame for intentions and actions: implications of the folk concept of intentionality. Unpublished manuscript.

Monson, T.C. and Snyder, M. (1977) Actors, observers, and the attribution process: a reconceptualization. *Journal of Experimental Social Psychology*, 13, 89–119.

Nisbett, R.E., Caputo, C., Legant, P. and Marecek, J. (1973) Behaviour as seen by the actor and as seen by the observer. *Journal of Social and Personality Psychology*, 27, 154–164.

Opotow, S. (1990) Moral exclusion and injustice: an introduction. *Journal of Social Issues*, 46, 1–20.

---

[25] Imagine the shift in the modern characterization of shared public space, if we employed new behaviors around the public use of mobile phones. (1) When we are observers of others' mobile phone use, we could concede that users are not intentionally being rude but are merely unaware of their behavior. We might then be able to approach them in a non-judgemental way to point out their social faux pas. (2) When we are mobile phone users we could indicate to our caller, our whereabouts. The caller's demands might then be measured against the social appropriateness of the conversation within one's physical context (not unlike if the person calling was actually present). (3) As users, our awareness of our surroundings is key. When we do receive a call, what would be the harm in reaching out to those in close proximity to ask them if they would mind if we took the call?

Ross, L. (1977) The intuitive psychologist and his shortcomings: distortions in the attribution process. In Berkowitz, L. (ed.), *Advances in Experimental Social Psychology*. Academic Press, New York.

Storms, M.D. (1973) Videotape and the attribution process: reversing the actors' and observers' point of view. *Journal of Social and Personality Psychology*, 27, 165–175.

Taylor, S.E. and Fiske, S. (1978) Salience, attention and attribution: top of the head phenomenon. In Berkowitz, L. (ed.), *Advances in Experimental Social Psychology*. Academic Press, New York.

16

Changing Learning and Teaching Cultures?

Louise Mifsud

16.1 Introduction

Computer Supported Collaborative Learning (CSCL[1]) is concerned with looking at how computers can be used to facilitate the learning process for students working together in a learning environment. These types of system are designed to augment face-to-face communication and not replace it. In a CSCL environment, the emphasis is placed on using collaborative systems (such as networked work stations) to foster group dynamics to aid the learning process.

From a theoretical perspective, there are several approaches that contribute to an understanding of CSCL. For example, Spiro et al. (1992) look at CSCL from a distributed cognition perspective with the emphasis being placed on investigating the interaction between individuals, environment and learning tools available. From this perspective the interest is on examining the "effect of" technology on the learning process and the "effect with" using technology to aid the learning process.

Another approach that has been adopted is to look at CSCL from a social–cultural perspective. Vygotsky (1978) stressed the importance of social facilitation in the learning process. Factors that contribute to this process in the classroom include face-to-face interaction, negotiation of tasks and coordination of activities and mobility. This theoretical perspective has also underpinned some recent research that looked at the effect of the introduction of handheld computers to the classroom environment. In their study, Zurita et al. (2003) found that school children who used a handheld computer to assist them in a maths-based collaborative

---

[1] Within the field of CSCL, and in the discussion of collaborative learning, the question of what is meant by collaborative learning has been a topic of debate as there does not appear to be unanimous agreement on what is meant by the term collaborative learning (Dillenbourg, 1999). However, for the purposes of this chapter I will not go into this discussion.

learning exercise performed better than those students who did not have technological support to help them with the task.

Following on from this, the study reported in this paper also looks at the affect of the introduction handheld devices on the classroom environment from a social–cultural perspective. In particular, this chapter is concerned with investigating if the introduction of handheld computers affects the group dynamics between teacher and pupils by challenging the teacher's ability to "control" the classroom environment.

Classrooms and their culture are ingrained in centuries of institutionalization. Control is a deeply rooted aspect of the learning and teaching cultures of classrooms. Enter new type of technology, which can be described as mobile, handheld, personalized, discrete and unobtrusive. Some educators are realizing the potential of handheld technology in education, and several research centers are both developing programs for educational purposes and also researching the implications of handheld technology in education. On the other hand, will the technology be perceived of as a potential disruption to the flow?

In this chapter, it is one of my objectives to focus on some of challenges that mobile and handheld learning technologies present to classrooms, and the impact they may have on the learning and teaching cultures. I explore an attempt which is made at implementing a mobile technology specifically for educational purposes, and argue that the learning and teaching cultures have to change or adapt to the new situation if the technology is going to contribute to learning and teaching cultures. Will the use of the technology be an add-on to the existing culture or will a new learning and teaching culture emerge?

16.2 The Case for Handhelds in Education

Mobile or handheld technologies are often referred to as PDAs (Personal Digital Assistants). They are just as often referred to as handhelds or even by their trade name such as Palm™ or iPaq™. According to Becta (undated),

> Any small, portable device that provides computing, information storage and retrieval can be described as a handheld computer.

Mobile technologies have been around for some years. If one looks at the example of the mobile telephone, the proportion of both Europeans and North Americans with mobile phones is rising fast, with the number of mobile phones being greater than the fixed type in Europe. Telephones with additional features such as built-in cameras, color screens, games and music are also increasing in popularity (*Economist*, 2002, 2003). Ling and Helmersen (2000) point out that mobile telephony studies show that the personalised, mobile, accessible and social technology is also wide-spread among children. Figures for Norway for 1999 show that 68% of 15-year-olds owned a mobile telephone, almost 40% of 13-year-olds and 82% of 20-year-olds possessed a mobile telephone in Norway (Ling and Helmersen, 2000), whereas figures for 2003 show that 99% of 16–24-year-olds, 88% of secondary school students (13–16-year-olds) and 96% of upper secondary school students (16–18-year-olds) have access to a mobile telephone (SSB, 2003). A UK survey in January 2001 by the NOP Research Group (NOP, 2001) found that 48% of children aged 7–16 owned a mobile phone.

Brown points out that although several features contribute to the popularity of the mobile phone, it is the 'mobile component' of the mobile phone that is its most important feature (Brown, 2002, p. 7). The mobile aspect involves for example the "instant availability" of information regardless of time and place (Cooper, 2002). One must also be clear over differences *within* handheld technologies, whether they are "networked" – i.e. Internet access available in a wireless mode – or whether access to the Internet has to go via a computer through synchronizing, and whether peer-to-peer communication can go over the Internet. Most PDAs have the option of infrared (IR) communication or "beaming", which is dependent on "face-to-face" contact between the users. The "personal" aspect must also be taken into consideration. Gant and Kiesler's (2002) study shows that people regard the phone as a personal possession. A text message or SMS, for example, is regarded as a personal message. The NOP Research Group study (NOP, 2001) also found that on average children aged 7–16 send 2.5 text messages per day. Inkpen points out that handheld computer technology for children is not a new idea – referring to the entertainment industry and hi-tech toys such as Sega Gameboy™, Nintendo™ and Tamagotchi™ (Inkpen, 1999). She further points out that one of the main advantages of these handheld electronic devices is their ease of integration into a child's world, and that the products themselves become a part of the children's culture.

Is the use of handheld technologies in an educational setting in a starting phase or is their use a "passing" phase? Trotter (2001) in his *Edweek* article, questions whether handheld technologies in school are "new best technology tool or just a fad". Vahey, for example, also questions whether

> ... handheld computers [are] just another flash-in-the pan technology that will have little to no impact on education? What makes this different from all the other technologies ... that claim to revolutionize education, but in the end have only marginal impact? (Vahey and Crawford, 2002, p. 5)

So what can personal and handheld technologies offer children? School scenarios have for example envisaged an "anywhere–anytime" learning (Brown, 2001).

One of the promises that handheld technology holds is that of "one-to-one" device-to-student ratio which many have argued is necessary for any sort of impact to be made in education (Soloway et al., 1999, 2001; Hennessy, 2000; Brown, 2001; Norris et al., 2002; Shields and Poftak, 2002). Another promise is that of *flexible access* to technology, which will provide tools to help children construct knowledge throughout their daily activities (Inkpen, 1999). She further points out that most computer technology available to children is segregated from other aspects of their lives.

The Palm Education Pioneers (PEP) Project Report summarizes the findings based on the evaluation of the PEP Classroom Teacher Awards. The study is based on 150 teachers across the USA. PEP teachers found that the key benefits to students were increased time using technology, increased student motivation, increased collaboration and communication and benefits from having a portable and accessible personal learning tool. They further report that the use of handheld computers has a motivating effect on students. The report further suggests that this is sustainable over time "… surprising because one would expect that as the novelty of handheld technology wanes, the motivational effects of using the technology would decline" (Vahey and Crawford, 2002, p. 14). One also has to keep in mind that the teachers themselves applied for the grant, which might also mean that the teachers were also motivated in the first place, eager to make changes in their learning culture and motivate their class.

Sharples (undated) points out that the assumption that "computer-mediated learning will occur in the classroom, managed by a teacher" is now being challenged by the growth in access of personal technologies. Handheld technologies can be viewed as a technology which can be used *in context* and *when needed*, perhaps as a result of the handhelds' unobtrusiveness and discreteness. The focus is moved from the technology to the task at hand. As Norman (1993) notes, the calculator as a technology is unobtrusive and undemanding, and that it is "we" who are in control as it supports our abilities to perform mathematical computations, but without getting in the way, an argument that can also be applied to handheld computers. The question that arises is what consequences can aspects such as "unobtrusive", "mobile" and "personal" have to the learning and teaching cultures?

16.3 Learning and Teaching Cultures

By culture I understand the way "we do things around here" – school and classroom practices – a "habitual way or mode of acting" (Simpson and Weiner, 1989). Säljö describes culture as a collection of ideas, attitudes, knowledge and other resources that we acquire through interaction with the rest of the world, including the physical tools and artefacts (Säljö, 2001, p. 30). In a traditional classroom culture and practice talking to one's friends, moving around in the classroom and changing activities on an individual basis are regarded "as signs of resistance to school practices" (Bergqvist and Säljö, in press, p. 118).

The learning and teaching culture of classroom has been described as characterized by memorization and reproduction of school texts, where lecturing and the question–answer method still seem to be the dominant forms of work in the classroom, and where teacher talk dominates and students' activity is largely limited to answering questions formulated by the teacher (Miettinen, 1999). Scardamalia and Bereiter, for example, point out that "question-asking in school is not really dialogic" (Scardamalia and Bereiter, 1993, p. 38). Resnick (1978) points out that the dominant form of learning, performance and judging in school is individual – success or failure at tasks are independent of what other students do, despite group activities. Ling (2000) argues that school, as an institution, is still Foucaultian in culture and practice, where the focus is on the individual and where the students are analyzed individually. A substantial body of research shows how teachers' actions represent institutional norms and imperative (Edwards and Mercer, 1987; Mercer, 1995). One can further argue that the Foucaultian culture argument can also be applied to the classroom and the learning-teaching culture. Edwards and Mercer (1987, p. 30) point out the teacher's need to maintain control of the content and destination of any discussion is a primary concern.

16.3.1 Frames of Control

The teacher's control of his or her classroom is embedded in what is often regarded as a "good" learning culture, characterized by an asymmetric, but legitimate, dialogue that flows between the teacher and the students. In this respect, control is regarded as a positive and legitimate aspect. According to Bernstein, "if classification regulates the voice of a category then framing regulates the form of its legitimate message" (Bernstein, 1990, p. 100, cited in Sadovnik, 2001). "Frame" refers to the "degree of control teacher and pupil possess over the selection, organization, pacing and timing of the knowledge transmitted and received in the pedagogical relationship" (Bernstein, 1973, p. 88). The manner of control, and the extent of the asymmetry are a part of this frame. However, control also has negative implications, where the teaching culture is characterised by the teacher's monopoly/control of all situations in the classroom. Mercer, for example, argues that the teacher controls the flow of the discourse, and is the sole evaluator of its content (Mercer, 1995, p. 35). *Flow* in the Csikszentmihalyi sense refers to a process of total involvement with life (Csikszentmihalyi, 1992, p. xi). Adapting the concept of *flow* to a school setting would thus mean a total involvement in the life at school, where the lines or boundaries between that which is related to "pleasure" – games, talking to friends, etc., and "business" or schoolwork are less clear-cut – a seamless flow of activities. This asymmetry can create tension and conflict and a disruption in the flow.

"Disruption" is defined by the *Oxford English Dictionary* (Simpson and Weiner, 1989) as a "dissolution of continuity" – which can be interpreted as a break in a continuous *flow*. In the world of teaching and learning cultures, it appears that the aspect of control is one which could create tension and conflict, where handheld, mobile and unobtrusive technology could conserve the asymmetry and take the role of a "disruptive technology". It seems that the handhelds release some form for defence of an established learning–teaching culture, a barrier because certain basic values, and the teacher's control, appear to be threatened.

Looking at the results of the PEP report, whereas the teachers found that the students felt

> ... empowered to find software and resources relevant to their school work, often contributing to improved class activities. (Vahey and Crawford, 2002, p. 42)

the PEP study reports that the teachers still wanted to be in control, as one of the teachers is quoted as saying

> One of the most important management issues that I have found to be associated with the use of the Palms [is] making sure that students are doing what they are supposed to be on the Palms, not playing games. (Vahey and Crawford, 2002)

A total of 56% of the teachers in this report indicated inappropriate uses, such as games played during class time, downloading inappropriate materials and inappropriate use of beaming (passing notes, cheating on tests and "copying" by handing in assignments beamed from other students) as a problem. These examples point in the direction of aspects of communication which the teacher no longer has the sole control of. Trotter (2001) points out that handheld devices have stirred debate in some schools. Administrators have banned their use, saying that some students use handheld computers to cheat in tests, play non-educational games or e-mail friends inside or outside the school. Trotter reports that in several schools, handhelds have either been banned, or it is up to the teacher to decide if they are allowed in class. Even in classrooms that allow the use of handheld computers, educators have had to be "vigilant" in preventing mischievous students from using the devices as mere toys. The question which comes up here is whether one can introduce a new technology without changing or adjusting the culture of teaching and learning, to make room for the aspects of the technology which are not included in the present culture. Is the technology merely an *add-on* to existing learning practices, or is the technology contributing to a new teaching and learning culture, becoming embedded in the context? A dilemma which crops up is that students can command an increasingly sophisticated set of communication and computing devices, which they are forbidden to use. The question which then comes up is of who controls the network for purpose of access, and what infrared technology is in relation to this? So far, the impression which Norwegian media has given is that the response of Norwegian educational institutions to such technologies appears to have been to treat them as a threat to be countered (Dagbladet, 2002; Dagsavisen, 2004; Sæter, 2004).

Sharples suggests that handhelds may become "a zone of conflict" between teachers and learners, with both trying to get control of the opportunities it affords for managing and monitoring learning (Sharples, undated). He further points out that this potential for confrontation needs to be recognized and addressed. Mifsud (2002) addresses the same topic, referring to personal experience of the classroom culture, where, for example, communication is mainly "controlled" by the teacher and punctuated by raised hands from students wishing to contribute to a school or classroom-related discussion. In this setting, other forms of communication, whether digital or analog, between friends or classmates, for example, are not legitimate. Even in the relatively open Norwegian learning and teaching culture of project and group work, communication is expected to be related to the tasks at hand. Sharples further argues that the assumption that "computer-mediated learning will occur in the classroom, managed by a teacher" is now being challenged by the growth in access to personal technologies (Sharples, undated).

Trotter (2001) describes a lesson where the class was instantly connected in an electronic network when they turned on their handheld devices. A cart that held a laptop coordinated the wireless network and ran software that let the teacher see what was on every student's handheld computer. The teacher told students to examine the painting which was visible on their screens: the interior of a crowded subway car. He instructed them to choose a person in the car and write a first-person monologue of that individual's thoughts. Trotter further describes how the teacher can monitor what each student is doing on his device. This may sound like an ideal situation for teachers, as it also means that the teacher does not have to let go of control over what the students are doing. It appears that this particular teacher actually has more control than a teacher in a non-wireless, "pen-and-paper", classroom would otherwise have. He can even make sure that the students are writing the story in the first person, thus also controlling on-task activities. Does this open up control of a different nature, which is close to surveillance? Foucault's thesis of technology (the telescope) to observe without being observed appears to be particularly suited here (Foucault, 1977, p. 171). It is the teacher who is in power, observing and correcting.

This leads the discussion to what kind of learning and teaching culture is predominant in the classroom and points to the way in which new technologies and wireless communications challenge the existing learning and teaching cultures. These are some challenges that schools face when implementing handheld and mobile technologies – challenges to timetable, curriculum, assessment, testing, … the backbone of the structure, and "grammar", of school (Tyack and Cuban, 1995) – the zone of conflict over control of different elements which make up the learning and teaching culture. Handheld technologies also appear to challenge the "image" of what is a good-quality learning–teaching culture.

16.4 Methodology

This chapter builds on a pilot study undertaken February 2003 at a school in the mid-west USA, Midlands Intermediate School. The classes are part of the University of Michigan's Center for Highly Interactive Computers in Education (Hi-CE) "Learning in the Palm of your Hands" project[2]. The students in this study were 12 years old (two sixth-grade classes, 24 students in each class, using Palm™ III). At the time of the study, the teachers had been using the handhelds in class for 3 years; the students started using their handhelds at the beginning of the school year in September 2002. The students took their handheld home if they wanted to. One of the teachers, Mrs S., is a Jason project[3] teacher for the ninth year – she can therefore be described as a teacher who is motivated and wants to make her teaching interesting. Mrs S. had received information about the Hi-CE project and had taken the initiative to contact the Hi-CE milieu. She reported that she has had excellent technical help from the Hi-CE team.

The aim of the pilot study was to ascertain where, how and why the students used their handheld computers and to describe the learning and teaching cultures when mobile and handheld technologies are introduced to the classroom culture.

The students and the teachers were observed in the classroom and the sessions were video-taped. There were informal conversations with the teachers, which were also video-recorded. This chapter focuses on observations of one social science lesson with the following language lesson.

Conversation analysis is a method for the analysis of naturally occurring interaction. Its key assumption is that language use is a site for social action (Wetherell et al., 2001). Jordan and Henderson point out that video analysis is a valuable tool for the study of learning activities as the video provides "optimal data if we are interested in what 'really' happened

---

[2] See http://www.handheld.hice-dev.org/.
[3] The JASON Project aims at offering an interdisciplinary, multimedia approach to enhance teaching and learning for students and teachers. http://www.jasonproject.org/jason_project/jason_project.htm.

rather than in accounts of what happened" (Jordan and Henderson, 1994, p. 13). A question which arises when using a video camera in classroom observations and interviews is the degree to which people are influenced by the presence of a camera. However, they continue by arguing that people get used to the camera quickly, especially if there is no operator behind it (Jordan and Henderson, 1994). Silverman (2001) emphasizes that the character of the data is critically affected by the positioning of the camera. In the pilot study the camera was mainly positioned at the back of the classroom (mainly without an operator). The videos were supplemented by observational notes. When the students gathered at one end of the classroom, the camera was moved accordingly.

The parts of the video that have been regarded as significant/relevant for this chapter have been transcribed. They contain an account of the participants' talk. They also contain annotations for non-verbal behavior, such as gaze, gesture and actions such as the manipulation of the handheld technologies where this was judged to be important to the whole context. Pauses in the conversation are indicated by [...]. The actual length of the pauses is not indicated. However, as Jordan and Henderson point out, it is "impossible to include all potentially relevant aspects of an interaction ... so the transcript emerges as an iteratively modified document that increasingly reflects the categories the analyst has found relevant to her or his categories" (Jordan and Henderson, 1994, p. 10).

The study has proceeded inductively, attempting, to use Geertz's terms, to stay close "to the ground" (Geertz, 2000, p. 24).

16.5 Preliminary Findings and Discussion

16.5.1 The Question of Control

In the classroom observed, the students appeared to work mainly in groups of four. The teacher and the students had agreed on a set of rules for use of the handheld, which specifically included the games aspect. The students were allowed to play games on their handheld provided that they had finished the task which was set to them.

From a memo, 15 February:
Mrs S. told me that in the beginning [when she was still not used to working with the handhelds] that she would define to the pupils which programs were to be used for which purposes and even specify the name for the document that the pupils had to hand in, something she said that she did not practice any more. She said that she had eventually let go of some of the control. Asked about playing games, Mrs S. answered that the pupils had agreed as to when and where game playing was allowed. Breaking the rules could mean having the Palm withdrawn for a period of

time and all the games deleted. She added, however, that it took next to no time for the pupils to get the games back – from the Internet or from other students.

Although the teacher herself commented that she had "eventually let go of some of the control", several questions can be raised here: control over what?; both input and output processes?; what kind of control has the teacher "let go of" and why?; in which way is this control "let go of"?; where is the teacher's area of control?; are we moving towards a "contract pedagogy", conditional freedom?; is this an example of "contract pedagogy" and is this different from a reward and sanction pedagogy, or is this an example of a different teaching culture where the students are allowed to play games, as long as they follow the rules; is one exchanging one form of control for another – the students still get sanctioned if the rules of the game are not followed? The sanctions appear to be tinged with a hint of "reality" – the teacher knows that the sanction – removing the handheld and the games – will only work for a short period of time.

In the classroom, the students appeared to move freely from one group to the other, exchanging information from one handheld to another (sometimes it appeared that they also exchanged games or results of games[4]). There was also infrared communication (beaming) between the students, where the students have to be physically close to each other, which technically means that the teacher can see that beaming is going on. Where infrared communication is the only form of networking the students can have, the classroom space is still the physical space. Games were mostly downloaded from the Internet and then shared around by beaming. The handhelds, although similar to start off with, had been personalized – the students had decorated the covers with colorful stickers. They were also personalized in the form of the information and games that the students chose to download on to their handhelds.

Circulating freely around the classroom when not engaged in whole class activities appeared to be part of this class's culture. The students did not appear to be bound by any format other than that they have to present their work – it is up to the students to choose the format for their output. In one of the lessons observed, part of their work was a presentation to the rest of the class, and the students were free to choose how to present their work – whether to use their Palms or not and programs which they chose. Some of the groups chose word processing, others chose a concept map, others an animation while one group chose to create a role play. The "unobtrusiveness" of the technology can be seen in this setting as it is the students who chose how and when to use the technology – the students are in control and the handheld technology appears to be supporting the

---

[4] Some of the conversations between the students are not clear enough as an external microphone was not used.

tasks the students have at hand. It seems that the students could, to a certain extent, decide in which context the handhelds were necessary and when and how to use them. It also appears that the handhelds contribute to the activity space becoming multi-dimensional.

The students, and the tasks they are set, are still bound by time constraints, however, as one can see from the next excerpt.

Memo and transcription, 17 February: from the "Freedom of Religion" lesson:
Passage 1:
Mrs S.: You have five questions that you have to answer and they are not the same five questions that you have in your book. Number one is right there and that is where you put in you cursor and start typing right there.

… [noise – students get their books from the cupboard at the end of the classroom]

…

Then it's your job to sync [synchronize] between now and Thursday cause the Internet is down. I know that the instructions say that you make sure that you sync it by Wednesday, but I've extended it to Thursday. … That way I can check your work. It's your job to make sure that it's in by Thursday and if it's not in by then I'll have to mark you down.

… [some students start opening books, others attaching their folding keyboard to their Palm]

…

Any questions?

…

Now open your social science books. … I'm going to give you until 12.30 to finish your presentation.

When the students synchronize, the students' work is downloaded on to a website to which both the teacher and the students have access when they are connected to the Internet. The teacher can then check the students' work, and give feedback through this website, and the students can either upload the feedback on to their handheld or read it directly from the website. The teacher still has control over the written work, the 'output' – the students' tasks – she is the one who evaluates the work. The pupils have been given a time limit – and their work has to be done between 'today' (which happened to be a Monday), and Thursday – and they were free to do their work on any day – provided that they finished it by Thursday. Failing to do so would result in their being "marked down". One can argue that this is an interesting choice of wording – Foucault argues that "marking down" is one option of sanctioning (Foucault, 1977, p. 175). Foucault argues that all the functions of surveillance were duplicated by a pedagogical role … and "marking down" trouble-makers was one of the tasks of the assistant teacher. On the other hand, the learning–teaching culture appears to give the *students* the responsibility of making sure that they handed in their work – conditional freedom. The teacher also explains to the students why she is extending the deadline – Internet problems. The dialogue in this particular culture appears to be less asymmetric, it appears to be based on one which flows between the teacher and the students – the teacher explains why the deadline has been moved to Thursday. As Bergqvist and Säljö point out, in another setting, the teacher operates by making her expectations clear about what is to be done, but this freedom is conditional, and there are tacit rules that limit the options (Bergqvist and Säljö, in press, p. 118). The "frame" for the teaching culture appears to be well structured, but with more autonomy for the students within this frame.

In both Mrs S.'s and Mrs G.'s classrooms, the teachers did not comment upon the pupils playing games or doing other off-task activities unless the teacher was addressing the whole classroom or, as in the excerpt below, where the pupils were supposed to be paying attention to the rest of the class's presentations. The teacher comments on the fact that she can hear a student using his/her handheld, and points out that she "shouldn't be hearing any Palm noises".

Passage 2: [During presentation of group work. Students are grouped close to the end of the class, close to the monitor]
Mrs S.: Group number 4 should be ready to go right away. [Peep] … and I shouldn't be hearing any Palm noises.

After their recess, the students had a spelling lesson with Mrs G.:

Passage 3:
The pupils in Mrs G.'s class were working on spelling. Mrs G. went around answering questions. When the pupils finished typing their task they went to the printer to print out their work. Some opened a book and started reading. Some played games. One girl had not started on her task at all. She played Tetris throughout the lesson (as far as I could see). Asked why she played Tetris "now", she answered that she liked Tetris and that she did not feel like writing now – she would do it later at home.

Some of the pupils were also beaming things to each other (too far away for the camera to catch what they were beaming).

In this context I asked Mrs G.:
LM: Is there any cheating?

Mrs G.: No. Most of them are pretty good about it. They take pride in most of their work. When they beam work, there are certain things I know. When they print out their names are already on it. If they were to beam it, I would know. Their names would be on it.

Mrs G. did not appear to be preoccupied that some of the students were not doing what they were asked to do. On the question of cheating there appears to be a mixture of trust and some measures taken in advance, as the program does not allow for cheating, copying of homework, etc. The culture can still be described as having control over some of the students' learning process. Contract pedagogy also appears to be at work here – as long as the students do their work and hand it in when they are supposed to, it does not matter where they do the work – whether they do it at home or at school. The dialogue appears to flow between teacher and students, without too much disruption. Hence the conflict over the zone of control appears to be minimised to few settings. The vigilance described by Trotter (2001) does not appear to be evident in this culture. The handhelds appeared to be integrated into the whole context of the classroom, and the unobtrusiveness of the handhelds appeared to contribute to this integration.

16.6 Closing Reflections

The discussion around the use of handhelds in education appears to be what kind of control the teachers have relinquished and why. The degree of control appears to make a difference to learning–teaching cultures, when handheld technologies are introduced. It appears that it is the *quality* of the control which is important, and not the *quantity* of control. Which control qualities are important to maintain flow? And what is freedom in a learning culture? The students were relatively free to roam around while they were working on their respective tasks, and could play games – provided that they followed the tacit classroom rules: conditional freedom. When the teacher is not addressing the whole class, then it seems as though there is a different culture which takes over. The students are then free to wander around, talk, beam information, etc., and the teacher does not try to control what they do. What kinds of activities occur in a different learning and teaching culture? The teacher no longer has a monopoly over communication, whether on- or off-task.

Do the handhelds contribute to changing the learning and teaching cultures? Do the cultures change or are they just different? What consequences on learning–teaching cultures do the mobile, personal, unobtrusive qualities of a handheld technology in the classroom have? It seems that the personal, mobile and unobtrusive aspects lead to the student and the individual being in the center and as such a variety in the teaching culture. The mobility of the handhelds appears to contribute to an uninterrupted flow. The teachers did not appear to be threatened by the handheld technology, but appeared to make use of it, integrate wholly into the learning–teaching culture. In addition, the results from this study and the Zurita study (Zurita et al., 2003) suggest that mobility has opened up another fascinating dimension for research in the area of CSCL.

16.7 Acknowledgments

I thank Professor Soloway and the Hi-CE (University of Michigan, USA) team for their support and for giving access to very interesting material. I also thank the school teachers involved for their patience in answering my questions. My thanks also go to the Institute of Information Systems at Agder University College, Kristiansand, for funding this research and InterMedia, University of Oslo, for their moral support. I also extend my thanks to Sigmund Lieberg (supervisor) for reading through and commenting on previous drafts.

16.8 References

Becta (undated) What is a handheld computer? Retrieved 24 March 2004 from http://www.ictadvice.org.uk/index.php?section=te&cat=001003&rid=437.

Bergqvist, K. and Säljö, R. (in press) Learning to plan. A study of reflexivity and discipline in modern pedagogy. In Renshaw, P. (ed.), *Dialogic Learning*. Kluwer, Dordrecht.

Bernstein, B. (1973) *Class, Codes and Control*, 2nd edn. Routledge and Kegan Paul, London.

Bernstein, B. (1990) *Class, Codes and Control. Vol. 4*. The Structuring of Pedagogic Discourse. Routledge, London.

Brown, B. (2002) Studying the use of mobile technology. In Brown, B., Green, N. and Harper, R. (eds), *Wireless World: Social and Interactional Aspects of the Mobile Age*. Springer, London.

Brown, M.D. (2001) Handhelds in the classroom. Retrieved 20 February 2004 from http://www.education-world.com/a-tech/tech083.shtml.

Cooper, G. (2002) The mutable mobility: social theory in the wireless world. In Brown, B., Green N. and Harper, R. (eds), *Wireless World: Social and Interactional Aspects of the Mobile Age*. Springer, London.

Csikszentmihalyi, M. (1992) *Flow: The Psychology of Optimal Experience*. Harper Perennial, New York.

Dagbladet (2002) Nektes mobil på skolen. Retrieved 24 March 2004 from http://www.dagbladet.no/nyheter/2002/06/13/337827.html.

Dagsavisen (2004). Nei til mobil på skolen. Retrieved 26 March 2004 from http://www.dagsavisen.no/innenriks/apor/2001/10/609423.shtml.

Dillenbourg, P. (1999) What do you mean by 'collborative learning'? In Dillenbourg, P. (ed.), *Collaborative Learning: Cognitive and Computational Approaches*. Pergamon, Amsterdam, pp. 31–63; http://tecfa.unige.ch/tecfa/publicat/dil-papers-2/Dil.7.1.14.pdf

Economist (2002) The fight for digital dominance. *The Economist*, 21 November.

Economist (2003) Mobile phones. *The Economist*, 9 January.

Edwards, D. and Mercer, N. (1987). *Common Knowledge: the Development of Understanding in the Classroom*. Routledge, London.

Foucault, M. (1977). *Discipline and Punish: the Birth of the Prison*. Allen Lane, London.

Gant, D. and Kiesler, S. (2002) Blurring the boundaries: cell phones, mobility, and the line between work and personal life. In Brown, B., Green, N. and Harper, R. (eds), *Wireless World: Social and Interactional Aspects of the Mobile Age*. Springer, London.

Geertz, C. (2000) *The Interpretation of Cultures: Selected Essays*. Basic Books, New York.

Hennessy, S. (2000) Graphing investigations using portable (palmtop) technology. *Journal of Computer Assisted Learning*, 16, 243–258.

Inkpen, K.M. (1999) Designing handheld technologies for kids. *Personal Technologies Journal*, 3(1&2), 81–89.

Jordan, B. and Henderson, A. (1994) Interaction analysis: foundations and practice. *The Journal of the Learning Sciences*, 4(1), 72s.

Ling, R. (2000) The impact of the mobile telephone on four established institutions. Paper presented at the ISSEI 2000 Conference of the International Society for the Study of European Ideas, Bergen, Norway, 14–18 August.

Ling, R. and Helmersen, P. (2000) It must be necessary, it has to cover a need: the adoption of mobile telephony among pre-adolescents and adolescents. Paper presented at the Sosiale Konsekvenser av Mobiltelefoni, Oslo.

Mercer, N. (1995) *The Guided Construction of Knowledge: Talk Amongst Teachers and Learners.* Multilingual Matters, Clevedon.

Miettinen, R. (1999) Transcending traditional school learning: teachers' work and networks of learning. In Engeström, Y., Miettinen, R. and Punamäki-Gitai, R.-L. (eds), *Perspectives on Activity Theory.* Cambridge University Press, Cambridge, pp. 325–344.

Mifsud, L. (2002) Alternative learning arenas – pedagogical challenges to mobile learning in education. Paper presented at the Wireless and Mobile Technologies in Education Conference, Växjo, Sweden.

NOP (2001) Half of 7–16s now have a mobile phone. Retrieved 24 March 2004 from http://www.nop.co.uk/news/news_survey_half_of_7-16s.shtml.

Norman, D.A. (1993) *Things That Make Us Smart: Defending Human Attributes in the Age of the Machine.* Addison-Wesley, Reading, MA.

Norris, C., Soloway, E. and Sullivan, T. (2002) Examining 25 years of technology in U.S. education. *Communications of the ACM*, 24(8).

Resnick, L (1978). Learning in school and out. *Educational Researcher*, 16(9), 13–20.

Sæter, K. (2004) Forby kameramobiler på skolen. *Aftenposten*, 29 March.

Sadovnik, A.R. (2001) Basil Bernstein. *Prospects: the Quarterly Review of Comparative Education*, 31, 687-703.

Säljö, R. (2001) *Læring i Praksis: et Sosiokulturelt Perspektiv.* Cappelen Akademisk, Oslo.

Scardamalia, M. and Bereiter, C. (1993) Technologies for knowledge-building discourse. *Communications of the ACM*, 36(5).

Sharples, M. (undated) Disruptive devices: mobile technology for conversational learning. Retrieved 10 October 2003 from http://www.eee.bham.ac.uk/sharplem/Papers/ijceell.pdf.

Shields, J. and Poftak, A. (2002) A report card on handheld computing. *Technology and Learning*, February.

Silverman, D. (2001). *Interpreting Qualitative Data: Methods for Analysing Talk, Text and Interaction*, 2nd edn. Sage, London.

Simpson, J.A. and Weiner, E.S.C. (eds) (1989). *The Oxford English Dictionary*, 2nd edn. Clarendon Press, Oxford:

Soloway, E., Grant, W., Tinker, R., Roschelle, J., Mills, M., Resnick, M., Berg, R. and Eisenberg, M. (1999). Science in the palms of their hands. *Communications of the ACM*, 42(8).

Soloway, E., Norris, C., Blumenfeld, P., Fishman, B., Krajcik, J. and Marx, R. (2001). Devices are ready-at-hand. *Communications of the ACM*, June.

Spiro, R.J., Feltovich, P.J., Jacobson, M.L. and Coulson, R.L. (1995) Cognitive flexibility, constructivism, and hypertext: Random access instruction for advanced knowledge acquistion in ill-structured domains. http://www.ilt.columbia.edu/ilt/papers/Spiro.html.

SSB (2003) Andel som har tilgang til diverse informasjon – og kommunikasjonsteknologi, etter familietype, husholdningsinntekt, kjønn, alder, utdannelse og arbeidssituasjon. 2. Kvartal 2003. Retrieved 24 March 2004 from http://www.ssb.no/ikthus/tab-2003-11-06-01.html.

Trotter, A. (2001) Handheld computing: new best tech tool or just a fad? Retrieved 2002 from http://www.edweek.org/ew/newstory.cfm?slug=04palm.h21.

Tyack, D.B. and Cuban, L. (1995) *Tinkering Toward Utopia: A Century of Public School Reform.* Harvard University Press, Cambridge, MA.

Vahey, P and Crawford, V. (2002) *Palm Education Pioneers Program: Final Evaluation Report.* SRI International.

Vygotsky, L. (1978) *Mind in Society: the Development of Higher Psychological Processes.* Harvard University Press, Cambridge, MA.

Wetherell, M., Yates, S.J. and Taylor, S. (2001) *Discourse as Data: a Guide for Analysis*. Sage, London.

Zurita, G., Nussbaum, M. and Sharples, M. (2003) Encouraging face-to-face collaborative learning through the use of handheld computers in the classroom. *In Chittaor, L. (ed.), Proceedings of Mobile HCI 2003*. Springer, London, pp. 193–208.

Mobile Phone Addiction

Woong Ki Park

17.1 Introduction

The use and the definition of mobile phone is undergoing reinterpretation as the mobile phone blurs the distinction between personal communicator and mass media. The mobile phone has become one of the most omnipresent communication devices within the past decade. According to Cohen and Lemish (2002), mobile phones used to be an esoteric device. Today, the mobile phone is certainly the most pervasive communicative device that people carry. The mobile phone can connect people "anytime", "anywhere" and with "any body", with the added benefit of mobility and portability.

Especially in Korea, mobile phone use has proliferated since the first service was launched 15 years ago. In fact, Korea's mobile phone penetration has reached at 66%[1] (Ministry of Information and Communication, 2002) and Korea is one of the leading countries in the world in the areas of mobile phone use and technology.

With the wide spread and high penetration of mobile phones, Koreans use the mobile phone literally at any time, anywhere and with anybody – from subways to restrooms. The rapid adoption of the mobile phone can be attributed to many factors, including increased portability, subsidized subscriptions, declining costs and added value-added services. No segment of Korean society is exempt from the mobile phone culture.

Furthermore, mobile phone technology is becoming more advanced every year, especially in the Korean market. For example, new mobile telephone models are offered every 3 months from numerous manufacturers. Also, mobile phone users now can watch movies, play video games, listen to music, pay for goods and services and so forth. Hence the mobile telephone is "more than meets the eye".

---

[1] Korea's total mobile phone users are estimated at 32,324,588 (about 66% of the total population of South Korea). The following is the breakdown of the three mobile telecommunication carriers in Korea. SK Telecom (SKT) has 17,149,621 (53%) users, Korea Telecom Freetel (KTF) has 10,496,588 (32.3%) users and LG Telecom (LGT) has 4,783,075 (14.7%) users Ministry of Information and Communication, 2002).

Despite the rapid acceptance of mobile phone use around the world, the social and cultural impacts of mobile phone have attracted little academic attention so far (Katz and Aakus, 2001), but there are a small number of research projects that deal with the social implications of mobile phone use from various countries (Leung and Wei, 2000; Townsend, 2000; Beaubrun and Pierre, 2001; Colombo, 2001; Fortunati, 2001; Katz and Aakhus, 2001; Licoppe and Heurtin, 2001; Ling, 2000, 2001; Palen et al., 2001; Wikle, 2001). In Korea, there are five published studies that dealt with the social uses of mobile phones (Bae, 2001, 2002; Kim, 2001; Na, 2001; Sung and Choi, 2002). However, previous studies were basically exploratory in nature and viewed the use of the mobile phone from functional perspectives. With the proliferation of mobile phone use, not a single study has dealt with the possible negative consequences of mobile phone use despite the fact that there is some evidence that mobile phone use can have negative effects. Some of the popularized negative consequences of mobile phone use include dangers of driving while on the mobile phone (Green, 2001) and harmful effects of radiation emitted from the mobile phone itself (Sandstrom et al., 2001). However, these reported ill-effects of mobile phone use result from either the medium's machinery itself (e.g. radiation, heat) or ill-coordination of physical movement. None of the previous studies has dealt with the possible negative psychological effects on the mobile phone user, especially dependence on the mobile phone due to heavy use of the medium. In a recent survey of Korean college students in the Seoul metropolitan area, 73% of the respondents reported that if they do not have access to mobile telephone, they feel uncomfortable and irritated (Lee, 2002) – which is a sign of a withdrawal symptom of addiction. In a similar way, American people have developed an "obsession" to carrying the mobile phone everywhere (Wikle, 2001) and show signs of heavy dependence on the use of mobile phones (Licoppe and Heurtin, 2001). Although the term "addiction" can be imprecise, it does grasp the notion of real-life problems. Addiction can cause detrimental damage both to individuals and to society because addicted people cannot work or study owing to their physical and psychological dependence on substance/media, disabling their functionality within society.

Hence, in the light of reported cases of college students and other similar cases, we can safely conclude that there is a possibility of addiction to mobile phone use. The study of addiction to the mobile phone is important for two reasons. First, it is important to define the nature of mobile phone behavior among people because what constitutes a problem use or not must be defined. Furthermore, the study of addiction can better understand the possible social and psychological effects and processes of media use because the notion of addiction includes high/over-use of the medium. For instance, cultivation studies argue that extensive watching of television correlates with the distorted perception of real life (Gerbner and Gross, 1976; Signorielli, 1986).

The mobile phone is an extraordinary medium. People can communicate with each other without time and space restrictions. However, the problem is that people sense that they cannot live a normal life without it. The knowledge of how the medium can affect people can help people who are in need of help.

Therefore, the main objective of this research is to provide some means of making an empirical distinction between normal and problem mobile phone use. This research attempts to assess whether problematic mobile phone use exists, and how it might happen. In an attempt to explain why and how certain people become highly dependent on (addicted to) the mobile phone, and what problematic use might be, this research adopts the notions of substance use/abuse.

17.2 Theoretical Conception and Literature Review

17.2.1 Definition of Addiction

Addiction can be described as an unusually high dependence on a particular medium. The term addiction can be used to denote all types of extreme behavior, such as an unusual dependence on drugs (e.g. alcohol, narcotics), food, exercise, gambling, gaming, television viewing and Internet use (Peele, 1985). Newer types of addictions that have been studied include on-line sexual addiction (Bingham and Piotrowski, 1996; Young, 1998; Stein et al., 2001), and addictive consumer behavior (Elliott et al., 1996; Faber et al., 1987).

In general, addiction is a psychological dependence that is life organizing and more important than other coping instruments (Peele, 1985). According to Peele (1985), any compulsive or overused activity should be considered as addiction. Chemical and biological models of the causes of addiction exist. For example, one theory of drug addiction says that the drug itself causes dependency, leading an individual to be under its control (Handdock and Beto, 1988). Other theorists have cited genetic predispositions or brain differences as causes of dependency (Schuckit, 1987). Some other theories of causes of addiction are: social (e.g. demographic factors cause perceived deficits for which an individual must compensate, or predispose individuals to addiction); lifestyle (e.g. peer groups pressure individuals into addictive behavior); and psychological (e.g. personality traits such as depression or hyperactivity increase motives to indulge in addictive behavior) (Haddock and Beto, 1988). According to Peele (1985), the major motives for addictive behavior are relief of pain, anxiety, and other negative emotional states (i.e. escape); enhanced control, power, and self-esteem (i.e. compensation); simplifying and making life seem more manageable (i.e. ritual); and as a mood modifier or way of feeling good (i.e. instrumental).

It is not surprising that previous research on addiction had confounding problems because addiction can be defined by many factors. For example, in the study of alcoholism, alcohol addiction has been defined as "getting drunk, heavy drinking, excessive drinking, deviant drinking and unpopular drinking" (Keller and Doria, 1991, p. 253). Many researchers now agree that heavy drinking does not constitute alcoholism unless other symptoms are evident.

17.2.1.1 Psychological Addiction

Sensation seeking has been linked to smoking addiction (Carton et al., 1994) and drug addiction (Zuckerman, 1979; Jaffee and Archer, 1987; Andrucci et al., 1989). According to Russell and Bond (1980), two-thirds of the alcoholics in their study sample believed that alcohol would balance out their unpleasant feelings. In another study, shopping-addicted people reported feeling motivated by escape. They reported that shopping helped them to stop feeling depressed (Elliott et al., 1996). These psychological factors play an important role in addiction research.

According to Akers (1991), the traditional concept of addiction included tolerance, dependence and withdrawal, and applied to the physiological demand for a drug. The term psychological dependence replaced addiction in the 1960s to refer to the craving for a drug without physical dependence (Horvath, 1999). The term psychological dependence is now used to describe habitual behavior in the absence of proof for physical addiction. Although popular opinion now favors traditional conceptualization of addiction, the term dependence remains in place (Akers, 1991).[2] Criteria for dependence (i.e. addiction) are outlined in the American Psychiatric Association's (APA) (1994) *Diagnostic and Statistical Manual of Mental Disorders* (DSM-IV), which includes spending a great deal of time using the substance; using it more often than one intends; thinking about reducing use or making repeated unsuccessful efforts to reduce use; giving up important social, family or occupational activities to use it; and reporting withdrawal symptoms when one stops using it.[3]

Substance abuse is characterized by severe impairment or distress, such as loss of employment, child neglect or putting one's life at risk due to the behavior. Interestingly, based on the above criteria, a person could be diagnosed as substance dependent solely based on psychological factors.

---

[2] Addiction refers to dependence on drugs but the term is not as widely used by researchers (Jung, 2001). In this research, addiction and dependence are used interchangeably.
[3] An individual only has to respond affirmatively to three or more symptoms during any 12-month period to be classified as substance dependent.

From these definitions of addiction in the literature, many questions emerge with regard to mobile phone use. First, can people be addicted to mobile phone use? Can causal states for addiction such as compensation, escape, ritual, instrumental or sensation/stimulation seeking also exist for addiction to mobile phone use? Although no research has been done on mobile phone use, the same principles and constructs can be applied.

17.2.2 Media Addiction

17.2.2.1 Uses and Gratifications Paradigm

According to Katz et al. (1974), people use media actively. They argued that people approach media based on perceived needs that emerge from psychological and social influences. Most research lists reasons for media use as information seeking, escape or relaxation and companionship. For example, McQuail et al. (1972) made a typology of media use that included escape, companionship/social utility, personal reference/value information and reality exploration/information surveillance. In another study, Katz et al (1973) created a typology of social and psychological media use that included information/cognitive, emotion/affect, cred-itability/status, social contacts and escape.

Research on the uses and gratifications paradigm has consistently argued that the gratifications sought motivate the use of a particular medium and hence users fulfil certain psychological needs. They have identified about six major motives for gratifying felt needs through tele-vision. These included learning, habit, pass time, companionship, escape, arousal and relaxation (Rubin, 1994) and these motives satisfy two core motive orientations: ritualized and instrumental (Horvath, 1999).

Also, studies on mobile phone from uses and gratifications perspec-tives list similar gratifications motives. Bae (2001, 2002) argues that mobile phone users' gratifications sought were entertainment, compan-ionship, immediacy and privacy. In another study of mobile phone user's usage pattern, Sung and Choi (2002) found that companionship, instru-ment, entertainment and self-actualization were the gratifications sought by the users. Also, previous studies on the land-line fixed telephone have found two types of telephone use (Keller, 1977; Noble, 1987): which included intrinsic (social-oriented) and instrumental (task-oriented).

Hence previously identified uses of media can be applied to the study of the mobile phone. To sum up, the reasons people give for using certain media can be reduced to two core factors: ritualistic and instrumental. Instrumental use of media is typically defined by motives for information-seeking or arousal-seeking behavior, whereas ritualistic use of media is defined by motives such as habit, passing time or companionship (Rubin, 1981b, 1984). People who have ritualized motives are habitual media users and have a high affinity with the medium, whereas instrumental viewers use media selectively and have an affinity with specific content (Rubin, 1981a, 1983). Habitual/ritualistic use of media implies dependency or addiction.

The most convincing argument that the media can act like a drug comes from so-called "deprivation studies" research. These studies argue that people actually experience withdrawal symptoms when they cut back on the use of a certain medium. Studies of deprivation show what needs are most frustrated when media are unavailable. Steiner (1963) found individual accounts of people whose television set was out of order, which included "the family walked around like a chicken without a head". In Winick's (1988) study of people who were deprived of television viewing owing to a broken or stolen television, people felt great anxiety and aggression. In another study, by deBock (1980), ritualistic motives were found to be frustrated the most by television viewers who had power failure problems in their residential areas.

A study by Foss and Alexander (1996) comparing heavy and non-television viewers found that non-viewers called television a drug or religion and that viewing caused less social interaction. In a similar study, Edgar (1977) found that people regarded television as a depressant drug that dulls the senses. Anderson et al. (1996) also found that people used television just like alcohol. McIlwraith (1988)'s extensive study on those who called themselves television addicts found that television addicts were more neurotic, introverted and easily bored than non-television addicts.

Kubey (1996) argued that a study using APA's DSM-IV criteria can identify television addiction. A few studies did actually used APA criteria to find action in video game playing and computer use (Fisher, 1994; Phillips et al., 1995).

17.2.3 A Model of Psychological Addiction to the Mobile Phone

Psychology/psychiatry studies on addiction have extensively used APA's DSM-IV criteria to find people who are addicts and communication studies on media dependence and uses and gratifications found major motives for using the media. A synthesis of these studies can shed light on the study of mobile phone addiction.

Finn (1992) argued that there are two types of media use: (1) social compensation type and (2) mood management type. Similarly, Keller (1977), Rubin (1984) and Noble (1987) have found media use patterns that are ritualistic or instrumental. Either route may result in addiction, although the literature suggests that social compensation is probably the primary route. The first route is compensatory in nature and the results reinforced motives for using the medium and symptoms of addiction. In this route, mobile telephone use is analogous to drug use as a depressant: to relax, escape or fulfil a habit. The second route is about mood manage ment. In this route, mobile phone use is a stimulant, fulfilling needs for cognition. Mobile phone addiction, like any other addiction, may be the result of dependence on its stimulant or depressant effects.

Based on these conceptual dimensions of mobile phone addiction, this study seeks to expand mobile phone usage research by addressing the following important question: what constitutes mobile phone addiction? Specifically, one research question (RQ) and three hypotheses (H) are raised:

RQ1: Do concepts comprising stimulant route or depressant route best predict mobile phone addiction?

The mobile phone's basic function is inherently to enable two distant parties to communicate at the same time, and basically eliminates the fundamental human anxiety regarding loneliness (Townsend, 2000). Hence it is conceivable that people who exhibit some degree of loneliness might be inclined to use the mobile phone more to eliminate such anxiety. In a similar way, people might need stimulation in order to compensate for unfulfilled needs. Griffiths and Dancaster (1995) argued that the need for sensation seeking can be a part of dependence. Finn (1997) found that extroverted people preferred talking rather than using media.

H1: Loneliness will be positively related to ritualistic motives and mobile phone addiction.

H2: Need for cognition will be positively related to instrumental motives and mobile phone addiction.

H3: Stronger relationships will result between ritualistic motives and mobile phone addiction than between instrumental motive and television addiction.

17.3 Method

A survey was administered to student mobile phone users[4] in Seoul metropolitan area colleges from December 2002 to January 2003. Participants were asked to report the minutes of telephone usage (number of incoming calls received, number of outgoing calls, average number of minutes spent per call and number of text messages sent) and subsequently these were divided into light user and heavy user categories for further analysis. The questionnaire included the following measures to assess various

---

[4] 99% of Korean college students own mobile phone (Lee, 2002).

dimensions that are discussed[5] in the study: demographic items, CAGE items,[6] mobile addition items, usage motive items (ritualistic and instrumental) and psychological items(loneliness and need for cognition).[7]

The sample consisted of 200 people who participated in the survey and 157 usable data resulted (Tables 17.1 and 17.2). To answer research questions, the following statistical procedures were done. First, Cronbach's alpha tests were performed to measure the reliability of the various measures that were used in the study. Second, factor analyses were carried out to find the factors. Third, Pearson correlation analysis was applied to see the relationships among different variables. Fourth, hierarchical regression equations were applied to see whether mobile phone addiction was comprised of the stimulant or depressant route.

Table 17.1 Descriptive data for the respondents

Demographic items		No. of respondents	Proportion (%)
Sex	Male	76	48.4
	Female	81	51.6
Age (years)	Under 21	6	3.8
	21–25	91	58.0
	25–32	37	33.1
	Over 32	8	5.1
Total		157	100.0

---

[5] Korean's average usage of the mobile phone (minute of use: MOU) is 9 minutes per day based on the data from Korea's three mobile phone carriers' (SK Telecom, Korea Telecom and LG Telecom) (official data). In this research, respondents who reported less than 9 MOU were considered "light" users and respondents who reported more than 9 MOU were considered "heavy" users. To aid the accuracy of MOU reported by the respondents, survey administrators were instructed to check with the respondents' mobile phone's incoming call, outgoing call and the number of text messages records.

[6] The CAGE questionnaire is a self-report alcoholism screening device which consists of direct questions about substance use and related problem (Ewing, 1984). The acronym stands for Cutting down, Annoyance by criticism, Guilty feeling and Eye opener. The measure has an accuracy rating of over 90% in screening out patients (Ewing, 1984). Item responses on the CAGE are scored for 0 for "no" and 1 for "yes" answers, with a higher score an indication of alcohol problems. While the normal cut-off for the CAGE is two positive answers, the consensus panel of the substance abuse and mental health services administration recommends that primary care physicians lower the threshold to one positive answer to cast a wider net and identify more patients who may have substance abuse disorders. In this study, the following modified versions of CAGE were used to screen possible mobile phone use addiction: "Have you ever felt you should Cut down on your mobile phone usage?" (83% responded "yes"); "Have people Annoyed you by criticizing your mobile phone usage?" (31% responded "yes"); "Have you ever felt bad or Guilty about your mobile phone usage?" (60% responded "yes"); and "Do you usually turn on the mobile phone first thing in the morning?"(70% responded "yes").

[7] The addiction studies have reported difficulty in finding people who call themselves addicts due to the nature of question that are raised (Mcllwrath, 1998; Smith, 1986). Hence, the best way to find addicts is to seek participants who identify themselves as addicts or who can be identified by others as addicts (Horvath, 1999).

Table 17.2 Descriptive data on mobile phone use[a]

Sex	Age (years)	Average calls per day	Average messages per day	Total average calls per day (calls + text)
Male	Mean	7	11	18
	N	76	76	76
	S.D.	10	6	7
Female	Mean	12	9	21
	N	81	81	81
	S.D.	3	7	5
Total	Mean	9	12	21
	N	157	157	157
	S.D.	2	9	11

[a] Frequency: total number of calls placed and received.

17.3.1 Measures

17.3.1.1 Mobile Phone Addiction

Mobile phone addiction was measured by the revised version of the Television Addiction Scale developed by Horvath (1999), which is based on the DSM-IV. The scale represented each of the seven criteria of dependence and consisted of a five-point Likert-type scale with 20 items: tolerance, withdrawal, unintended use, cutting down, time spent, displacement of other activities and continued use.

In order to study the nature of mobile phone addiction, the validity of the revised measure is central to the study. Hence three types of validity test were performed (content, criterion and construct) first. First, the mobile phone addiction scale's content represents the dimensions of addiction since its original scale is based on the DSM-IV scale itself. Second, the mobile phone addiction scale was compared with the CAGE instrument to test criterion validity and it was found that two measures had high correlations ($r = 0.52, p < 0.05$). Third, after factor analysis of the scale, the result indicates the theoretical constructs of addiction that the study tried to measure (Table 17.3).[8] The first factor was labeled "problem use" and accounted for 29% of the total variance after rotation (eigenvalue = 5.51). Its 14 loadings described mobile phone use as problematic. The second factor was labeled "guilty use" and accounted for 15.3% of the total variance after rotation (eigenvalue = 2.91) and its six loadings described mobile phone use as resulting in a feeling of guilt. To check the

---

[8] Used principal axis factor analysis with oblique rotation. A factor needs a minimum eigenvalue of 1.0 and at least three loadings meet a 60/40 rule to be retained.

Table 17.3 Factor analysis of the mobile phone addiction scale

Addiction item	Mean	S.D.	Factor					
			1	2	3	4	5	6
1. Problem use								
Family members get angry	3.32	1.84	0.85	0.29	0.09	0.09	−0.11	0.07
Loved ones can't stand it	3.77	0.88	0.84	0.26	0.04	0.11	0.08	0.04
Created real problems for me	3.22	1.03	0.83	0.26	0.09	0.09	0.11	0.03
You can call it withdrawal	3.63	1.03	0.82	0.23	0.18	0.02	0.00	0.03
Feel bad but cannot stop using it	3.94	0.93	0.81	0.22	0.22	0.20	0.02	0.18
Causing serious problems	3.49	0.89	0.79	0.26	0.20	0.09	0.09	0.10
Use it because I missed it so much	3.52	0.97	0.74	0.40	0.28	0.20	0.05	0.10
Whole life revolves around mobile phone	3.05	0.91	0.72	0.32	0.17	0.10	0.03	0.07
Embarrassed to tell people	3.07	0.89	0.71	0.24	0.25	0.25	−0.04	0.03
Alienating people around me	3.36	1.05	0.70	0.20	0.03	0.04	0.07	0.04
Spend more time than anything	3.52	1.18	0.69	0.33	0.16	−0.03	0.02	0.03
Use more	3.65	1.12	0.67	0.48	0.22	0.10	−0.06	0.02
Spend great deal of time using it	3.00	0.89	0.64	0.30	0.18	0.10	0.07	0.09
Use more to feel the same	3.07	1.22	0.61	0.14	0.17	0.20	0.01	0.18
2. Guilty use								
More productive if I didn't use it	3.75	1.00	0.30	0.82	0.18	0.02	−0.02	0.01
I think I should cut down	3.65	1.20	0.34	0.68	0.16	0.05	−0.03	0.05
Spending many hours	3.01	1.37	0.30	0.61	0.34	0.10	0.06	0.02
Feel guilty about over-using it	3.77	1.02	0.30	0.58	0.15	0.03	0.08	0.20
Should concentrate on study/work	3.57	0.94	0.32	0.55	0.17	0.07	0.06	0.20
Should spend more on other activities	3.45	0.97	0.30	0.51	0.25	0.20	0.03	0.30
Eigenvalue			5.51	2.91	2.00	1.49	1.27	1.06
Variance explained (%)			29.0	15.3	10.0	7.84	6.19	4.98

reliability of the measure, Cronbach's alpha was measured and found to be reliable (0.85) for further analysis in the study.

17.3.1.2 Ritualized/Instrumental Using Motives

Ritualized and instrumental use motives were measured with the revised version of the Television Viewing Motives Scale developed by Rubin (1983) to fit with the nature of the study. The measure consisted of a five-point Likert-type scale with 27 items and measured the user's underlying motives for using a specific medium, such as relaxation, habit, pass time, entertainment, social interaction, information, arousal and escape.

Before the main analysis of the study, factor analysis was performed to see the underlying dimensions of the mobile phone use among the respondents (Table 17.4).

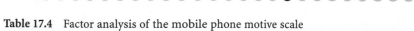
Table 17.4 Factor analysis of the mobile phone motive scale

Item	Mean (5-point)	S.D.	Factor 1	2	3	4	5	6
1. Habit/pass time								
Passes the time when bored	3.03	1.17	0.83	0.11	0.17	0.21	0.10	0.03
Something to do	3.65	1.13	0.78	0.28	0.21	0.24	0.29	0.11
Habit	3.46	1.00	0.76	0.17	0.11	0.16	0.14	0.14
Nothing else to do	3.24	1.28	0.74	0.10	0.07	0.16	0.12	0.23
Because it's there	3.05	1.01	0.62	0.21	0.14	0.15	0.14	0.18
No one else to talk to	3.43	0.91	0.58	0.16	0.20	0.26	0.30	0.12
Don't have to be alone	3.14	0.95	0.55	0.22	0.09	0.03	0.25	0.20
Makes me feel less lonely	3.44	0.84	0.54	0.01	0.14	0.04	0.14	0.17
Like to use	3.07	0.90	0.54	0.17	0.28	0.11	0.12	0.18
2. Escape								
Forget about school/work	3.38	0.89	0.37	0.83	0.23	0.19	18	0.19
Can get away from present	3.37	0.7	0.13	0.71	0.20	0.16	0.34	0.14
Can get away from others	3.13	0.94	0.11	0.69	0.12	0.18	0.29	0.23
3. Entertainment								
Amuses me	3.24	0.97	0.28	0.34	−0.67	0.28	0.25	0.36
Enjoyable	3.33	1.00	0.39	0.20	−0.63	0.10	0.34	0.35
Entertains me	3.35	1.02	0.21	0.27	−0.58	0.21	0.16	0.24
4. Relaxation								
It relaxes me	3.01	1.00	0.17	0.21	0.24	0.66	0.29	0.27
Pleasant rest	2.56	0.75	0.31	0.22	0.19	0.56	0.36	0.21
5. Arousal								
Thrilling	2.36	1.00	0.17	0.15	0.27	0.15	0.58	0.30
Exciting	3.90	0.68	0.21	0.40	0.32	0.14	0.54	0.20
6. Companionship								
Talk to others	3.94	0.67	0.40	0.10	0.11	0.11	0.12	0.79
Talk to friends	3.49	0.93	0.12	0.37	0.14	0.14	0.12	0.53
Non-loading items								
Allows me to unwind	2.54	0.80	0.32	0.36	0.22	0.38	0.17	0.26
Can be with others	2.30	0.88	0.21	0.28	0.41	0.29	0.22	0.17
Obtain much information	2.01	0.85	0.37	0.10	0.34	0.34	0.31	0.28
Learn new information	3.00	0.87	0.28	0.38	0.20	0.15	0.16	0.30
Learn around the world	2.66	0.79	0.12	0.19	0.38	0.33	0.10	0.31
Peps me up	2.57	0.99	0.35	0.36	0.34	0.23	0.20	0.21
Eigenvalue			5.86	4.78	2.84	2.37	2.01	1.84
Variance explained (%)			21.7	17.7	10.5	8.78	7.59	6.84

The first factor was labeled "habit/pass time" and had nine loadings that represented mobile phone use to pass time. The second factor was labeled "escape" and had three loadings that represented mobile phone use as an escape route from the present settings. The third factor was labeled "entertainment" and had three loadings that represented mobile phone use as (non-)entertainment. The fourth factor was labeled "relaxation" and had two loadings that represented mobile phone use as a way to relax. The fifth factor was labeled "arousal" and had two loadings that represented mobile phone use as a stimulating device. The sixth factor was labeled "companionship" and had two loadings that represented the mobile phone use as a network communicating device with others.

To check the reliability of the measure, Cronbach's alpha was measured and found to be reliable (0.81) for further analysis in the study.

17.3.1.3 Loneliness

To see whether loneliness was related to mobile phone use and addiction, respondents' loneliness was measured. Loneliness was measured by using the revised UCLA loneliness scale developed by Russell and Cutrona (1988). The scale consisted of 20 items and the reliability of the scale was 0.79.

17.3.1.4 Mobile Phone as Stimulant

To test whether respondents were groups of people who need a high degree of stimulation, the Need for Cognition Scale developed by Cacioppo et al. (1984) was used to measure mental stimulation. The scale consisted of 18 items and the reliability of the scale was 0.89.

17.4 Results

Hypothesis 1 predicted that loneliness will be positively related to ritualistic motives and mobile phone addiction. Pearson correlation analysis (Table 17.5) indicated that addiction, loneliness, habit/pass time and escape had positive correlations. Hence the hypothesis was partially supported. Loneliness was positively correlated with addiction ($r = 0.10$, $p < 0.01$). Loneliness was also positively correlated with ritualistic motive habit/pass time ($r = 0.20$, $p < 0.05$) and with ritualistic motive escape ($r = 0.10$, $p < 0.05$). Hence the results suggest that lonely students were more likely to use a mobile phone as a matter of habit but also to get away from current situations/settings in which he/she is involved.

Hypothesis 2 predicted that need for cognition will be positively related to instrumental motives and mobile phone addiction. Pearson

Table 17.5 Pearson Correlations among Major Variables

	1	2	3	4	5	6	7	8	9	10	11	12
1 Addiction	–											
2 Sex	0.27**	–										
3 CAGE	0.52*	0.15*	–									
4 MOU	0.55***	0.17**	0.29**	–								
5 Cognition	-0.10**	0.02	-0.10*	-0.09*	–							
6 Loneliness	0.10**	-0.43**	0.14**	0.06	-0.11	–						
7 Habit/pass time	0.20*	0.08	0.17	0.20**	0.47	0.24*	–					
8 Escape	0.10*	-0.36**	-0.13**	-0.21**	0.11	0.14*	0.09*	–				
9 Entertainment	-0.03	-0.26**	0.20	0.22	-0.03	-0.02	0.11*	0.14	–			
10 Relaxation	-0.06	0.03	0.16**	0.17*	0.07	0.08	0.07	0.02	0.05	–		
11 Arousal	0.09	0.03	0.10	0.09	0.14*	0.02	0.02	0.12*	0.10	0.18	–	
12 Companionship	0.07	-0.30**	0.12**	0.43**	-0.04	0.19	0.04	0.05	0.09	0.18*	0.04	–

($N = 157$): *$p < 0.05$; **$p < 0.01$; ***$p < 0.001$.

correlation analysis indicated that need for cognition was not positively related to any instrumental motives such as entertainment, arousal and companionship. In fact, need for cognition was negatively related to addiction, suggesting that people who are addicted to mobile phone use do not need a high degree of mental stimulation.

Hypothesis 3 predicted that stronger relationships will result between ritualistic motives and mobile phone addiction. The mobile phone addiction scale was positively correlated with ritualistic motive habit/pass time ($r = 0.20$) and escape ($r = 0.10$), whereas the mobile phone addiction scale was negatively correlated with instrumental motive entertainment ($r = -0.03$) and was positively correlated with instrumental motive arousal ($r = 0.09$). To see whether the ritualistic motive correlations were significantly different from the instrumental motive correlations, a t-test was applied and the result, $t(314) = 3.24$, $p < 0.05$, indicated that the differences were statistically significant. Hence hypothesis 3 was supported.

Research question 1 asked whether the stimulant or depressant model would best predict mobile phone addiction. Hierarchical regression analysis was applied to both models (Tables 17.6 and 17.7) and the results indicated that the depressant model (explaining 38% of the addiction versus 20% of the addiction) was the better predictor of addiction than the stimulant model.

Sex, loneliness, habit/pass time, companionship, relaxation and escape variables explained 38% of the total addiction variance in the depressant model. Loneliness and habit/pass time were significant predictors of the addiction.

Table 17.6   Hierarchical regression analysis of the depressant model of mobile phone addiction

Independent variable		Dependent variable			Mobile phone addiction	
		B	SE B	Beta	T	Sig. T
1 step	Sex	0.559	0.159	0.271	3.509	0.001*
		$R^2 = 0.07$	$F(6.383)1$	Sig. $= 0.000$		
2 step	Sex	0.423	0.175	0.205	2.410	0.017
	Loneliness	0.160	0.089	0.154	1.805	0.053
		$R^2 = 0.21$	$F(7.874)$	Sig. $= 0.000$		
3 step	Sex	0.178	0.184	0.086	0.963	0.337
	Lonely	0.294	0.091	0.282	3.240	0.001*
	Habit	0.391	0.091	0.362	4.288	0.000*
	Companionship	-0.110	0.092	-0.100	-1.200	0.184
	Relaxation	0.317	0.076	0.310	4.170	0.136
	Escape	0.135	0.094	0.118	1.443	0.232
		$R^2 = 0.38$	$F(12.313)$	Sig. $= 0.000$	$N = 157$	

*$p < 0.05$.

Table 17.7 Hierarchical regression analysis of the stimulant model of mobile phone addiction

Independent variable		Dependent variable			Mobile phone addiction	
		B	SE B	Beta	T	Sig. T
1 step	Sex	0.559	0.159	0.271	3.509	0.001*
		$R^2 = 0.07$	$F(11.186)$	Sig. = 0.000		
2 step	Sex	0.713	0.150	0.346	4.743	0.000*
	Cognition	0.521	0.100	0.381	5.222	0.000*
		$R^2 = 0.09$	$F(13.895)$	Sig. = 0.000		
3 step	Sex	0.606	0.163	0.294	3.727	0.000*
	Cognition	0.524	0.108	0.383	4.873	0.000*
	Arousal	0.291	0.115	0.144	1.576	0.005*
	Entertainment	0.131	0.081	0.128	1.618	0.108
		$R^2 = 0.20$	$F(20.836)$	Sig. = 0.000	$N = 157$	

*$p < 0.05$.

Sex, cognition, arousal and entertainment variables explained 20% of the total addiction variance in the stimulant model. Cognition and arousal were significant predictors of the addiction.

17.5 Discussion

The purpose of this study was to discover a means of making an empirical distinction between normal and problematic mobile phone use. The study was based on substance dependence and television viewing dependence, and the study proposed two models: depressant and stimulant models. The study results indicate that mobile phone users use their telephones as depressants. Hence mobile phone users satisfy the needs through habitual use and to pass time.

The study results are similar to those of earlier studies on mass media. Rubin (1984) and Finn (1992) argued that people use media either to compensate for or to manage their mood. People who use the mobile phone as a depressant are people who are lonely and habitual users. Also, people who use the mobile phone as a stimulant are people who need to fulfil a high degree of stimulation such as arousal.

Effects of addiction such as tolerance and withdrawal accompany both depressant and stimulant models. For example, mobile phone users grow tolerant of mobile phones despite the fact that they may cause some problems such as high phone bills and public annoyance. If the mobile phone is unavailable for a time, people become highly anxious and irritated by the absence of the mobile phone. These kinds of behaviors can go on despite the fact that these are troubling signs of addiction.

Consistent with the proposed model, students responded to measures of loneliness, habit/pass time and need for cognition, enabling this research to conclude the followings.

The mobile phone (and also other communication devices) is an inherently interpersonal technology – its sole function is basically to allow two parties at a distance to communicate with each other. However, the study results indicated that respondents reported significant correlations with loneliness and mobile phone addiction. Previous research findings on mobile phone use indicate that it strengthens existing social networks (e.g. friends and family members) but, at the same time, this study found that it can also mean that it might help inhibit growth of new social networks. Thus, heavy mobile phone users might experience more loneliness than light mobile phone users because they render themselves inaccessible to others (than the existing social networks) since they rely on existing social networks. Hence having access to a mobile phone does not necessarily make people more accessible to others and feel less lonely.

Ritualistic motives were stronger than instrumental motives in explaining mobile phone addiction. Although both motives are related to addiction, instrumental motives were better met by use of an alternative medium by the respondents. For example, if one wants to fulfil information needs, one can use newspapers or television, or if one wants to fulfil entertainment needs, one can simply find a video arcade room or play computer games on home desktop computers.

Instrumental motives were not correlated with the need for cognition. Furthermore, the need for cognition was unrelated to all the instrumental motives such as arousal and entertainment. Hence people who are in high need of stimulation do not necessarily use the mobile phone as their main medium for fulfilling their stimulation needs. Moreover, the need for cognition was negatively related to mobile phone addiction. Conceptually it does make coherent sense with regard to the study results since lonely people used the mobile phone habitually – indicating that habitual use does not require a high degree of stimulation.

To sum up this research, it is not likely that people use mobile phones to fulfil stimulation needs since there is an abundance of alternative media that can be used to meet such needs. Also, need for cognition may not be an adequate construct to grasp the notion of addiction. Furthermore, entertainment and information seeking were not significant variables in all analyses, suggesting that students are basically using the mobile phone as a communication device, not a multimedia tool.

17.5.1 Limitations

Several important limitations of this study must be mentioned.

First, consistent with other survey research, the findings from this research represented only a small sample of the population. Furthermore, the sample population was limited to college students, thus disabling the applicability of the findings to a wider population. Compounding the research problem was that college students may not be the best sample for representing "loneliness" because these students are at the peak of their life, filled with many exciting activities and work. Although many college students can be psychologically lonely in spite of many activities and studies, this variable must be treated with special care and attention in the future studies.

Second, consistent with the cultivation studies, the validity of categorizing overuse of the media must be considered in future studies. In fact, there is no consensus on what constitutes heavy use and light use, although the MOU is increasing every year.

Third, because mobile phone technology is rapidly evolving into something else, future studies can be divided into a number of sections dependent upon its functionality. For example, mobile phones can have the function of obtaining information via the Internet. Mobile phones can be used to capture photographs and video images and serve as a utility function. Also, mobile phones can play video games and serve as an entertainment function. Furthermore, a mobile phone can now act as a credit card and thus serve as a commerce function.

Keeping control over one's mobile phone use is more of a challenge today than ever before owing to the increase in its various functions. Its basic functions serve as an important asset to everyday life in modern society, but when it interferes with people's lives and creates problems, then it must be taken seriously and identified as an addiction.

17.6 Acknowledgment

This work was supported by the Soongil University Research Fund.

17.7 References

Akers, R.L. (1991) Addiction: the troublesome concept. *Journal of Drug Issues*, **21**, 777–793.

American Psychiatric Association (1994) *Diagnostic and Statistical Manual of Mental Disorders*, 4th edn. American Psychiatric Association, Washington, DC.

Anderson, D.R., Collins, P.A., Schmitt, K.L. and Jacobvitz, R.S. (1996) Stressful life events and television viewing. *Communication Research*, **23**, 243–260.

Andrucci, G.L., Archer, R.P., Pancoast, D.L. and Gordon, R.A. (1989) The relationship of MMPI and sensation seeking scales to adolescent drug us. *Journal of Personality Assessment*, **53**, 253–256.

Bae, J.H. (2001) Generation effect on gratification obtained from mobile phone use and perception as interpersonal communication medium. *Korean Journalism and Communication Studies*, **45**, 160–188.

Bae, J.H. (2002) Uses and gratifications and characteristics of telephone: a comparative study between telephone and mobile telephone's face-to-face channel. *Korean Journalism and Information Journal*, 18, 128–160.

Beaubrun, R. and Pierre, S. (2001) Technological developments and socio-economic issues of wireless mobile communications. *Telematics and Informatics*, 18, 143–158.

Bingham, J.E. and Piotrowski, C. (1996) On-line sexual addition: a contemporary enigma. *Psychological Reports*, 79, 257–258.

Cacioppo, J.T. and Petty, R.E. (1982) The need for cognition. *Journal of Personality and Social Psychology*, 42, 116–131.

Cacioppo, J.T., Petty, R.E. and Kao, C.F. (1984) The efficient assessment of need for cognition. *Journal of Personality Assessment*, 48, 306–307.

Carton, S., Jouvent, R. and Widlocher, D. (1994) Sensation seeking, nicotine dependence, and smoking motivation in female and male smokers. *Addictive Behaviors*, 19, 219–227.

Cohen, A. and Lemish, D. (2002) Real time versus survey measures in research on mobile use. Paper presented at the annual International Communication Association Conference, Chunchon, Korea.

Colombo, F. (2001) Mobile telephone use in Italy in the 1990s: interpretive models. *Modern Italy*, 6, 141–151.

deBock, H. (1980) Gratification frustration during a newspaper strike and a TV blackout. *Journalism Quarterly*, 57, 61–66, 78.

Edgar, P. (1977) Families without television. *Journal of Communication*, 27, 73–77.

Elliott, R., Eccles, S. and Gournay, K. (1996). Revenge, existential choice, and addictive consumption. *Psychology and Marketing*, 13, 753–768.

Ewing, J.A. (1984) Detecting alcoholism: the CAGE questionnaire. *Journal of the American Medical Association*, 252, 1905–1907.

Faber, R.J., O'Guinn, T.C. and Krych, R. (1987) Compulsive consumption. *Advances in Consumer Research*, 19, 459–469.

Finn, S. (1992) Television "addiction?": an evaluation of competing media use models. *Journalism Quarterly*, 69, 422–435.

Finn, S. (1997) Origins of media exposure: linking personality traits to TV, radio, print, and film use. *Communication Research*, 24, 507–529.

Fisher, S. (1994) Identifying video game addiction in children and adolescents. *Addictive Behaviors*, 19, 545–553.

Fortunati, L. (2001) The mobile phone: an identity on the move. *Personal and Ubiquitous Computing*, 5(2), 85-98.

Foss, K.A. and Alexander, A.F. (1996) Exploring the margins of television viewing. *Communication Reports*, 9, 61–68.

Gerbner, G. and Gross, L. (1976) Living with television: the violence profile. *Journal of Communication*, 26, 173–199.

Green, M. (2001) Do mobile phones pose an unacceptable risk? Adequacy of the evidence. *Risk Management*, 48(11), 40–48.

Griffiths, M.D. and Dancaster, I. (1995) The effect of Type A personality on physiological arousal while playing computer games. *Addictive Behaviors*, 20, 543–548.

Haddock, B.D. and Beto, D.R. (1988) Assessment of drug and alcohol problems: a probational model. *Federal Probation*, 52, 10–16.

Hill, C.A. (1987) Affiliation motivation: people who need people … but in different ways. *Journal of Personality and Social Psychology*, 52, 1008–1018.

Horvath, C.L. (1999) Psychological addiction to television: scale development and validation. Unpublished doctoral dissertation, Kent State University.

Jaffee, L.T. and Archer, R.P. (1987) The prediction of drug use among college students from MMPI, MCMI, and sensation seeking scales. *Journal of Personality Assessment*, 51, 243–253.

Jung, J. (2001) Psychology of Alcohol and Other Drugs: a Research Perspective. Sage, Thousands Oaks, CA.

Katz, E.J. and Aakhus, M. (2001) *Perpetual Contact: Mobile Communication, Private Talk, Public Performance*. Cambridge University Press, Cambridge.

Katz, E., Gurevitch, M. and Haas, H. (1973) On the use of the mass media for important things. *American Sociological Review*, 38, 164–181.

Katz, E., Blumler, J.G. and Gurevitch, M. (1974) Utilization of mass communication by the individual. In Blumler, J.G. and Katz, E. (eds), *The Uses of Mass Communication: Current Perspectives on Gratifications Research*. Sage, Beverly Hills, CA, pp. 19–32.

Keller, M. and Doria, J. (1991) On defining alcoholism. *Alcohol Health and Research World*, 15, 253–259.

Keller, S. (1977) The telephone is new and old communities. In Poole, I.S. (ed.), *The Social Impact of the Telephone*. MIT Press, Cambridge, MA, pp. 281–298.

Kim, S.D. (2001) The socio-cultural factors of mobile telephone proliferation. *Korean Journalism and Communication Studies*, 45, 189–228.

Kubey, R.W. (1996) Television dependence, diagnosis, and prevention. In MacBeth, T.M. (ed.), *Tuning in to Young Viewers*. Sage, Thousand Oaks, CA, pp. 221–160.

Lee, D.J. (2002) College students' hand-phone usage culture survey. *University Culture Newspaper*, 31 October.

Leung, L. and Wei, R. (2000) More than just talk on the move: uses and gratifications of the cellular phone. *Journalism and Mass Communication Quarterly*, 77, 308–320.

Licoppe, C. and Heurtin, J.P. (2001) Managing one's availability to telephone communication through mobile phones: a French case study of development dynamics of mobile phone use. *Personal and Ubiquitous Computing*, 5, 99–108.

Ling, R. (2000) "We will be reached": the use of mobile telephony among Norwegian youth. *Information Technology and People*, 13, 102–120.

Ling, R. (2001) "We release them little by little": maturation and gender identity as seen in the use of mobile telephony. *Personal and Ubiquitous Computing*, 5, 123–136.

McIlwraith, R.D. (1998) "I'm addicted to television": the personality, imagination, and TV watching patterns of self-identified TV addicts. *Journal of Broadcasting and Electronic Media*, 42, 371–386.

McQuail, D., Blumler, J.G. and Brown, J.R. (1972) The television audience: a revised perspective. In McQuail, D. (ed.), *Sociology of Mass Communications*. Penguin, Harmondsworth, pp. 166–194.

Ministry of Information and Communication (2002) Homepage: http://www.mic.go.kr/jsp/mic_d/d700-0002-1.jsp?code=Handm_code.

Na, E.Y. (2001) Consistency between media and user characteristics in individual-centeredness, immediacy, and directness influencing the selection of mobile phones as interpersonal communication tool. *Korean Journal of Journalism and Communication Studies*, 45, 189–288.

Noble, G. (1987) Discriminating between the intrinsic and instrumental domestic telephone user. *Australia Journal of Communication*, 11, 63–85

Palen, L., Salzman, M. and Youngs, E. (2001) Discovery and integration of mobile communication in everyday life. *Personal and Ubiquitous Computing*, 5, 109–122.

Peele, S. (1985) *The Meaning of Addiction*. Lexington Books, Lexington, MA.

Phillips, C.A., Rolls, S., Rouse, A. and Griffiths, M.D. (1995) Home video game playing in school children: a study of incidence and patterns of play. *Journal of Adolescence*, 18, 687–691.

Rubin, A.M. (1981a) An examination of television viewing motivations. *Communication Research*, 8, 141–165.

Rubin, A.M. (1981b) A multivariate analysis of "60 minutes" viewing motivations. *Journalism Quarterly*, 58, 529–534.

Rubin, A.M. (1983) Television uses and gratifications: the interactions of viewing patterns and motivations. *Journal of Broadcasting*, 27, 37–51.

Rubin, A.M. (1984) Ritualized and instrumental television viewing. *Journal of Communication*, 34(3), 67–77.

Rubin, A.M. (1994) Media uses and effects: a uses-and-gratifications perspective. In Bryant, J. and Zillmann, D. (eds), *Media Effects: Advances in Theory and Research*. Lawrence Erlbaum, Mahwah, NJ, pp. 417–436.

Russell, D.W. and Cutrona, C.E. (1988) Development and evolution of the UCLA Loneliness Scale. Unpublished manuscript, Center for Health Services Research. College of Medicine, University of Iowa, Iowa City, IA.

Russell, J.A. and Bond, C.R. (1980) Individual differences in beliefs concerning emotions conducive to alcohol use. *Journal of Studies on Alcohol*, 41, 753–759.

Sandstrom, M., Wilen, J., Oftdeal, G. and Mild, K.H. (2001) Mobile phone use and subjective symptoms: Comparison of symptoms experienced by users of analogue and digital mobile phones. *Occupational Medicine*, 51, 25–35.

Scherhorn, G. (1990) The addiction trait in buying behavior. *Journal of Consumer Policy*, 13, 33–51.

Schuckit, M.A. (1987) Biological vulnerability to alcoholism. *Journal of Consulting and Clinical Psychology*, 55, 301–309.

Signorielli, N. (1986) Selective television viewing: a limited possibility. *Journal of Communication*, 36(3), 64–76.

Smith, R. (1986) Television addiction. In Bryant, J. and Zillmann, D. (eds), *Perspectives on Media Effects*. Lawrence Erlbaum, Mahwah, NJ, pp. 109–128.

Stein, D.J., Black, D.W., Shapira, N.A. and Spitzer, R.L. (2001) Hypersexual disorder and preoccupation with Internet pornography. *American Journal of Psychiatry*, 158, 1590–1594.

Steiner, G.A. (1963) *The People Look at Television*. Knopf, New York.

Sung, D.K. (2002) A study on communication behavior using mobile Internet. Paper presented at the Annual International Communication Association Conference, Chunchon, Korea.

Sung, D.K. and Choi, Y.K. (2002) The difference of using pattern according to the mobile phone user group's characters. *Korean Journal of Journalism and Communication Studies*, 46, 153–190.

Townsend, A.M. (2000) Life in the real-time city: mobile telephones and urban metabolism. *Journal of Urban Technology*, 7, 85–104.

Wikle, T.A. (2001) America's cellular telephone obsession: new geographies of personal communication. *Journal of American and Comparative Cultures*, 24(12), 123–128.

Winick, C. (1988) The functions of television: life without the big box. In Oskamp, S. (ed.), *Television as a Social Issue*. Sage, Beverly Hills, CA, pp. 217–238.

Young, K.S. (1998) *Caught in the Net: How to Recognize the Signs of Internet Addiction and a Winning Strategy for Recovery*. Wiley, New York.

Zuckerman, M. (1979) *Sensation Seeking: Beyond the Optimal Level of Arousal*. Lawrence Erlbaum, Mahwah, NJ.

18

Does Personality Affect Peoples' Attitude Towards Mobile Phone Use in Public Places?

Steve Love and Joanne Kewley

18.1 Introduction

Mobile communication in its various forms such as voice telephony, SMS, WAP, WLAN technology and PDAs is being used by broader and broader sections of society. As a result of this increased usage (particularly for mobile phone use), there is a small but growing body of research that indicates that the use of mobile communications is influencing how we go about our daily lives from both a social and economic perspective. For example, as mobile phone usage increases, it is no longer unusual to see mobile phones being used in a wide variety of contexts (e.g. social, business) in various locations (e.g. trains, cafes). Wei and Leung (1999) found that the majority of calls being made by mobile phone users take place on the streets and in public transport, shops and restaurants.

Mobile phones now occupy concurrent social spaces, spaces with norms that sometimes conflict such as the space of the mobile phone user and the virtual space where the conversation takes place (Palen et al., 2000). There is a lively public debate concerning whether or not it is acceptable to use mobile phones in restaurants, streets and parks or on public transport. In relation to this, Ling (1997) highlights the fact that how mobile phones are used in public has become an element in the definition of socially appropriate/inappropriate behavior. In his study he found that people perceived mobile phone use in places such as restaurants as unacceptable, partly because people tend to talk louder than usual when using mobile phones and, as a result, individuals located near mobile phone users felt coerced into eavesdropping into their conversation. Even mobile phone companies are issuing guides on "mobile etiquette", encouraging sensible and responsible mobile phone behavior in public places (BT Cellnet, 2001).

In another study, Murtagh (2000) presented findings from an observational study of the non-verbal aspects of mobile phone use in a train carriage. Murtagh found that changing the direction of one's gaze, turning one's head and upper body away from those co-present was common feature of mobile phone behavior on trains. These behavioral responses were seen as being indicative of the subtle complexities involved when using mobile phones in public locations.

In terms of a theoretical perspective, the work of the groups of Murtagh, Ling and Palen has been influenced by the work of Erving Goffman. Goffman (1959) suggested that people have specific "public faces" and personas for different social locations. The idea behind this is that individuals have rules that determine their behavior in public places, or what Burns (1992) refers to as the "observance of social propriety".

One can see the relevance of Goffman's theory when considering Murtagh's study. Goffman talks about civil inattention in public places and when a person engages in a mobile phone conversation on a train, other individuals in close proximity may be drawn, unwillingly ("coerced into eavesdropping" to use Goffman's terms), into what is essentially a private conversation. Hence the mobile phone medium interacts with the social status quo and individuals' change their behavior as a result of this.

In terms of research methodology, the underlying approach to most of the work carried out in this area has been qualitative. This approach has been used in a variety of domains, such as information visualization (Martin and Bowers, 1999) and organizational change (Harper et al., 2000). Adopting a qualitative research approach is regarded as being particularly useful in looking at the effects of technology on social behavior as it can provide a detailed understanding of patterns of user behavior when using technologies such as mobile phones.

However, there is a need to complement and augment the research carried out in this area by investigating the affects of individual characteristics (such as personality, age and gender) on peoples' perception of the social usability of mobile phone applications in public places. By investigating the effects of the individual characteristics–situation interaction, the salient factors influencing individuals' perception and behavior to mobile phone use in public places can be identified.

Research into the interaction between the individual and the situation has been well documented in psychology. Argyle et al. (1981), for example, looked at the personality–situation interaction (personality × situation) and found that personality traits such as extraversion and introversion had an impact on how individuals would behave in particular situations. They found that introverts, for example, would use avoidance behavior in order to get out of a situation they would find (or would potentially find) uncomfortable. This implies that there is some form of

Does Personality Affect Peoples' Attitude Towards Mobile Phone Use in Public Places?

18

social perception taking place on the part of the individual. This social perception is also known as a "social schema" and acts as a guide for individuals to gauge their feelings, thoughts and actions in social situations (Fiske, 1995). This clearly is an important factor to consider when it comes to explaining the attitudes (and behavior) of individuals towards mobile phone use in public places.

Another factor that is relevant to this area of work (individual characteristics × situation interaction) is the issue of interpersonal space. The majority of studies investigating interpersonal space issues have been carried out in public places such as shops, libraries and workplaces. For example, Wollman et al. (1994) focused their study on the intrusion of an individual's personal space in the workplace. Others, such as Veitch and Arkkelin (1995), investigated the relationship between individuals who did not know each other.

One major underlying factor emerging from this work has been the idea of an individual feeling crowded when they perceive their personal space to have been invaded. Sears et al. (1988) define crowding as the feelings of discomfort and stress related to spatial aspects of the environment an individual is currently in. The idea of personal space being related to some measurement of interpersonal distance can be traced back to the work of Hall (1966).

In his well-known analysis, Hall (1966) stated that personal space could be divided into a series of zones:

- *Intimate zone*: a distance of up to 45 cm from the individual. Only close relatives or close friends are normally allowed into this zone (e.g. girlfriend or boyfriend).
- *Personal zone*: a distance of up to 1.2 m from the individual. Usually family and friends are allowed in this zone.
- *Social zone*: a distance between about 1 and 3 m from the individual. An example of the social zone is the typical space between work colleagues who are engaged in conversation.
- *Public zone*: a distance between about 3 and 8 m from the individual. An example of the public zone is a lecturer delivering a lecture to a group of students in a lecture theater.

However, there are several factors that can have an effect on the interpersonal distance preferred by individuals. First there is the cross-cultural dimension. Shortly after Hall published his work, Watson and Graves (1966) published an account of the difference in personal space preferred between individuals from America and some Mediterranean countries (who preferred a shorter distance between the speakers) when it came to having normal conversation. A violation of this space by either party led to a feeling of discomfort by the other. Personal characteristics can also have an impact. Studies have suggested that females interacting with other females tend to have a smaller distance between them than males interacting with other males (Gifford, 1987). In addition, Kaya and Erkíp (1999) suggest that females prefer a greater distance between themselves and males.

Another important factor in determining interpersonal distance is the situation the interaction takes place in. For example, in high-density situations (e.g. traveling on an increasingly crowded underground train), people can experience feelings of discomfort. In situations like this, limited physical resources have to be shared between greater numbers of people and, at the same time, there is a concomitant increase in physical contact between individuals that can lead to a decrease in the individual's feelings of privacy. As a result of situations like such as this, Hall (1966) reported that people tend to experience more negative feelings towards others in high-density situations than in other lower density situations.

Taking the interaction between the individual and the situation as a starting point, Love (2001) carried out an experiment to investigate the idea that participants perceived themselves being drawn, short-term, into the personal zone of an individual who engaged in a mobile phone conversation while seated in a waiting room. The results obtained from this study suggest that participants react and feel differently towards mobile phones being used in public places. However, the results also suggested that personality could have an effect on an individual's reaction to being in close proximity to a mobile phone conversation in a public place, or whether people make mobile phone calls in public places or not. For example, some individuals who took part in the study stated that they felt as if they were being drawn unwillingly into the personal space of the individual receiving the call, causing them to feel uncomfortable. In addition, these individuals reported that they would be reluctant to use their mobile phones in a public place, as they would not want to draw attention to themselves. On the other hand, there was another group of participants in this study who appeared to be more relaxed and stated that they were comfortable about being in close proximity to a mobile phone conversation in a public place. This group of participants also stated that they did not feel uneasy about using their mobile phones in a public place.

18.2 Current Study

A review of the research literature indicates that use of mobile communication technologies, such as mobile phones, is having an impact on our behaviour in public places (e.g. Ling, 1997; Murtagh, 2000). There is, however, a need to identify the salient individual characteristics that have an affect on the social usability of mobile phone use. Therefore, the specific aims of this study were to:

Does Personality Affect Peoples' Attitude Towards Mobile Phone Use in Public Places?

18

1. Conduct an initial investigation into the possible affects of individual characteristics (e.g. personality) × situation interaction when mediated by mobile phone use.
2. Begin to describe the process that may explain peoples' behavior in relation to mobile phone use in public places.

18.3 Methodology

18.3.1 Participants

Forty-two participants took part in this study. There were 22 males and 20 females. All participants had been mobile phone users for at least 12 months before the start of the study.

18.3.2 Data Gathering Techniques

This study adopted a cross-sectional survey design approach. This type of design allows for the collection of information from a group of people in a number of different conditions that are expected to be significant to change. In this case of this study, the conditions are participants' responses to the specific scenarios that are explained below. In addition, this type of design allows for a comparison of results between specific subgroups of the participant sample. In terms of the study reported here, this refers to the personality traits of the individuals.

Therefore, the data for this study were obtained from the following two sources:

Participants responses to scenarios
The experiment was based, partly, on Hall's theory of personal space. Participants were asked to indicate how they would feel about receiving or making social and personal calls in the following locations: café, walking down the street or on a train. Participants were also asked how they would feel about someone else making or receiving a personal or social call when they were next to them in the following locations: walking down the street, in a café or on a train. These scenarios were presented to the participants at different interpersonal distances (these were based on Hall's original zones of personal space):

- 0–0.5 m away from someone else (*intimate zone*)
- 0.5–1 m away from someone else (*personal zone*)
- 1–3 m away from someone else (*social zone*)
- 3–8 m away from someone else (*public zone*).

Table 18.1 Example of scenarios used in the experiment

You receive an "intimate call" on your mobile phone from a loved one when you are in a café. How comfortable or uncomfortable would you feel taking this call?

	VC	C	N	U	VU
At less than 0.5 m from other people round about you?	O	O	O	O	O
Between 0.5 m and 1 m from other people round about you?	O	O	O	O	O
Between 1 and 3 m from other people round about you?	O	O	O	O	O
Between 3 and 8 m from other people round about you?	O	O	O	O	O

VC = very comfortable; C = comfortable; N = neutral; U = uncomfortable;
VU = very uncomfortable.

An example of a location and context scenario given to participants is shown in Table 18.1.

Eysenck's Personality Questionnaire (short form)
The trait model of personality forms the background to personality aspects of this research study. The fundamental issue facing trait-based approaches to the study of personality concerns the classification and number of basic traits that can be used effectively to describe the basic structure of personality. For the purposes of the study reported here, the Eysenck Personality Questionnaire (short form) (Eysenck, 1978) was considered to be the most appropriate psychometric test to use. This personality inventory was chosen as it offers a concise measure of the personality dimensions (e.g. extroversion–introversion, neurotic–calm, agreeable–disagreeable) that could have an effect on individuals' attitudes to mobile phone usage in public places. This personality inventory is self-administered, with each individual answering each statement with either "yes" or "no". The Eysenck Personality Questionnaire obtains scores on three continuously distributed trait dimensions, which can be summarized in the form of a three-dimensional personality profile of the scorer.

18.3.3 Procedure

Participants were chosen from an opportunistic sample and asked to complete the two questionnaires and return them to the experimenters once they had completed both. The participants were informed that the results would be anonymous and they would be able to get the results of the experiment from the experimenters on request. Participants were also paid for taking part in the study and a follow-up call was made to allow the participants to ask questions about the study in which they had just taken part.

Does Personality Affect Peoples' Attitude Towards Mobile Phone Use in Public Places?

18.4 Results

18.4.1 Participants' Responses to Scenarios

For the purposes of carrying out statistical analysis, participants' scores on each of the three personality dimensions were converted into a binary distinction between high and low, the former being defined as the upper half of the distribution and the latter as the lower half of the distribution. It should be remembered that these scenario ratings represent how comfortable or uncomfortable individuals feel in a particular context and location.

The first personality trait to be looked at was extroversion. The results of the analysis produced significant differences between high and low on the extroversion scale in the contexts and locations shown in Table 18.2.

The second personality trait to be looked at was neuroticism. The results of the analysis produced significant differences between high and low on the neuroticism scale in the contexts and locations shown in Table 18.3.

Table 18.2 High and low extroversion mean scores for context and location

You are next to someone receiving a call on their mobile phone from a friend when you are on a train.

Distance	High extroversion	Low extroversion	p-Value
Less than 0.5 m from other people round about you	3.04	2.21	<0.05
Between 0.5 and 1 m from other people round about you	2.78	2.11	<0.05
Between 1 and 3 m from other people round about you	2.91	1.84	<0.05
Between 3 and 8 m from other people round about you	2.30	1.74	<0.05

Table 18.3 High and low neuroticism mean scores for context and location

You want to make a mobile phone call to a friend when you are on a train.

Distance	High neuroticism	Low neuroticism	p-Value
Between 0.5 and 1 m from other people round about you	3.21	2.35	<0.01
Between 1 and 3 m from other people round about you	3.00	2.09	<0.01
Between 3 and 8 m from other people round about you	2.74	1.78	<0.001

Table 18.4 High and low psychoticism mean scores for non-significant contexts and locations

Distance	Context
Between 3 and 8 m from other people round about you	You are next to someone receiving a call on their mobile phone from a friend when you are in a café
Between 1 and 3 m from other people round about you	You are next to someone receiving a call on their mobile phone from a friend when you are in a café
Between 0 and 0.5 m from other people round about you	You are next to someone making a call on their mobile phone to a friend when you are walking down the street
Between 0.5 and 1 m from other people round about you	You are next to someone making a call on their mobile phone to a friend when you are walking down the street
Between 1 and 3 m from other people round about you	You are next to someone making a call on their mobile phone to a friend when you are walking down the street
Between 3 and 8 m from other people round about you	You are next to someone making a call on their mobile phone to a friend when you are walking down the street
Between 1 and 3 m from other people round about you	You are next to someone making a call on their mobile phone to a friend when you are in a café
Between 3 and 8 m from other people round about you	You are next to someone making a call on their mobile phone to a friend when you are in a café

The final personality trait to be looked at was psychoticism. The results of the analysis produced significant findings between those who scored high and low on the neuroticism scale in virtually all of the contexts and locations. Table 18.4 indicates the contexts and locations where there were no significant differences between these two groups of participants.

18.5 Discussion

When investigating how an individual's personality affects their behavior in any given situation, it is important to consider social, physical and temporal parameters. This study focused on two of these: physical and social, mediated by mobile phone use. In relation to the results reported here, it is important to add a caveat about the sample size: the sample size was relatively small and any conclusions must be treated with caution.

Bearing this in mind, the first personality trait to be looked at was extroversion. The results obtained from this study indicate that extroversion had only a limited effect on peoples' attitude to mobile phone use in a specific context and location. In this case, extroverts felt more comfortable than introverts being next to someone on a train making a mobile phone call. Argyle et al. (1981) found that introverts would use avoidance

Does Personality Affect Peoples' Attitude Towards Mobile Phone Use in Public Places?

18

behavior to get out of a situation they would find (or would potentially find) uncomfortable. Taking on board this explanation, it appears that in this instance, introverts do feel uncomfortable owing to the physical and social parameters of the situation in which they find themselves (e.g. sitting next to, or close to, an individual making a call on their mobile phone on the train).

Neuroticism (the second personality dimension that the Eysenck Personality Questionaire measures) was also found to have a significant, but limited, effect on peoples' attitude towards mobile phone use in a specific context and location. Once again this was in relation to being on a train, although in this case, it was the individual who would be uncomfortable about making a mobile phone call to a friend within this context and location.

These results appear to suggest that although it is now commonplace to see people using mobile phones on the train, there are a group of people who may find this situation uncomfortable. Previous results (Love and Kewley, 2002) appear to suggest that these individuals' do not like the idea of going back stage with people they do not know in this particular context and location. The reason offered by these participants was that they felt they could not discuss "private matters" with strangers sitting so close to them, as they would be listening to what they were saying. In addition, they felt that, as bystanders, they would not like to be drawn into the back stage environment of the individuals they were sitting next to.

Pyschoticism (the final personality dimension in the Eysenck Personality Questionnaire) was also found to have a significant effect on peoples' attitudes towards mobile phone use and behavior in public places. This dimension has been studied less than the other two dimensions in the Eysenck trait model of personality and therefore less is known about it.

Eysenck described individuals obtaining a high score on this dimension as being insensitive and unconcerned about other peoples' welfare or feelings. Another way to describe this dimension could be along an agreeable–disagreeable dimension. Digman (1990) described the agreeableness personality dimension as ranging from those people who are friendly, warm and likeable to those individuals who are antagonistic and hostile towards other people.

Adopting this definition, the results obtained in this study are interesting. It appears that when considering the affects of personality × situation interaction in the specific scenarios used in this experiment, the most dominant and significant personality trait appears to be agreeableness. The results suggest that there are significant differences between agreeable and non-agreeable in terms of how comfortable or uncomfortable they feel in relation to most of the physical and social contexts and physical locations in which they find themselves. An interesting avenue to explore here would be to explore for any perceived differences in attitude when these individuals are the users of the mobile phone and when they are the bystander.

The implications of these findings can be related to the second aim of this study: the explanation of the factors and processes that affect peoples' behavior in public places when mediated by mobile phone use. We would like to put forward a potential model to explain peoples' behaviour in relation to mobile phone use in public places (Figure 18.1).

In the first instance, an individual finds himself or herself in a situation where a mobile phone is being used, for example, sitting next to someone on a train (location) and they want to make a short call (time duration) to a friend (context). The model suggests that the environmental stimulation interacts with individual user characteristics (in the case of the current study this would be personality traits) to create a behavioral response. For example, the person making the call could adopt back-stage behavior (e.g. gestures, dialogue, posture) and talk to their friend about the good time they had the last time they met up for a drink. The results obtained from this study suggest that personality traits can have a significant affect on peoples' behavior in this context and location.

As mobile phone usage has increased in public places, there has been a concomitant increase in relation to the social information available about this phenomenon (i.e. behavioral styles and attitudes towards mobile phone use in public places). Meyrowitz (1985) stated that individuals con-

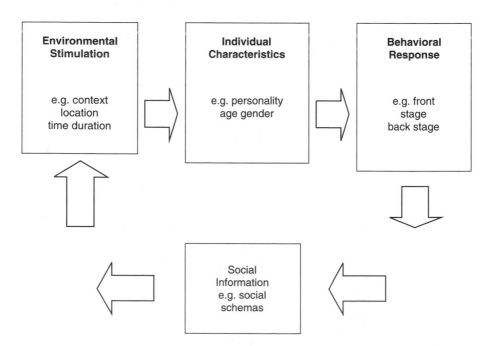

Figure 18.1   Model of peoples' behaviour in relation to mobile phone use in public places.

Does Personality Affect Peoples' Attitude Towards Mobile Phone Use in Public Places?

18

struct social situations. This implies, from a psychological perspective, that there is some form of social perception taking place on the part of the individual. This can be developed by the individual into a "social schema" that will influence their feelings, thoughts and actions in social situations. Fiske (1995) has shown that our schemas affect our perception of the behavior of others. Therefore, social information is an important factor in the model described above. In addition, Meyrowitz (1985) also talks about how situations are often defined by the behavior of individuals in physical locations. In relation to telephone behaviour he states:

> When two teenagers speak to each other on the telephone, they override the physical distance and create a backstage area apart from the adults with whom they live. (Meyrowitz, 1985, p. 37)

In the long term this can lead to the development of new behaviors (e.g. gestures, dialogue, posture) and, in relation to mobile phone use in public areas, this appears to be the case.

This model, as stated, is only in its rudimentary stages. The next stage of the work will be to test formally the validity of this process of behavior change. In addition, there are other components whose role has yet to be assessed. The first of these could be to look at how other individual characteristics such as gender and age interact with the temporal, context and location parameters associated with mobile phone use in order to create new behavioral responses that, in turn, feed into our social information (or social schemas). There is also the temporal aspect to consider: in what ways will our social schemas for this phenomenon change over time? In addition, further research will ultimately lead to the model being refined with new components being added and others, perhaps, dropping out.

Overall, this research also provides an opportunity to combine a psychological and sociological methodological approach in order to produce a comprehensive model that explains peoples' behavior in relation to mobile phone use in public places. In addition, this model and the related research can be seen as an attempt to develop and adopt more of an ecological approach to human–computer interaction research where the emphasis is on context and location.

18.6 References

Argyle, M., Furnham, A. and Graham, J.A. (1981) *Social Situations*. Cambridge University Press, Cambridge

BT Cellnet (2001) Mobile etiquette: changing the way we use our mobiles [online]. http://www.btcellnet.net/cgi-bin; accessed 15 October 2001.

Burns, T. (1992) *Erving Goffman*. Routledge, London.

Digman, J.M. (1990) Personality structure: emergence of the five factor model. *Annual Review of Psychology*, **41**, 417–440.

Eysenck, H.J. (1978) Superfactors P, E and N in a comprehensive factor space. *Multivariate*

Behavioural Research, **13**, 475–482; cited in Eysenck, H.J. and Eysenck, M.W. (1992). *Personality and Individual Differences*. Plenum Press, New York.

Fiske, S. (1995) Social cognition. In Tesser, A. (ed.), *Advanced Social Psychology*. McGraw-Hill, New York, pp. 149–194.

Gifford, R. (1987) *Environmental Psychology: Principles and Practice*. Allyn, Boston.

Goffman, E. (1959) *The Presentation of Self in Everyday Life*. Doubleday, New York.

Hall, E.T. (1966) *The Hidden Dimension: Man's Use of Space in Public and Private*. London, Bodley Head.

Harper, R., Randall, D. and Rouncefield, M. (2000) *Organisational Change and Retail Finance: an Ethnographic Perspective*. Routledge, London.

Kaya, N. and Erkíp, F. (1999) Invasion of personal space under the condition of short-term crowding: a case study on an automatic teller machine. *Journal of Environmental Psychology*, **19**, 183–189.

Ling, R. (1997) 'One can talk about common manners'!: the use of mobile telephones in inappropriate situations. In *Haddon, L. (ed.), Themes in Mobile Telephony. Final Report of the COST 248 Home and Work Group*.

Love, S. (2001) Space invaders: do mobile phone conversations invade peoples' personal space?. In *Proceedings of the18th International Human Factors in Telecommunications Symposium*, Bergen, 5–7 November 2001, pp. 125–131.

Love, S. and Kewley, J. (2002) Does personality affect peoples' attitude towards mobile phone conversations in public places? In *Proceedings of the 3rd Wireless World Conference, University of Surrey*, 17–18 July 2002, pp. 205–218.

Martin, D. and Bowers, J. (1999) Informing collaborative information visualisation through an ethnography of ambulance control. In Bodker, S., Kyng, M. and Schmidt, K. (eds), *Proceedings of the Sixth European Conference on Computer Supported Collaborative Work*, Copenhagen, 12–16 September 1999, pp. 311–330.

Meyrowitz, J. (1985) *No Sense of Place: the Impact of Electronic Media on Social Behaviour*. Oxford University Press, Oxford.

Morgan, K., Morris, R.L., Macleod, H. and Gibbs, S. (1992) Experiment III: comparing the performance and preferences of naive computer users on functionally equivalent graphical computer interfaces and command line interfaces. Presented at the 6th International Conference on Systems Research, Informatics and Cybernetics, August 1992, Baden-Baden.

Murtagh, G. (2000) Seeing the "rules": preliminary observations of action, interaction and mobile phone use. In Brown, B., Green, N. and Harper, R. (eds), *Wireless World: Social and Interactional Aspects of the Mobile Age*. Springer, London.

Palen, L., Salzman, M. and Youngs, E. (2000) Going wireless: behaviour and practice of new mobile phone users. In *Proceeding of the ACM 2000 Conference on Computer Supported Cooperative Work*, Philadelphia, PA. ACM Press, New York, pp. 201–210.

Sears, O.D., Peplau, A. and Freedman, J. (1988) *Social Psychology*. Prentice Hall, New York.

Veitch, R. and Arkkelin, D. (1995) *Environmental Psychology: An Interdisciplinary Perspective*. Prentice Hall, New York.

Watson, O.M. and Graves, T.D. (1966) Quantitative research in proxemic behaviour. *American Anthropology*, **68**, 971–985;. cited in Eysenck, M.W. (2000) *Psychology: a Student's Handbook*. Psychology Press.

Wei, R. and Leung, L. (1999) Blurring public and private behaviours in public space: policy challenges in the use and improper use of the cell phone. *Telematics and Informatics*, **16**, 11–26.

Wollman, N., Kelly, B.M. and Bordens, K.S. (1994) Environmental and intrapersonalpredictors of reactions to potential territorial intrusions in the workplace. *Environment and Behaviour*, **26**, 179–194.

Part 4

Language and Mobile Communication

19

Introduction

Naomi S. Baron

As a college freshman, the first essay I was assigned in my obligatory English Composition course was a description of how to tie a pair of shoes. The class quickly learned how difficult it can be to deconstruct a set of behaviors that have become automatic after years of practice. Human language is somewhat analogous to shoe-tying. By the time we become fluent speakers and writers, we tend to give little thought to how we utter sounds, construct sentences or, these days, write e-mails or text messages.

Linguists endeavor to take language use off automatic pilot by identifying the building-blocks and combination procedures that result in spoken and written practices. Complementing this structural analysis is a functional component, which dissects relevant social variables. Among the factors that may correlate with particular linguistic behaviors are age, gender, education, cultural background, knowledge of multiple languages and the specific context in which the linguistic event is taking place.

Technology is another domain potentially affecting language. Some of these effects are obvious. One hundred years ago, both telegrams and long-distance phone calls were expensive, so our words were few. Today's 160 character limit on standard mobile phones militates against lengthy prose. Yet if you query average users of instant messaging (IM) on personal computers or of texting on mobile phones how the medium affects the message, most would be hard pressed to produce much beyond superficial answers about abbreviations or spelling.

There is a growing literature analyzing the use of interactive computer-based media for composing and transmitting linguistic messages [see Herring (2002) for an overview]. The study of computer-mediated communication (CMC) has largely emphasized the social side of language use: what is the role of humor in e-mail?; do males or females dominate mixed-gender chat rooms?; do students prefer to send one another instant messages or to speak face-to-face? A smaller number of studies have concentrated on issues of linguistic structure: does e-mail style look more like speech or writing?; are the linguistic conventions found in IM the same as in SMS?; are there significant age or gender differences in the construction of IMs or SMSs?

In the USA, where personal computers abound in homes and offices, the dominant forms of written cybertalk have been e-mail and IM, typically sent from and received at computers with full keyboards. As of 2004, only a relatively small proportion of Americans used portable texting devices, including Blackberries (among business people) for e-mail, and mobile phones (among teenagers and young adults) for e-mail, IM or text messaging. In Europe and Asia, where mobile phone penetration has long outstripped that in the USA, texting on mobiles is the usual form of written message exchange. Thus, whereas Americans (on average) have been somewhat tethered in their written communication – it is hard to lug your computer everywhere you go – Europeans and Asians have been more mobile, given the proliferation of SMS.

When talking about mobile phones, we also need to distinguish between form and function. Those in the fashion business or automotive industry have long understood that the functionality of a piece of clothing (e.g. resists stains) or an automobile (is highly fuel efficient) is often less important to a consumer than its external form (Britney Spears wears shirts like this one, or macho men drive this kind of car). As communication technologies become increasingly "domesticated" into our everyday activities (Silverstone and Haddon, 1996), users have the luxury of concentrating on external form (for example, smaller and smaller mobile phones) or features that have marginal functionality (such as funky prerecorded voice mail messages you can download to your mobile for a fee).

The chapters in this part enrich our understanding of the linguistic aspects of computer-mediated communication, as practiced on both personal computers and mobile phones. The four contributions represent a diverse swath of linguistic and cultural practices: IM from "tethered" PCs in the USA; SMS from "mobile" phones in Sweden and Norway; print media references to mobiles in Hong Kong.

The chapter by Naomi Baron, Lauren Squires, Sara Tench and Marshall Thompson ("Use of Away Messages in Instant Messaging by American College Students") examines "away messages", which are a feature of America Online's instant messaging program known as AIM. Away messages enable AIM users who were still logged on to AIM but temporarily away from their machines to alert possible interlocutors not to expect an immediate response to an IM. As computers are increasingly left on all day (thanks to broadband or cable connections), away messages enable IM users to establish a sense of social presence, even when they are not physically at their computers (e.g. having supper, in the bathroom, at a concert).

Over time, users have transformed the away message function of IM into a medium supporting a myriad of social functions, from entertaining IM friends with song lyrics or witticisms to monitoring which IMs to respond to and which to ignore. (A surprising number of AIM users now post away messages while sitting at their computers.) The chapter examines a corpus of 190 away messages and also explores the extent to which a cohort of American undergraduates feel "tethered" or "mobile" in their communication via personal computers or mobile phones.

The second and third chapters (Ylva Hård af Segerstad's "Language Use in Swedish Mobile Text Messaging" and Rich Ling's "The Sociolinguistics of SMS: an Analysis of SMS Use by a Random Sample of Norwegians") analyze sets of SMS messages. Building upon a small group of empirical studies of German SMS (Schlobinski et al., 2001; Androutsopoulos and Schmidt, 2002; Doering, 2002), Ylva Hård af Segerstad and Ling provide important new perspectives in their studies of SMS as a communication medium, in addition to expanding the base of languages (and cultures) for which we have empirical data.

Hård af Segerstad approaches SMS with a trained linguist's eye, interested in how the form of text messages resembles or differs from other types of CMC (including e-mail and IM) or from traditional written language. The data presented here are part of a larger comparative analysis (Hård af Segerstad, 2002). Drawing upon a corpus of 1152 SMS messages, Hård af Segerstad carefully scrutinizes the texts with respect to punctuation, spelling, grammar and non-alphabetic graphical elements such as emoticons. The chapter reveals a number of textual strategies that users employ (e.g. acronyms, omission of subject pronouns) to adapt messages to the limitations of the medium (e.g. small screen, cumbersome input, restricted number of characters per message). It also highlights user creativity and the role that English plays in Swedish SMS.

Ling's chapter explores the intersection of linguistic and social aspects of SMS. Bringing to bear his broad sociological perspective on mobile telephony (Ling, 2004), Ling collected and analyzed 867 SMS messages as part of a larger telephone survey of Norwegians regarding attitudes towards and usage of mobile phones for voice and texting functions. Although the messages came from users across the age spectrum, Ling is especially interested in the SMS texting culture that has emerged among Norwegian teens, particularly females. Ling reports that teenage girls send more text messages, use more complex syntax, include more salutations and closings and even employ better punctuation than their male or their older counterparts. Ling's conclusion that females are more sophisticated users of the medium is consonant with other research findings that female writing (and also speech) tends to approach normative standards more than that of men (Labov, 1991; National Center for Educational Statistics, 2002; Baron, 2004).

The last chapter in this part moves us from questions concerning the linguistic content of SMS messages to the language people use to talk about mobile phones themselves. In "The Construction of Symbolic Values of the Mobile Phone in the Hong Kong Chinese Print Media",

Vicki Yung analyzes three popular Chinese-language publications in Hong Kong to see how news headlines, descriptions in the technology and business sections, entertainment news and advertisements talk about mobile phones. Mobiles have become highly domesticated in Hong Kong culture – averaging more than one phone subscription per person in 2003. Yung argues that the local print media create linguistically symbolic values associated with mobile phones by drawing upon ordinary language practices with which their readership is familiar. For example, the large-sized phones of the 1980s are described as *dai go dai* (which, in Cantonese, means "bigger than big brother"). Readers of these publications know that such phones were expensive, and that only political honchos or gang members (both commonly referred to in Hong Kong as *dai go*, i.e. "big brother") could afford to buy them. Yung's chapter highlights the important symbolic value of mobile phones to the people who carry them and those who judge us by our physical appearance and technological accoutrements (Fortunati et al., 2003; Katz and Sugiyama, this volume, Chapter 5).

The popular press continues to be enamored with computer-mediated communication and with the devices upon which such messages are produced. As a result, we have been become familiar with superficial descriptions of how these media actually work – descriptions which, often as not, fail to identify such critical variables as age, gender, genre or nationality. The chapters in this part make us pause to examine real-life linguistic practices with these new communication media. In the process of unraveling the stands of language used to construct computer-mediated messages or to talk about the devices that convey them, we come away with a heightened appreciation of the complexity of human linguistic and social behavior.

19.1 References

Androutsopoulos, J. and Schmidt, G. (2002) SMS-Kommunikation: Etnografische Gattungsanalyse am Beispiel einer Kleingruppe. *Zeitschrift für Angewandte Linguistik*, 36, 49–80.

Baron, N.S. (2004) See you online: gender issues in college student use of instant messaging. *Journal of Language and Social Psychology*.

Doering, N. (2002) 'Kurzm. wird gesendet' – Abkürzungen und Akronyme in der SMS-Kommunikation. *Muttersprache. Vierteljahresschrift für Deutsche Sprache*, Heft 2.

Fortunati, L., Katz, J.E. and Riccini, R. (eds) (2003) *Mediating the Human Body: Technology, Communication, and Fashion*. Lawrence Erlbaum, Mahwah, NJ.

Hård af Segerstad, Y. (2002) Use and adaptation of written language to the conditions of computer-mediated communication. Doctoral Dissertation, Department of Linguistics, Göteborg University.

Herring, S. (2002) Computer-mediated communication and the Internet. In Cronin, B. (ed.), *Annual Review of Information Science and Technology 36*. American Society for Information Science and Technology, Medford, NJ, pp. 109–168.

Labov, W. (1991) The intersection of sex and social class in the course of linguistic change. *Language Variation and Change*, 2, 205–254.

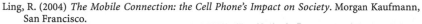

Ling, R. (2004) *The Mobile Connection: the Cell Phone's Impact on Society*. Morgan Kaufmann, San Francisco.

National Center for Educational Statistics (2002) *The Nation's Report Card: Writing 2002*. National Assessment of Educational Progress. U.S. Department of Education, Institute of Educational Sciences, Washington, DC.

Schlobinski, P., Fortmann, N., Gross, O., Hogg, F., Horstmann, F. and Theel, R. (2001) Simsen: Eine Pilotstudie zu sprachlichen und kommunikativen Aspekten in der SMS-Kommunikation. Networx Nr. 22, Hannover.

Silverstone, R. and Haddon, L. (1996) Design and domestication of information communication technologies: technical change and everyday life. In Silverstone, R. and Mansell, R. (eds), *Communication by Design: the Politics of Information and Communication Technologies*. Oxford University Press, Oxford, pp. 44–74.

20

Tethered or Mobile? Use of Away Messages in Instant Messaging by American College Students

Naomi S. Baron, Lauren Squires, Sara Tench and Marshall Thompson

20.1 Defining the Problem

Since the appearance of the telegraph and the telephone, interlocutors have had options about how to communicate with one another at a distance. Typically, there is a settling-in period for new language technologies, as people gradually work out what medium is most appropriate to use with which interlocutor, and how messages should be formulated (Baron, 2002). The kinds of usage patterns that emerge with new technologies are rarely monochromatic. Differences may reflect economic considerations, age, gender, education level and cultural habits. For example, telephone usage was far more pervasive in the USA than in many other countries until well after World War II (Baron, 2000). E-mail was largely restricted to the university community until the 1990s, and teenage use of mobile phones to send short text messages dwarfs SMS usage by older cohorts (Ling, 2004).

The aim of this chapter is to contribute to a discussion of cultural diversity in information and communication technologies (ICTs) used to convey writing at a distance, where choices include listservs, chat, newsgroups, e-mail, instant messaging (IM) from laptop computers or PCs and text messaging on mobile phones. Our particular focus is the USA, which currently relies far more heavily upon computers than mobile phones for such written communication. Specifically, we will explore how American college students use away messages (one component of America Online's instant messaging system) to negotiate social activities and relationships.

Like texting on mobile phones, away messages in IM enable users to establish a continuing sense of social "presence". That is, the away message function of IM enables those posting messages to be physically mobile (even when not using a mobile communication device) because the user's social presence is maintained in his or her physical absence from the computer. Before looking at away messages themselves and the role they play in fostering both social presence and mobility, we need to understand the American computing milieu out of which they grew.

20.2 ICTs American Style

The USA is filled with computers that can access the Internet. By the end of 2002, more than 72% of American adults reported they had gone online within the past 30 days.[1] Of all the uses of these networked computers, e-mail continues to top the list.[2] Among American teenagers, socializing over the Internet is commonly done through IM rather than e-mail.[3]

America is also increasingly a country of mobile phone users. While current market penetration had moved to roughly 54% in 2003, the USA lagged behind countries such as Norway (with roughly 91% penetration) and Taiwan (with 111% penetration).[4] Equally importantly, when Americans pick up their mobile phones, overwhelming they do so to place or receive voice calls, not to send or receive an SMS. In countries such as Norway and Japan, at least among younger users, mobiles are more frequently used as texting devices than for speech.[5]

ICT usage patterns in the USA are not homogeneous across the population. As the Pew Internet and American Life studies have shown, to understand American computer-based communication patterns, one needs to look separately at, for example, teenagers, college students, family groups and older citizens. The research and discussion that follow focus on American college students.

For transmitting personal written messages via computer, Americans have two main options: e-mail or IM.[6] A number of different IM systems are available, including ICQ, MSN Messenger and Yahoo! Messenger. However, most college instant messaging in the USA is done through America Online's freely available program known as AIM (AOL Instant Messenger). The current AIM system encompasses a suite of functions that enable users not only to send synchronous messages to individuals

---

[1] www.ipsos-reid.com.

[2] www.pewinternet.org/reports/chart.asp?img=Daily_A6.htm.

[3] Pew Internet and American Life Project, 20 June 2001, "Teenage Life Online: the Rise of the Instant-Message Generation and the Internet's Impact on Friendships and Family Relationships; www.pewinternet.org/reports/toc.asp?Report=36.

[4] International Telecommunication Union; www.itu.int/ITU-D/ict/statistics/at_glance/cellular03.pdf.

[5] Raban et al. (2002); Ling et al. (2002); Hashimoto (2002).

[6] Such multi-user forms of computer-mediated communication as chat or listservs are excluded from the present discussion.

Tethered or Mobile? Use of Away Messages in Instant Messaging by American College Students

20

but also to "present" themselves to members of their buddy list or anyone knowing their screen name.[7] These forms of presentation include screen names, profiles, buddy icons, expressions, emoticons, fonts and colors and away messages (Squires and Stacey, 2002).

The first three of these functions – screen names, profiles, and buddy icons – tend to be reasonably stable over time. Sometimes selection of screen names is seen as a creative act (e.g. the choice of "Swissmiss" – also the name of a hot chocolate mix – by an American who had lived in Switzerland). Profiles and buddy icons enable users to create a persona (real or imagined) that they wish to reveal (or project) to others. A profile might contain information such as date of birth, hobbies, favorite movies, contact information, links to web sites or quotations. Buddy icons (which are projected in the lower left-hand corner of an instant messaging box during an IM conversational exchange) can be created independently or selected from a set of graphics that AOL provides. AIM Expressions are essentially themed electronic stationery. The "theme" you select (e.g. basketball, a pop star) shows up as a sidebar (or a light overall image) on many of the IM functions you are using. AIM offers a variety of preformed emoticons that can be inserted into an IM, a profile screen or an away message. Similarly, users can customize fonts and colors when constructing text for IMs, profiles or away messages.

20.3 AIM Away Messages

20.3.1 How Away Messages Work

Away messages were originally designed to enable AIM users who were still logged on to their computers but not physically sitting at their machines to alert possible interlocutors not to expect immediate replies to instant messages. For example, the user might have gone off to the bathroom, to get something to eat or to attend class. During a user's absence, an away message creates a social link with other members of the user's messaging circle. As one female undergraduate put it, "Even if they are not chatting [on IM], you can still know all about someone's life by reading their away messages".

We can think of away messages as a form of "on-stage" behavior in contrast to IM conversations, which may be seen as "back-stage" activity (Goffman, 1959). Jacobs argues that among American teenage girls, "the back-stage conversations [i.e. IM] are where alliances are formed,

---

[7] Individual users can create their own AIM social circle by constructing a buddy list, which compiles the screen names of people with whom they regularly wish to communicate (or whose whereabouts they wish to track). By looking at their buddy list, IM users can see which "buddies" are currently logged on to AIM. Users can also check one another's profiles and away messages.

problems are discussed and solved, and plans are made beyond the hearing of others ... [while] the on-stage places [i.e. away messages] are where alliances are declared and social positions and presence are established" (Jacobs, 2003, p. 13).

AIM users know that a member of their online social circle has posted an away message by looking at the buddy list that appears on their screen. This list indicates not only which members of the list are currently online but also which ones (i.e. of those logged on) have posted away messages. A (virtual) yellow piece of paper next to a buddy's screen name indicates that that person has posted an away message. By clicking on the piece of paper, you can view the message. AIM provides a default away message ("I'm away from my computer right now"), and hundreds of public access web sites (e.g. www.awaymessages.com) list thousands of sample messages.

20.3.2 Presence and Mobility in Away Messages and SMS[8]

We have suggested that away messages enable users to create a sense of social presence even when they are, ostensibly, absent. That is, users posting away messages can be physically mobile (i.e. absent from a networked computer) while remaining in persistent contact with their social circle through the use of away messages.

Like SMS on mobile phones, away messages provide a textual medium both for constructing indicators of presence and for managing incoming messages from constituents in your social circle. Unlike IM and voice functions on mobile phones,[9] both away messages and SMS are asynchronous forms of communication. Users of away messages and SMS can screen incoming messages, determining when – or whether – to respond. Both sets of users also have the option of constructing texts of their own choosing, which may or may not reflect reality. For example, a user can post an away message declaring absence from the computer (when he or she is actually sitting in front of it) or send an SMS announcing an afternoon of shopping (when the sender is actually in the library doing homework).

Yet the analogy between away messages and SMS is imprecise. Whereas there is a cost for doing texting on mobile phones, away messages are "free", perhaps leading to divergent language or usage patterns in the two media. Moreover, although asynchronous, SMS tends to be more interactive, frequently generating conversational exchanges more akin to instant messaging than to away messages. Third, mobile phones enable users to view and send SMS during more times of the day and in more physical

---

[8] This comparison benefited from discussion with Rich Ling.
[9] We set aside here issues of caller ID and voice mail on mobile phones.

Tethered or Mobile? Use of Away Messages in Instant Messaging by American College Students

20

locations than do computers (via which away messages are posted and read). This issue of accessibility raises the question of whether users of away messages and IM more generally perceive themselves to be "tethered" in comparison with those doing texting on "mobile" phones.

20.4 Case Study of Away Message Usage on an American College Campus

To understand better how away messages function in the USA, we gathered and analyzed a corpus of away messages posted by American University undergraduates or by age-mates on their AIM buddy list. American University is a selective undergraduate and graduate institution located in a residential setting in the upper-northwest corner of Washington, DC. The school has approximately 5350 full-time undergraduates, about 55% of whom live in university housing. Campus dormitories are all wired via a high-speed network. Additionally, there are open computing labs on campus, and numerous locations on campus, including dormitories, offer wireless network connections.

20.4.1 Design of Study

The corpus was gathered in Fall 2002 by 19 undergraduate students enrolled in an Honors Colloquium entitled "Language in the New Millennium".[10] Each student collected data from two subjects, one male and one female, who were on the student's AIM buddy list, generating a total of 38 subjects. Five away messages were collected from each subject, yielding a corpus of 190 away messages. A randomization process was used for selecting the screen names on the buddy list from which away messages were collected. Data collection took place over a 2-week period.

In addition to collecting the away messages themselves, researchers calculated the average number of words per message for each of their subjects, did a brief content and stylistic analysis of the messages and offered their own snapshot view of the subjects' offline personalities. Several researchers also interviewed their subjects, eliciting the writers' rationales for the content and style of their away messages. The present study only formally reports on issues relating to message length, gender and content, although it also taps into the interviews done with some of the 38 subjects.

---

[10] The senior author is grateful to the entire class for its role in gathering and analyzing data, and to Michael Mass, Director of the University Honors Program at the American University, for initially bringing the phenomenon of away messages to the senior author's attention.

20.4.2 Data and Analysis

The corpus of 190 away messages yielded an array of styles and moods, a good deal of humor and a substantial display of personal information. While subjects sometimes recycled their own away messages (since they can be saved), no one resorted to AIM's default away message or to public away message sites, Our analysis also includes several examples that did not appear in the formal corpus but that student researchers had used or otherwise encountered in their prior experience with away messages. Such examples are indicated with the notation "{not in corpus}".

20.4.3 Message Length and Gender

We began by calculating the mean length of the away messages. Message length varied enormously across individual subjects. Summing across each subject's five-message sample, individual means ranged from 1 to 49.2 words per message. Summing across female subjects, the average length of away message was 12.3 words. The average length for male subjects was 13.3 words.

Compared with IM conversations, where messages sent by college students average 5–6 words per turn (Baron, 2004), away messages are fairly lengthy. Clearly, technology presents no barrier to producing substantial messages in either venue, given the ease with which text can be generated on a full computer keyboard. Moreover, students in the survey were accomplished computer users, with years of experience in writing papers, collecting research materials and composing e-mail and/or instant messages). Nonetheless, an obvious question (which we address below) is why so many subjects in the study sent lengthy away messages.

20.4.4 Content Analysis

A preliminary content analysis suggested separating the away messages into two major categories: Informational/Discursive and Entertainment. Each of these categories was then divisible into subcategories:

Informational/Discursive:
   "I'm away".
   Initiate discussion or social encounter.
   Convey personal information (about self, opinions, sense of humor).
   Convey personal information to selected other(s).
Entertainment:
   Humorous comments (e.g. jokes, plays on words).
   Quotations (from authors, songs, movies, TV shows, friends).
   Links to web sites.

Tethered or Mobile? Use of Away Messages in Instant Messaging by American College Students

20

Because the sample size was small (190 messages), we did not undertake a statistical analysis of the data by type. Rather, we focused on understanding the spectrum of message types and, equally importantly, the communicative or social role messages appeared to be serving.

Tables 20.1–20.4 summarize the subcategories of Informational/ Discursive messages, distinguishing in each case between the overt function of the message type and the social or communicative functions that the messages appear to be filling. Table 20.5 summarizes the analysis of Entertainment away messages (combining together the three subcategories). A number of away messages overlap categories (especially where humor or a quotation is involved). We have included such messages under the particular category we wish to illustrate, though mindful of their multi-functionality.

20.4.4.1 Overt Function: "I'm Away"

The first subcategory of Informational/Discursive messages contains away messages that overtly declare that their authors are away from their computers and therefore not available to respond to instant messages, despite still being logged on to AIM. Table 20.1 summarizes the social and communicative functions of these messages.

Table 20.1 "I'm away" messages

Overt form: "I'm away"	
Social/communicative functions	Examples
I really am away	Out
	In the bowels of hell … or what some would call the library.
	Sleeping … don't bother
Itinerary	Voter registration, peace corps meeting, class, choir, dinner, dorm council
	Class till 1235, study for test until 205, test at 210, work from 5 to 1130
Randomly selected or generated message	Cleaning my room [Note: sender has actually left the room] {not in corpus}
Remaining in the loop	Not here … Please leave me a message! Thanks.
	Asleep … leave one
Lurking/filtering	Sleeping … or am I ☺
	Maybe I'm doing work … maybe I'm not … the question of the night
Intentional misrepresentation	Dinner with Mark and dancing all night [Note: sender actually in dorm alone watching TV] {not in corpus}

The purely communicative function of "I'm away" messages is to alert members of your buddy list that even though you are logged on to AIM, you cannot (or will not) be responding immediately to instant messages they might send. Subjects in the study used three forms of away messages to fill this function. The simplest ("I really am away") states that the user is unavailable (e.g. "out") or engaged in a specific activity or at a specific location (e.g. "the library"), thus accounting for the absence. A variant on this message type ("Itinerary") spells out the sequence of activities in which the sender will be engaged. Although the level of detail appearing in some of these messages may seem unnecessary, student researchers noted the social usefulness of informing friends how you are spending your day. Such detail enables members of an online social circle to continue a conversation stream (e.g. "So how was your test?") rather than needing to begin the encounter from scratch (e.g. "What did you do today?") when they resume IM or encounter one another face-to-face.

Some AIM users care less about laying out their agendas than conveying the essential information that they will be unavailable. For this purpose, users can grab whatever message from their saved arsenal they happen upon first ("Randomly selected message"). As one student researcher explained, it is irrelevant if you are actually cleaning your room or off at a class, since in either event, you are unavailable. However, the reader of such an away message generally has no way of knowing whether the literal content of the message is true.

Three variants of overt "I'm away" messages appear to be serving predominantly social functions. "Remaining in the loop" messages constitute requests for a message to be waiting when the individual posting the away message returns to active use of AIM. A more pronounced form of "remaining in the loop" is to request immediate communication through another ICT (see Table 20.2). This "remaining in the loop" function is similar to telephone voice mail (e.g. "I can't take your call now. Please leave a message."). Both media invite all comers to leave messages. However, in the case of voice mail, the caller has a particular interlocutor in mind. With away messages, the equivalent of the caller (i.e. the person checking his or her buddy list to see which members of the list are online and who has posted an away message) may or may not be seeking to communicate with a specific individual. "Callers" can access the away messages of anyone online, thus using the system as a social checkpoint for assessing the status of a collection of people.

A second socially motivated function of "I'm away" messages ("Lurking/filtering") is monitoring the incoming traffic of IMs, allowing senders to decide which messages to respond to and which to ignore. An away message such as "Sleeping ... or am I ☺" signals buddies there is some chance their IMs will be read (and responded to) immediately, but the recipient of such an IM is not obligated to do so. Apparently many college students post "I really am away" messages (e.g. "out") when they are

Tethered or Mobile? Use of Away Messages in Instant Messaging by American College Students

⟨20⟩

actually sitting at their computers. This ruse allows them to ignore whatever IMs might come in but also to commence IM if someone on the buddy list should post an interesting away message or send an interesting IM.

The final group of socially driven "I'm away" messages ("Intentional misrepresentation") enables senders to construct a self-image through use of creative license. Although we could not confirm examples of such usage in the database, class researchers reported instances in which friends posted away messages detailing socially impressive activities (e.g. an elaborate date with a desirable partner) when in fact the sender was sitting at home watching TV. Since computer-mediated communication more generally allows for invention of new identities (e.g. regarding age, gender, personality, nationality), it is hardly surprising to find fabrication of activities in away messages.

Given the ostensible function of away messages (namely, to say that the writer cannot be reached on IM), why don't people who will not be responding to IMs simply sign off from AIM? Part of the problem seems to be technological. The default setting of AIM triggers a sound whenever you log on or off, alerting everyone who has you on his or her buddy list. Although the default sound can be turned off, a visual icon still appears, showing a door opening or closing. The researchers suggested that such an intrusion was socially "too loud". Not only are you noisily announcing your presence (when you log on), but you are inviting a deluge of IMs. Commenting on the "Lurking/filtering" function (or use of an "I really am away" message when you are not), students again noted the importance of social politeness. If you are on AIM but only want to reply selectively to messages, posting an away message and then only responding selectively to IMs you receive is a way of not hurting the feelings of people whose IMs you ignore.[11]

20.4.4.2 Overt Function: Initiate Discussion or Social Encounter

The second Informational/Discursive subcategory of away messages invites communication in the immediate or near future by another ICT (generally the mobile phone or an IM message) or face-to-face. Table 20.2 summarizes the social and communicative functions of this group of messages.

The first cluster of these away messages ("Reach me through a different ICT") entreats readers to initiate communication immediately. Sometimes the primary benefactor is the person posting the away message (e.g.

---

[11] SMS on mobile phones also allows users to manage their "presence" (i.e. their availability to respond to an incoming message). However, as a technologically asynchronous ICT, SMS affords its users more "presence" anonymity than does IM.

Table 20.2 Messages inviting virtual and face-to-face contact

Overt form: initiate discussion or social encounter	
Social/communicative functions	Examples
Reach me through a different ICT	Call my cell [phone] Since I am never around, try TEXT MESSAGING me! Send it to [phone number]. I wanna feel the love!
Let's chat online	Please distract me, I'm not accomplishing anything sitting at my desk … im me … we'll talk …
Invite face-to-face contact	Attempting to get all of this work done … if anyone is feeling productive please feel free to come and help me out ☹ WANTED: One date to the DTD Grab-a-Date on Friday. Applications now being accepted. Preference given to candidates who are hot, smart and outgoing. :-D

"I wanna feel the love!"), whereas at other times posters report a sense of obligation to their buddies. In the words of one subject in the study, "I feel like I should be accessible. My cell phone is always attached. I don't want my friends to think there's a time when they can't reach me."

A second cluster of messages ("Let's chat online") invites buddies to IM the person posting the away message when he or she is working at the computer (e.g. "Please distract me"). This "boredom" function of away messages looks like an oxymoron, but it turns out to be an efficient means of broadcasting a message to a large number of possible readers in the hope that someone will respond. As in the case of "Lurking/filtering" messages (Table 20.1), individuals posting away messages that solicit online chatting can select the incoming IMs to which they wish to respond. The third set of away messages in this cluster ("Invite face-to-face contact") constitutes a similar sort of broadcast request (e.g. "WANTED: One date") for someone's physical rather than virtual presence.

20.4.4.3 Overt Function: Messages Conveying Personal Information (About Self, Opinions, Humor)

Users sometimes create away messages to convey personal information – about themselves, their opinions, or their facility with humor.[12] Table 20.3 summarizes the social and communicative functions of this category of messages.

---

[12] CMC researchers have often commented on the importance of humor in the medium (e.g. Baym, 1995; Danet, 2001).

Table 20.3 Messages conveying personal information (about self, opinions, humor)

Overt form: convey personal information (about self, opinions, sense of humor)	
Social/communicative functions	Examples
Current state, activity or judgment of sender	Reading for once ... the joy of being an English major is sooo overwhelming right now ... (the sarcasm is very much intended) Yes, I am <u>that</u> good
Opinions	You have very little say in your fate or what will eventually befall you, but don't let that keep you from voting Most girls would rather be beaten to death by a pretty boy, than touched by an ugly one. [Note: sent by female]
Sense of humor	I could easily be replaced with a dancing chimp ... and at times I believe people would prefer the chimp (but then again, so would I) This chik needs filla [Note: Chick-Fil-A is one of the fast food shops on campus]

Many users of AIM see the away message function as a useful venue for presenting how they define themselves (either in general or at the moment) and how they wish others to perceive them. A number of the messages in this category serve multiple functions. For example, "Reading for once" indicates the writer's attitude towards having a lot to read in addition to implying unavailability ("I really am away") without specifying location. "This chik needs filla" conveys unavailability (the author has left her computer to go and eat), but does so in a humorous way, thereby revealing something of the writer's personality.

20.4.4.4 Overt Function: Convey Personal Information to Selected Other(s)

The last Informational/Discursive subcategory is broadcasting to one's entire online social circle personal information that is intended for a specific person (or persons). Table 20.4 summarizes the social and communicative functions of this group of messages.

Why do some AIM users choose the public forum of away messages to convey information seemingly targeted to selective members of their buddy list? One explanation is rooted in AIM technology. As of Fall 2002, AIM did not enable people to send IMs to users who were not signed on to the system. The only way to communicate via IM with someone logged off was through an away message, which could be accessed when the intended reader returned online. Responding to a question about privacy

Table 20.4 Messages conveying personal information to selected other(s)

Overt form: convey personal information to selected other(s)	
Social/communicative functions	Examples
Communiqué intended for single individual	Working at the multicultural bilingual center … Sam, we will hang later tonight!!! J I promise! "Let's see what DC [= Washington, DC] can do for us" – Jane (out exploring the city) … get well soon hot stuff :::sending good vibes:::: :-P
Communiqué intended for group of "insiders"	Back in D.C. missing Houston very much … Suz, Jan and Rick, thank you for an amazing weekend!

in away messages, one interviewee bluntly explained that he writes what he feels like writing and does not care who sees it. In other instances, students explained that they generally communicated through IM with a tight circle of friends, all of whom would likely know the individuals named in an away message. Therefore, away messages referring to specific people constituted intentional sharing of personal information with friends.

However, the rationale behind some postings of "private" messages may also be public display. By addressing or referring to a significant other or particular friend in an away message, authors publicize their personal relationships, reminding members of their buddy list (or potentially anyone else with access to their screen name) that they are the sort of people who have such friends. This public display function is similar to the use of "I'm away" messages containing intentional misrepresentation in order to impress others (e.g. claiming to be on a date when actually at home).

20.4.4.5 Overt Function: Entertainment (Humor, Quotes, Links)

The final group of away messages in the corpus encompasses postings primarily designed to entertain. Table 20.5 presents examples, clustered by type.

Why do AIM users post away messages designed essentially to entertain? The answer is grounded partly in the technology and partly in the social goals and expectations of users in this age cohort.

Experienced users of e-mail are familiar with "signature files", which allow senders automatically to post at the end of their e-mails not only professional contact information but also pithy sayings or quotations. Although instant messages themselves have no signature files, "Entertainment" away messages may serve comparable functions, as may IM profiles.

Tethered or Mobile? Use of Away Messages in Instant Messaging by American College Students

20

Table 20.5 Messages for entertainment (humor, quotations, links to websites)

Overt form: entertainment (humor, quotations, links)

Type	Examples
Humor	Work rhymes with beserk and jerk. i lurk in the murk and do my work, ya big jerk. ok, I gotta go do some work now
	Prepositions are not words to end sentences with
Quotations	"Good breeding consists of concealing how much we think of ourselves and how little we think of other people" – Mark Twain
	"Make a career out of humanity" – Dr Martin Luther King Jr
Poems, song lyrics	I sleep just to dream her
	I beg the night just to see her
	That my only love should be her
	Just to lie in her arms
	:-D
Links	Link to the website of a band, an actor or actress, a movie {not in corpus}

Both student researchers and subjects reported viewing their selections of entertainment clips as a form of self-expression. However, a number of users also commented that they perceived entertainment to be an essential component of the away messages genre. One subject noted, "I like to make people happy with my messages". Another indicated that since she enjoyed other people's away messages that made her laugh, she tried to make her own messages funny. A third said he likes to entertain people. Several others felt they had to justify themselves when their away messages weren't funny or creative, typically explaining that they lacked time or energy to craft amusing messages.

20.4.5 How Senders View Away Messages

Away messages clearly have multiple functions, not all of which are revealed by the overt form of the messages themselves. Interviews with some of the 38 subjects offered insights regarding their senders' intentions. We have already noted how important students felt humor was in creating away messages. Other factors include personal motivations for posting away messages and the ways in which users can creatively manipulate the use for which the medium was originally designed.

20.4.5.1 Personal Motivations

There is a lack of consensus among American college students as to how much away messages should reveal or justify their whereabouts. Some users deemed it important to let potential interlocutors know why the person posting an away message was absent and how to locate him or her. Others felt strongly that specifying their precise location was an invasion of privacy.

Similarly, away message users held differing opinions regarding the appropriate length of an away message. Whereas some advocated – and sent – one-word messages, others scoffed at the "laziness" of such writers. In the words of one subject, "I don't appreciate/agree with people whose away messages consist of one word (such as 'away', 'sleeping' or 'work'). I know these people are more interesting than that, and away messages can be indicative of your mood, your state of mind, and what you're doing at the time. The best ones can do all three."

20.4.5.2 Creative Manipulation of Away Messages

A number of users consciously manipulated the away message genre to serve individual needs. For example, one interviewee noted that she only posts away messages when she is in her room, working at her computer. (Her messages included the likes of "Eating the souls of my fellow man" and "*sigh*".) For her, away messages were a way of expressing personal information (sometimes humorously) about her current state (see Table 20.3), perhaps to generate conversation with people viewing her messages.

The flip side of using away messages to express your feelings is to craft messages that intentionally camouflage your state of mind. One subject commented that she posts quotations when she does not feel like talking or giving away too much information. The same individual reported using self-deprecation ("I could easily be replaced with a dancing chimp" – Table 20.3) when "it has been a long day" and she doesn't want to go into the details of why. Another subject revealed that she uses humor to mask her stress level in order not to bother friends with her troubles, but at the same time hinting that not all is well.

20.4.5.3 Prescriptions for Constructing Away Messages

Several of the people interviewed offered overall assessments of how away messages should be constructed. One asserted, "I figure away messages should either make you think, laugh or tell you where I am ... and sometimes they're just random." Another said, "I believe away messages should be funny, informative and reflect who you are." A comprehensive summary of how to construct an away message was posted in the AIM profile of someone on the buddy list of a member of the research group:[13]

---

[13] The summary (slightly edited) was constructed by Brian Ayers and Rob Berrey.

Tethered or Mobile? Use of Away Messages in Instant Messaging by American College Students

20

it is important not to underestimate the value of a good away message. too much internet time is wasted by people reading mediocre/poor away messages. a few rules to go by:

1. no one word away messages – EVER
2. quotes/lyrics, unless appropriately timely, are a poor excuse for away messages and make the writer look like a hack
3. humor is the only way to go – i'm not looking for a deeper understanding of life, or a little tug on the heart strings from my instant messenger
4. keep inside jokes inside
5. announcements are alright, however they should be followed by something humorous
6. if you're using colors, do so tastefully
7. don't leave your cell phone number. people aren't looking at your away message to contact you, they're looking at it cause they're bored out of their mind writing some paper
8. no freakin emoticons

20.5 Tethered or Mobile: American Perceptions

Contemporary American college students effectively use the away message function of AOL's Instant Messenger to help manage their non-face-to-face social relationships. Outside the USA, many teens and young adults use the SMS function on mobile phones as the main written means by which these social ties are maintained.[14] From the perspective of a mobile phone/SMS culture, Americans may seem tethered to their computers (either networked at stationary locations or lugged along to function wirelessly) in order to engage in such communicative activity. But do Americans perceive themselves to be tethered?

20.5.1 ICT Usage Patterns and Perceptions of American College Students

We have been arguing that among American college students, the away message function of IM is perceived as a means of staying in the social loop, even when not physically tethered to one's computer. In looking at the issue of mobility, we need to keep in mind that many contemporary American college students are accustomed to carrying laptops around with them (to use in class, in the library or in public spaces). Moreover, public access computers are commonly available (e.g. in libraries or computer labs),[15] and the number of public network drops – along with areas

---

[14] See, e.g., Raban et al. (2002); Katz and Aakhus (2002). Hashimoto (2001) indicates that in Japan, students favor SMS over e-mail for this function, even when e-mail is available.

[15] An abbreviated version of AIM ("AIM Express") can easily be downloaded to public-access machines.

having wireless access – continues to increase on college campuses, in coffee shops and in airports. Thus, use of such Internet functions as AIM is no longer restricted to a student's place of residence. As a result, contemporary American college students have multiple venues for engaging in IM conversations (and for checking and posting away messages), potentially contributing to a feeling of mobility.

In a focus group with American University undergraduates,[16] we probed whether students felt "tethered" or "mobile" in using networked computers and mobile phones. For both forms of ICT, students reported that the terms "tethered" and "mobile" could be interpreted in multiple ways. One meaning of "tethered" with respect to computers was the need to stay near a computer, e.g. bringing your laptop to class or seeking out computers in public spaces to check instant messages (and related functions such as away messages). Focus group participants did not report feeling tethered in this way. They generally did not carry laptops around (noting they did not have wireless connections), and they rarely sought out public computers. Several commented on their sense of "time out" relief when not at their computers. When away from their computers during school vacations, students in the groups did not report feeling the need to get online to do e-mailing, initiate IM conversations or check away messages.

Some students noted they felt "tethered" when they returned to their computers after a period of time away, facing hundreds of e-mail messages that needed responses. One student complained about being "tethered" to a family member by AIM. The student had recently taught his mother how to use IM, and now she wished to keep in frequent IM communication with her son. The son began "blocking" her messages (an option of AIM) when he wanted to remain logged on to IM (for communicating with friends) and especially when he wished to post off-color away messages he did not wish his mother to see.

With regard to mobile phones, students commented on feeling two senses of mobility. First, they noted they could always be reached (either directly or via voice mail). Second, students reported feeling mobile in that they were free to ignore calls. By looking at caller ID, they could decide whether to take the call or let it go to voice mail. We wondered whether students in the group ever felt tethered to their mobile phones (the way some American adults do) because they were always "on call". The students' answer was uniformly "no".

Why do students feel they can walk away from their computers for periods of time without feeling a need to check e-mail, engage in IM

---

[16] The group consisted of five males and three females, including six sophomores, one freshman and one junior. All eight were frequent users of instant messaging. All had mobile phones, but only one had actively used SMS, and that was in connection with her job.

Tethered or Mobile? Use of Away Messages in Instant Messaging by American College Students

20

conversations or view away messages? For many American college students, the voice functions of mobile phones complement the written functions of computer-mediated communication. When not at a computer, students simply switch modes (from writing to voice). As we saw in Table 20.2, some students even post away messages requesting that buddies call the poster's mobile phone.

20.5.2 The Future of ICTs in America

Will current computer-based habits of American college-aged students persist as mobile phones (and the availability of texting functions) continue to penetrate the American market? The answer will be driven in part by the proliferation and technological development of portable ICT devices (especially PDAs and mobile phones that handle not just SMSs or brief IMs but also longer text files such as attachments, AIM profiles and away messages). However, some of the answers may reflect cultural rather than technological factors.

ICT cultures on the two sides of the Atlantic are both converging and (at least as of now) maintaining independence. On the one hand, personal computers and use of the Internet (technologies already entrenched in American culture) are making considerable headway in Europe.[17] On the other hand, growth of SMS in the USA does not nearly approach usage in Europe.[18] That is, the USA is not presently an SMS "culture". From the perspective of SMS cultures, the USA may appear at best slow and at worst misguided with regard to the adoption of mobile written communication devices in place of more tethered computer-based messaging through e-mail or IM. However, like most issues of cultural adoption, the ICT story is the product of many variables.

A plethora of factors can influence individual and social decisions to adopt a new communication medium. Among the technological issues are: how convenient are the new ICT devices (e.g. how much do they weigh; can they be worn)?; how efficient are the devices (e.g. number of characters on a screen, availability of predictive input software)?; if, for example, users are accustomed to full keyboards (the case in the USA), will they willingly shift to tiny mobile phone keypads?

---

[17] Raban et al. (2002).

[18] For example, in the third quarter of 2003, 1.7 billion text messages were sent in the USA (http://www.nytimes.com/2004/06/26/technology/26ALIB.html?ex=1089280032&ei=1&en=09043265157 765e3) whereas that same number were sent in the UK in the month of November 2003 alone (http://www.mda-mobiledata.org/resource/hottopics/sms.asp). Note that in 2004, the UK population was only 59 million compared with a US population of 293 million.

A second factor is market issues: how much does a device (or use of the device) cost?; can devices made by different companies communicate with one another?[19] which devices are successfully marketed?; are technologies available for free (such as AIM, which clearly encourages the use of instant messaging in the USA)?

Third, there are issues of cultural entrenchment: how much re-education would be needed to switch from one ICT base to another?; what is the user – and usage – base for a particular ICT?

Fourth, changing fashion can influence adoption – or rejection – of new ICTs: will mobile phone users get tired of inputting SMSs?; will senders of SMSs, e-mail and IMs get bored with emoticons?

Finally, there is the issue of physical praxis, which may play a critical role in determining the future of texting in America. The USA is a car-based culture. We drive to work, drive to the grocery store, even drive around the corner. We use mobile phones in the car, but for talking rather than texting. In much of the rest of the world, people use far more public transportation and also walk more. You can send an SMS while on a bus or tram, while waiting for a train or while walking down the street. If Americans are tethered, it is largely to their cars, and cars make text messaging difficult or impossible.[20]

20.5.3 The Future of IM and Away Messages in America

Since away messages are part of the larger system of instant messaging, part of the fate of away messages is tied to the future of instant messaging more generally. Among the factors that may influence future use of away messages in the USA to maintain social ties are these: will AOL continue to make AIM available for free?; will IM become a dominant communication medium for a broad spectrum of age cohorts, as American businesses increasingly replace e-mail with instant messaging?; as college graduates make the transition into the world of work, will they lose interest in the features of AIM designed to foster self-expression (including away messages)?

In this chapter, we have attempted to demonstrate how away messages in IM are used by American college-aged students to help manage their social spheres and have considered the extent to which these students feel "tethered" or "mobile" as a result of their ICT usage. We have also suggested that the choice of alternative ICT systems reflects cultural pat-

---

[19] As Raban et al. (2002) explain, SMS usage in Israel was low because, until recently, the four mobile phone providers in the country employed different technologies that did not permit customers on one network to send SMSs to those on another.

[20] On college campuses where few students drive, we might predict higher use of SMS.

Tethered or Mobile? Use of Away Messages in Instant Messaging by American College Students

terns, not simply technological affordances. Both computer-based communication (including IM conversations and away messages) and mobile phones (both as texting and voice devices) enable users to manage their social spheres. The particular forms of ICT that college-aged students choose to manage interpersonal relationships are likely to continue evolving in perhaps unpredictable ways for the foreseeable future.

20.6 References

Baron, N.S. (2000) *Alphabet to Email: How Written English Evolved and Where It's Heading.* Routledge, London.

Baron, N.S. (2002) Who sets e-mail style: prescriptivism, coping strategies, and democratizing communication access. *The Information Society*, 18, 403–413.

Baron, N.S. (2004) See you online: gender issues in college student use of instant messaging. *Journal of Language and Social Psychology.*

Baym, N. (1995) The performance of humor in computer-mediated communication. *Journal of Computer-Mediated Communication*, 1(2); www.ascusc.org/jcmc/vol1/issue2/baym.html.

Danet, B. (2001) *Cyberpl@y: Community Online.* Berg, London.

Goffman, E. (1959). *The Presentation of Self in Everyday Life.* Doubleday, Garden City, NY.

Hashimoto, Y. (2001) How do the Japanese youth use cellular messaging services? Comparison with the usage of e-mails on personal computer. *Japanology*, 20, September.

Hashimoto, Y. (2002) The spread of cellular phones and their influence on young people in Japan. In Kim, S.D. (ed.), *The Social and Cultural Impact/Meaning of Mobile Communication.* School of Communication, Hallym University, Chunchon, Korea, pp. 101–112.

Jacobs, G. (2003) Breaking down virtual walls: understanding the real space/cyberspace connections of language and literacy in adolescents' use of instant messaging. Paper presented at the 2003 Annual Conference of the American Educational Research Association, Chicago.

Katz, J.E. and Aakhus, M. (2002) *Perpetual Contact.* Cambridge University Press, Cambridge.

Ling, R. (2004) *The Mobile Connection: the Cell Phone's Impact on Society.* Morgan Kaufmann, San Francisco.

Ling, R., Yttri, B., Anderson, B. and Diduca, D. (2002) e-living D7.4: age, gender, and social capital: a cross-sectional analysis. *e-living: Life in a Digital Europe, an EU Fifth Framework Project* (IST-2000-25409); www.eurescom.de/e-living/.

Raban, Y., Soffer, T., Mihnev, P. and Ganev, K. (2002) e-living D7.1: ICT uptake and usage: a cross-sectional analysis. *e-Living: Life in a Digital Europe, an EU Fifth Framework Project* (IST-2000-25409); www.eurescom.de/e-living/.

Squires, L. and Stacey, Z. (2002) AOL Instant Messenger and self-presentation. Unpublished class project for HNRS 302.002 Language in the New Millennium, Fall 2002, American University, Washington, DC.

21

Language Use in Swedish Mobile Text Messaging

Ylva Hård af Segerstad

21.1 Introduction

This chapter presents some of the findings of a linguistic analysis of SMS messages written by Swedish users. The material presented here is part of a larger analysis that analysed and compared data from e-mail, web chat, instant messaging (IM) and mobile text messaging (Hård af Segerstad, 2002). The main focus of the larger work was on factors that influence written language in these particular settings, that is, what characteristics the resulting texts show in comparison with traditionally written language and spoken language.

The introduction and popularity of mobile phones and mobile text messaging have come to evoke excessive public discussions about the kinds of cultural, social and psychological impacts that the new technology is likely to have, just as with many earlier communication technologies (Turkle, 1995). It seems that popular media representations about new communication technologies are greatly concerned about the way that standard varieties and conventional linguistic and communicative practices are affected (cf. Kasesniemi 2003; Thurlow, 2003).

The aim of the present study was to analyze SMS messages in order to answer the following main questions: how is Swedish written language used in SMS?; what are the conditions that influence mobile text messaging?; how is written language adapted to suit the conditions of text messaging?; are syntactical and lexical reductions used?; are these reductions and short forms based on Swedish or some other language?

The chapter opens with a brief account of the communicative setting of mobile text messaging and findings of previous studies of SMS. A description of data and methods for the present study follows. Results are then presented and exemplified. The chapter closes with a discussion of the results. Note that the terms SMS, mobile text messaging and texting are used interchangeably.

21.2 Background

21.2.1 The Communicative Setting of Mobile Text Messaging

Compared with other modes of CMC such as instant messaging and web chat, text communication via mobile phones is constrained regarding production and perception conditions. Most commonly, messages are produced on the tiny keypad of the phone. They may also be typed on a computer keyboard when sending messages from one of a multitude of web-based SMS services. Messages are read on the small screen of the phone, which commonly does not display the whole message in the same screen view. Moreover, there is an upper limit of about 160 characters per message.

Typing letters on the phone is achieved either by using "multi-tap" or some form of predictive text entry. With the multi-tap method, the user presses each key one or more times to specify the input character. For example, the 2 key is pressed once for the character A, twice for B, three times for C, four times for the number 2, five times for the Swedish vowel Å and six times for the Swedish vowel Ä. Multi-tap can generate problems of character sequence segmentation when a new character is on the same key as the previous character (e.g. in the word "ON" where both O and N are on the 6 key).

With predictive text input, linguistic knowledge about how words are spelled is added to the system. MacKenzie and Soukoreff (2002) call this technique "one-key with disambiguation." An example of software for predictive text input is T9 by Tegic Communications. The user needs to press each number key only once for each letter. The phone software tries to identify what the user intends based on a dictionary stored in the phone's memory. Since multiple words sometimes have the same key sequence, disambiguation may be difficult. The resulting text in the message may therefore be "software driven", and not reflect what the writer intended, as we will see below. In these cases the most frequent word is the default. The predictive text input system does not accept slang or dialect variants unless taught to do so by the user. (New words can be entered into the phone's memory, using the multi-tap technique.) As Kasesniemi and Rautiainen (2002) argue, the mobile of a "lazy" user "speaks" only standard language with no personal tones.

Another means of adding personal tones, or even help disambiguating text, in SMS is the use of emoticons. Emoticons are used in text messaging in the same manner as in IM or chat, for example, to enhance alphabetic writing by conveying moods or emotions that are normally expressed with extralinguistic cues such as facial expressions and tone of voice in spoken interaction. Symbols created in imitation of face-to-face facial expressions may help to make it easier to interpret text-only communication. It is likely that the experienced communicator seems to know that some messages might need additional information to disambiguate text-only communication. Inserting non-alphabetic symbols in mobile text messages can be troublesome on most types of mobile phones, owing to the limited keypad. However, some phone models (especially targeted for young users) include several preformatted emoticons, e.g. the happy face [:-)], the sad [:-( ] and the winking emoticon [;-)]. These emoticons do not require insertion of three separate symbols; rather, the whole emoticon may be selected and inserted in one go, but still claims the space of three characters.

SMS is an asynchronous mode of communication, i.e., it does not require the communicators to be online simultaneously. As SMS employs writing as means of expression, it is monomodal and therefore can only take advantage of what can be conveyed through the single channel of the visible writing system. Studies have shown that most messages are sent between friends who share considerable amounts of background knowledge (Hård af Segerstad, 2002; Telia/Temo, 2002).

In sum, compared with speaking or with typing on a keyboard, texting is rather cumbersome and time consuming. The sender is constrained by the limited number of characters per message and software-driven spelling problems, and also the asynchronicity of the medium when producing messages. Such conditions are ripe for users devising strategies to save time, effort and space in a corpus of SMS messages.

21.3 Previous Studies of Mobile Text Messaging

Not very many studies of language use in SMS are available as yet, but a few studies have been published on the use and function of texting by German users (Androutsopoulos and Schmidt 2001; Schlobinski et al., 2001) Norwegian users (Ling, this volume, Chapter 22) and British users (Eldridge and Grinter, 2001; Grinter and Eldridge, 2001). However, the studies have shown that language usage in SMS messages overwhelmingly ends up shortening the text. The formation of text messages seems to be driven to a large degree by the demands for efficient writing, but textual shortenings may also serve multiple functions (cf. Ling, this volume, Chapter 23). Users employ various devices to color and tone their messages, and while saving time and effort by using certain words and phrases, for example, also may be seen as latching on to mutual knowledge and experience that exist between sender and receiver, which we will see examples of below.

Schlobinski et al. (2001) regarded language use in their corpus to be a hybridization of written and spoken language, judging from the use of colloquial expressions, reductions and assimilations. They also remarked on unconventional ways of writing, such as writing consistently in lower-case. They argued that syntactical reductions are caused by the medium, and that abbreviations and short forms of words were frequently used. Eldridge and Grinter (2001) and Grinter and Eldridge (2001) report that the reasons for teenagers preferring to text one another are that it is quicker, cheaper and easy to use/more convenient than other communicative methods. They found minimal support for predictive typing technologies. The teenagers' common use of abbreviations and shorthand made it barely usable in practice. Döring (2002) analyzed types, frequencies, and functions of short forms in text messages using a corpus of 1000 authentic text messages and questionnaire data. Like Thurlow (2003), she found that the widespread claims about the linguistic exclusivity appear greatly exaggerated. She found that almost no SMS-specific short forms exist, which could manifest a collective identity, and that abbreviations and acronyms are only seldom used in text messaging. Nevertheless, Döring categorized lexical and syntactical reductions. Her findings of syntactic reductions confirm the findings of Androutsopoulos and Schmidt (2001).

Table 21.1 summarizes common syntactic reductions found in German SMS corpora, as reported by Döring (2002), Androutsopoulos and Schmidt (2001) and Schlobinski et al. (2001).

Androutsopoulos and Schmidt suggest that the most common type of syntactical reduction is the deletion of subject pronoun. The study of Schlobinski et al. shows the same pattern. Döring divides the category "acronyms" into letter acronyms and phonetic value acronyms. The category "abbreviations" is divided into conventional abbreviations and unconventional, ad hoc abbreviations. Letter acronyms and ad hoc abbre-

Table 21.1 Common syntactical reductions in German SMS (deleted elements in square brackets)

Deletion of subject (especially subject pronoun)	[Ich] Komme später Heim ... ! ([I] will be home late ... !)
Deletion of preposition, article and possessive pronoun	Weißt du was [der] Eintritt kostet? (Do you know what [the] entrance fee is?)
Deletion of copula, auxiliary or modal verbs (+XP)	[Bist du] Schon wieder zurück aus [Ø] Urlaub? ([Are you] home from work already?)
Deletion of verb and subject pronoun; telegram style	[Hast du] Lust, dann komm vorbei? ([Do you] feel like coming over then?])

Table 21.2   Acronyms and abbreviations in German SMS

Acronyms	Abbreviations
Letter acronyms *Examples:* SMS, CD, PS, HDL, I.L.D.	Conventional abbreviations *Examples:* inkl. (inklusive = incl.: including, included, inclusive of), bzw. (beziehungsweise = resp.: respectively), Nr. (Nummer = No., no.: number)
Phonetic value acronyms *Examples:* RAM, SIM	Unconventional, ad hoc abbreviations (or ad hoc formations) *Examples:* he (heute = today), i (ich = I) e (ein =a, an), f (für = for), ü-nächste (übernächste = the next but one), zw.durch (zwischendurch = in between)

viations proved to be the most commonly used of the four types. Schlobinski et al. dub the latter type of abbreviations ad hoc formations. Table 21.2 uses Döring's taxonomy of acronyms and abbreviations to summarize findings by Döring and Schlobinski et al. regarding acronyms and abbreviations in German SMS.

Studies have shown that text messaging is mainly used for maintaining relations between people and for coordinating social activities. It is used mostly for private communication between partners, friends, family and loved ones (cf. e.g. Hård af Segerstad, 2002; Kasesniemi 2003, Ling, this volume, Chapter 23). In their work on SMS, Eldridge and Grinter (2001) and Grinter and Eldridge (2001) argue that texting allows teenagers to forego some spoken conversational conventions by reducing the overall time spent on the interaction. This conclusion is supported by Döring (2002), who concluded that with SMS, users can be brief without fear of being perceived as abrupt or rude. Logically, one has to be brief in order not to go beyond the scope of the limited number of characters allowable per message, because each message sent is charged at a relatively expensive rate (in contrast, for example, to IM or e-mail, where there is no "per message" cost.) As Döring appropriately observes, another important reason for being brief is that inputting text on mobile phones is cumbersome. Finally, because most SMS communication is interpersonal communication between people who know each other, shared background knowledge makes brevity pragmatically plausible. Since both interlocutors know the character limit of each message, users find this terse communicative behavior to be completely acceptable rather than rude. The physical structures of text messaging can prevent users from going "off topic" and making a conversation longer than planned.

Judging from the factors illustrated by Döring, production and perception conditions, in combination with situational parameters (i.e. the technical restriction of 160 characters per message, each message sent carries a charge, text input is cumbersome) both permit and force people to express themselves concisely in mobile text messaging. Previous studies of SMS have shown that there is frequent use of lexical shortenings and syntactic reductions that save keystrokes compared with standard writing by hand or with a full keyboard. However, economizing with time and effort is not always an issue. It appears that people are willing to labor over the text carefully, perhaps to achieve a certain effect in the appearance of the text and the choices of words and phrases, as we will see examples of below. The content of a message, and also its wording and formulation – or indeed the fact that the interaction takes place via SMS – may be regarded as implicitly communicating information too. Just like everyday spoken interaction, text messages often seem to be doing several tasks at the same time, serving multiple functions such as social grooming and information sharing (cf. Hård af Segerstad, 2002; Ling, this volume, Chapter 23).

21.4 Data and Methods

Three methods for collecting data were used in this study: a web-based questionnaire; user diaries and messages forwarded by formal study participants; and messages gathered from the researcher's friends and family. Each method has its merits and drawbacks. The aim of the present study was to draw upon complementary data sources. Despite the popularity of mobile text messaging, it proved difficult to gather data. The difficulties are due both to ethical aspects of conducting research, which does not allow for analyzing messages taken from mobile phone operators' logs, and to the private nature of texting. One of the reasons for why people feel it is hard to part with their messages may be that SMS is experienced as being even more private than e-mail, for example. Most messages are sent between friends who know each other well; they communicate about things that they do not want any one else to read, even though the content of most messages is seemingly trivial about everyday matters. Mobile phones are also experienced as highly personal gadgets. They are typically carried in a pocket or a bag, most of the time close to the body, which also may enhance the feeling of privacy. The resulting corpus is large enough, however, to offer insights into how written language is being used in SMS.

21.4.1 Data Collection

21.4.1.1 Web-based Questionnaire

Informants were recruited to the web-based questionnaire portion of the study by being referred to a link on the researcher's web site in a daily newspaper in Sweden, in which the researcher had previously been interviewed. Subjects were also invited through the researcher's web site. All entries in the web-based data questionnaire were anonymous, and all informants remained completely anonymous to the researcher. Given this data collection process, one had to rely on the respondents' honesty in giving their true gender and age, while at the same time keep in mind that it might be tempting to play with the freedom that electronic anonymity allows for.

A web page consisting of a questionnaire was constructed, which informants accessed to answer questions and share a text message that they themselves had sent either by mobile phone or a web site that provides SMS service. Subjects were requested to copy an SMS from their mobile phone on to the project web site, character by character, exactly as it appeared on their phone. The data from this web-based form were then entered automatically into a database. The affordances of the two different input devices may end up in differences in message formulation. Web site SMS services enable users to input SMS with full keyboards, and so one would expect these messages not to use some of the reduction techniques of mobile SMS transmissions. There are data suggesting that web SMS messages are longer than full keyboard messages, which we will see below. However, the data reported here is not broken down into reporting messages created on mobile phone and web site separately.

There are pitfalls to using both the copying and the free selection methods. Copying requires retyping on a keyboard a text that is read from the small screen of a mobile phone. In the course of doing such copying, it is very easy to make typographical errors or otherwise to copy the original incorrectly. Moreover, since the informants were invited to select a message that they wanted to share, their selections might have biased the resulting corpus in terms of content, language or both.

21.4.1.2 User Diaries

Four informants (two male and two female, between 12 and 25 years of age) were recruited to participate in the research project. They were asked to keep a user diary in which they recorded information about each text message that they sent and received during a period of 1 week, or longer if they so wished, as well as to forward messages directly to the researcher's mobile phone. Not all messages sent or received during this period were actually recorded, the informants only sent those which they felt confident sharing. This meant that the selection problem remained, but the reasons why some messages were withheld could be discussed in follow-up interviews. The messages were transferred into a database on a computer (equipped with appropriate software) by connecting the researcher's mobile phone to the computer by a serial cable. This method helped to reduce copying and transcription errors. Each of the informants was interviewed individually when he or she considered the forwarding and diary-keeping period to be finished. Participants were informed that all messages that were forwarded, and everything that the participants told the researcher in the interviews or wrote in the diaries, would be kept strictly confidential, and that the participants would remain anonymous to everyone but the researcher.

21.4.1.3 Friends and Family

The third method of data collection was to ask friends and family to part with their messages during a period of time. Sixteen informants (eight female, eight male) contributed to the corpus by sending 788 messages. Messages were either sent directly to the researcher's mobile phone or copied character-by-character into a text document in a word processor. Messages that concerned the researcher directly were omitted.

Combining together SMS messages data collected via the three collection methods yielded a total corpus of 1152 messages (112 from the questionnaire, 252 from the four informants and 788 from family and friends). The 1152 messages were made up of a total of 17024 words.

21.4.2 Methods of Analysis

The corpus was stored in machine-readable format, and analyzed automatically with the help of TraSA (Transcription Statistics with Automation),[1] a computer tool used in corpus linguistics to calculate measurements such as the number of lexical tokens, utterances, turns, pauses, overlaps and vocabulary richness. Frequency lists generated by the software were manually checked for occurrence of abbreviations (both established and new), complex punctuation expressions, emoticons and asterisks.

The corpus of messages was also analyzed manually for occurrences of syntactical and lexical reductions. Such reductions included deletion of subject pronoun and deletion of verb phrase elements (e.g. auxiliary verb, copula verb, prepositions). Such omissions are features that are typically associated with spoken interaction (cf. Allwood, 2000) and not common in standard traditional writing, which suggest that text messaging shows similarities with spoken interaction rather than written structures. The present chapter presents results from both the computer-generated and the manual analyses.

---

[1] TraSA was developed by Leif Grönqvist at the Department of Linguistics, Göteborg University, Sweden (Grönqvist, (2000).

Table 21.3 Linguistic features of Swedish SMS

Category	Feature
Punctuation	Omitting punctuation
	Unconventional punctuation
	Omitting blank space
Spelling	Mispredictions
	Speech-like spelling
	Split compounds
	Consonant writing
	Conventional abbreviations used conventionally
	Conventional abbreviations used unconventionally
	Unconventional abbreviations
	All capitals or all lower-case
	Replace long word with shorter word
Grammar	Omission of subject pronoun
	Omission of VP element (copula, auxiliary, or modal verb + preposition)
	Omission of article, preposition, possessive pronoun
Graphical (non-alphabetic) means	Emoticons
	Asterisks
	Symbol replacing word

21.5 Results

The message analysis revealed that Swedish SMS communicators use the same basic types of syntactical and lexical reductions as reported in analyses of German SMS by Döring (2002) and Schlobinski et al. (2001). Table 21.3 summarizes the most common linguistic features of the Swedish SMS corpus.

In the present chapter, only some of the features in the above table will be exemplified and discussed. For an extended discussion of these linguistic features, see Hård af Segerstad (2002).

21.5.1 Punctuation

21.5.1.1 Omitting Punctuation

By omitting punctuation, a user saves the time and effort it takes to type those characters (period, comma, etc.). Omitting punctuation also saves space, which could be important when the message size is restricted by technical limitations. However, the 160-character limit does not actually always seem to be an issue. The mean length of message in this corpus was no more than was 14.8 words, or 64 characters, i.e. well below the upper limit of 160 characters, which leads one to assume mostly "time and effort" are the real issues. The results show that the strategy of saving space and effort by omitting punctuation is also used in messages that are short enough not to threaten to exceed the restricted number of characters. Example 1 gives an illustration of such a case. Periods, commas and capital letters for new sentences have been omitted. The sentence in the example below may, of course, look slightly different with "correct" punctuation depending on what the writer's intentions were. The punctuation in brackets is intended to show a possible version following normative punctuation and orthorgraphical conventions.

Example 1
Ge mig ditt nummer [,] har det inte längre [.] s[S]å ringer ja ikväll [,] ska spela fotboll [.]
(Give me your number [,] don't have it anymore [.] t[T]hen I'll call tonight [,] will play football [.])

21.5.1.2 Omitting Space Between Words

Similar to the strategy of omitting punctuation, by omitting space between words the user also saves time and effort. In some cases it was not necessary to save space, but omitting the gap between words renders a rather playful design to the message. Example 2 was typed on the keypad of a mobile phone.

Example 2
KOMINIAFFÄREN
(COMEINTOTHESTORE)

Example 3 was typed on a computer keyboard and sent via a web-based SMS service. The latter utilized all 160 characters, and seems to be a good example of linguistic awareness: by capitalizing each word readability was increased. Moreover, given its content, lack of spaces between words seems to render a sort of pleading touch to the message. This technique to produce text would be very difficult and time consuming if typed on a mobile phone, and no equivalents were found that were produced on a keypad.

Example 3
UtanBaraÖkatOchJagÄlskarDigMer&MerFastänDuÄrSurPåMig.MenVillDuInteHa
MigSå..SåBlirJagJätteledsen.DuBehöverInteFlyttaUppHitKäraVän.DuÄrNogBaraR
ädd.
(ButJustExpandedAndILoveYouMore&MoreEvenThoughYouAreMadAtMe.ButIf
YouDon'tWantMeThen..ThenI'llBeVerySad.YouDon'tHaveToMoveUppHereMyD
ear.You'reProbablyJustAfraid.)

21.5.2 Spelling

21.5.2.1 Mispredictions and Typos

Spelling in mobile text messaging seemed to be very much dependent on whether the sender used predictive text entry or not. However, in most cases the user did not state whether or not the phone he or she was using had this software or whether it was actually employed. The predictive function is not available on web-based SMS. As indicated above, predictive text entry compares the entered sequence of characters with a lexicon stored in the software. It predicts the most likely word or words with that particular sequence, and the most frequent word in Swedish bearing those characters (as determined by the software program) will be the one that appears on the screen. The software does not consider syntactic or semantic context, and is not tailored to everyday informal conversation, which results prove to be the most common in SMS messages. Thus, many strange software-driven "typos" may appear in mobile text messaging. Example 4 illustrates a misprediction in which the predictive software judged på [on] to be more frequent than så [so, so that]. However, given the interlocutors' shared common knowledge of Swedish, this misprediction probably did not hinder comprehension.

> *Example 4*
> De e på [så] ibland bara.
> (Mispredicted output: It just is on sometimes. Intended output: It just is like that sometimes)

Example 5 is an illustration of a misprediction of a colloquial phrase "Gott mos", (the literal meaning of which is "Good mash" and the intended meaning is something like "Good stuff"). Strangely enough, the predictive software appears to judge the following output to be the best match.

> *Example 5*
> Gott öms

The word "öms" does not mean anything in Swedish, which renders it somewhat untranslatable. The predictive software does not complete the spelling of a word, it only speeds up typing by eliminating multiple pushing of keys. It seems that the software had a word in the dictionary that started with "öms" higher on its priority list and the user might have simply spelled out the word and probably did not bother to page down through other alternatives with that letter combination, perhaps for special comic effect. Anecdotal evidence from informants suggested that as a result of mispredictions of a common phrase such as the one in example 5, as a joke people start using the misprediction both in speech and writing in place of the original. Yet another anecdotal illustration of this is the case of the boyfriend writing a message to his girlfriend, intending to type "Jag saknar dig" ("I miss you"), but the predictive software suggests should be "Jag räknar dig" ("I count you").

21.5.2.2 Speech-like or Unconventional Spelling

In many cases unconventional spelling, or spelling which imitates the phonetic value of speech, saves space and time and effort. Again, most of the motivation for saving keystrokes is probably to reduce time and effort, not to avoid the 160 character limit. Sometimes, though, unconventional spelling results in the same number of keystrokes as the normative spelling, sometimes in even more keystrokes or more effort spent. Clearly, the economy principle is not absolute, and what is considered "rational" behavior may instead be instrumental, i.e. rational for the purpose it serves (cf. Allwood, 2000). The word åxå [också = too], in example 6, saved two keystrokes compared with normative spelling. Imitating the spoken rendering of the word, at the same time it brings a sort of informal and colloquial tone to the message.

> *Example 6*
> Tjena! tillsammans e rätt tuff! ska åxå köpa den nån gång .... snart!! jo ja har d åxå trevligt spelar golf d går bra! men du får ha d så bra!! kramas!!
> (Hi! together is pretty cool! will also buy it some time .... soon!! yes I'm having a good time too playing golf it's going fine! but you'll have to take care!! hugging!!)

Colloquial words are often included and dialectal pronunciation is often rendered in SMS texts, as opposed to standard norms for texts for more formal purposes. Choice of words and how to render them in writing seem to be strongly dependent on the relationship between communicators. Such choices possibly act as in-group markers and connect with the interlocutors' shared background knowledge. They might also function as a pragmatic strategy. The message in example 7 might well be interpreted as a way of reminding the receiver that he or she ought to get back with some information soon.

> *Example 7*
> Höllö höllö [hallå hallå]! Bler [blir = going to be] det nön [någon = any] beachvolley elle [eller = or]?
> (**Hullo Hullo**[Hello Hello]! Is there **gunna** [going to be] be any beach volley **or what?**)

Another type of unconventional spelling, or rather a feature that is unconventional to spell out in writing in the first place, is found when features, such as **eh, öh, hmm,**[2] common in spoken language appear in

---

[2] These features are sometimes called "disfluencies" (cf. Clark and Wasow, 1998). Allwood uses the term "Own Communication Management" or OCM (Allwood, 2000).

mobile text messaging. Example 8 was opened with an example of this and closed with a conscious marking of a word rendered in its spoken form.

Example 8
Hmm, kanske det!? Vi hörs i morr'n [morgon = morning] :-)
(**Hmm,** may be!? Let's get in touch tmorro (tomorrow) :-))

21.5.2.3 Split Compounds

The SMS log showed a few cases in which the predictive text entry software seemed to be responsible for splitting compound words into two separate lexical items. Splitting compound nouns is a tendency in progress in Swedish, which, similar to so many other changes, has started a public discussion[3] (cf. similar discussions in Kasesniemi, 2003; Thurlow, 2003). Whether this might expedite a tendency which is already in progress remains to be seen. Splitting compounds in many cases entails a semantic change. An example is when the Swedish phrase "en brunhårig sjuksköterska" ("a brown-haired nurse") is split into "en brun hårig sjuk sköterska", which then means " a brown hairy sick nurse". The split in the example below is not as severe, though.

Example 9
HAJ! JAG HAR KÖPT ENA NY LUR. **JÄTTE ROLIG** [JÄTTEROLIG]!
(HI! I'VE BOUGHT A NEW PHONE. **REALLY GREAT!**)

21.5.2.4 Replace Long Words With Shorter Words

Yet another strategy to save space was to replace a long word with a shorter one, even though the shorter word is not as frequently used as the longer one. For example, the word "ej" has a slightly archaic stylistic tone, and is not used as often as the everyday "inte". Both are equivalents to the English "not".

Example 10
japp men det syns **ej** [inte = not] vem det är ifrån. X
(yepp but it **doesn't** show who sent it. X)

In some instances, shorter replacement words were taken from English. In contemporary informal speech, young Swedes commonly interject the

---

[3] See, for example, the popular TV show on language: Värsta Språket (= the Worst Language, meaning cool language or something to that effect) (see www.svt.se). There is also a web community, Skrivihop.nu, dedicated to fighting the splitting of compounds (http://www.skrivihop.nu/). Note that the web address means, in translation, "write together.now" or "compound.now".

odd word or phrase in English into their conversations, rendering the same playing tone. Inserting English words or phrases probably also serves the double function of informality marker and saver of time, effort and space. The English word "KIDS" used in example 11 is just one character shorter than the Swedish word "UNGAR" which it replaced, suggesting that the motivation is using the informal English term.

Example 11
KATTEN HAR FÅTT **KIDS** [ungar]
(THE CAT GOT **KIDS** [young, babies])

21.5.2.5 Consonant Writing

By sometimes omitting the vowels in a word, an SMS message can be rendered entirely in consonants. The intended meaning seems to come through even without the original vowels. The examples below did not have to be shortened in order to save space, but appear, rather, to be expressions of language play. However, it is difficult to tell when the goal is language play and when it is saving keystrokes to reduce time and effort. In many cases the use probably carries multiple functions.

Example 12
Har du **prgmrt** [programmerat] videon?
(Have you **prgrmd** [programmed] the video?)

Some cases look like display of a kind of linguistic awareness and the joy of language play. In example 13, the more or less "consonant written" word SVNGLSKA (which means "Swenglish") points out the construction "blada" [a shortened form of an English loan term (roller blading), adapted to reflect the Swedish inflectional system]. It seems that the writer became aware how strange or funny the word looked just as he or she was writing it, and cheerfully pointed that out!

Example 13
BLADA? **SVNGLSKA** [SVENGELSKA]!
(BLADING? **SWNGLSH** [SWENGLISH]!)

21.5.2.6 Conventional Abbreviations

Conventional, or established, Swedish abbreviations were found in the SMS corpus. They are used for the same reasons as they are used in other contexts of written communication: saving time and space. Conventional use of conventional abbreviations is when established abbreviations are used in their normative way. An example of this is the abbreviation "t.ex." ("till exempel" = "for example"). This was also the case in the SMS corpus.

Conventional methods of abbreviation may be used unconventionally which would require additional explicit information if it had occurred in an autonomous text. The initial letter in a word may stand for the whole word, as in example 14.

Example 14
SKITSNACK! VILKEN **T** [tid]?
(BULLSHIT! WHAT **T** (TIME)?)

21.5.2.7 Unconventional Abbreviations

The Messages illustrated in examples 15 and 16 show innovative new types of abbreviations based on Swedish words. The abbreviations are unconventional and not yet established in more formal types of written communication. Time will tell whether they will become established and accepted in SMS, in non-CMC forms of writing or both.

Example 15
X! På väg till y. inga pengar kvar på **tfn:en** [telefonen]. ringer i e.m. **CS** [ses]!
(X! On my way to y. no money left in the **phn** [the phone]. will call this p.m. **CU** [see you]!)

Example 16
Va **QL**[kul] det ska bli på LÖR!
(How **FN** [fun] it'll be on SAT!)

Abbreviations such as **CS** [ses = see you], **CU** [see you] and **QL** [kul = fun] were found across message users and also across modes of CMC (Hård af Segerstad, 2002).

Another type of unconventional abbreviations are those based on non-Swedish words. These were also found in the SMS data. Words and whole phrases in English may appear in the midst of contemporary Swedish informal conversation. These might be in the form of snippets of song lyrics to popular songs, poems or fixed phrases. Examples of the latter are illustrated in examples 17 and 18. It seems likely that this usage has been observed and picked up from chat room norms.

Example 17
YES YES, MASTER OF BARBECUE RETURNS... **BTW** [by the way = förresten]: LYCKADES INTE FÅ MED EXPEDITEN PÅ SYSTEMET...
(YES YES, MASTER OF BARBECUE RETURNS ... **BTW** [by the way = förresten]: DIDN'T MANAGE TO BRING THE SALES PERSON AT THE LIQUOR STORE WITH ME...)

Example 18
Det var ju inte ett kryptisk meddelande... man skulle kunna tro att du hade X. på besök:) Lev Väl! Miss **U** [you = du]!
(As if that wasn't a cryptic message... one could well believe that you had x. visiting:) Take care! Miss **U** [you = du]!)

21.5.3 Grammar

Grammatical reductions were found to be used in mobile text messaging, in many cases presumably in order to save time, effort and space. The results suggest that written language in SMS shows characteristics of informal spoken interaction (cf. Allwood, 2000). Informants in the present study confirmed that most SMS messages in the corpus were sent between people who knew each other already, supporting results from a survey of SMS use in Sweden (Telia/Temo, 2002). The senders were thus able to rely on the receiver's ability to make pragmatic inferences when decoding abbreviated messages. Such inferences allowed senders to omit elements that the receiver could reconstruct. SMS texting does not have to be as grammatically explicit as traditional writing, which often must be autonomous and interpreted by readers unknown to the sender.

21.5.3.1 Omission of Subject Pronoun

The results support the findings of Döring (2002) and Androutsopoulos and Schmidt (2001) that subject pronouns were frequently omitted. If the sender's phone number is stored in the receiver's phone book, the sender's name will appear above the message on the screen, so it is clear who sent the message. As illustrated by example 19, it is therefore often obvious to the receiver whom the deictic expression "jag" ("I") refers to, and it may be left out.

Example 19
[Jag] kan inte ikväll. [Jag] måste jobba. [Jag] gillar dig i alla fall. KRAM x
([I] can't tonight. [Jag] have to work. [Jag] like you anyway. HUGS x)

As noted above, the omission of a subject pronoun is a feature that is normally characteristic of spoken informal interaction in Swedish (cf. Allwood, 2000), but is not associated with standard writing in which "full sentences" are required.

21.5.3.2 Omission of Verb Phrase Element

Other grammatical reductions appeared in the SMS corpus, and resulted in a savings of time and space. Leaving out a copula, auxiliary or modal verb in combination with omission of a preposition may save several keystrokes. Several interpretations of the message in example 20 are possible (e.g. blir det på … [is it going to be at …], sa vi på … [did we say at …]), but the receiver, who presumably had all the necessary background information, probably had no trouble decoding it.

Example 20
[Ska vi ses på] cafe japan [kl]19?
(Can we meet at] cafe japan [at] 19 [o'clock]?)

21.5.3.3 Omission of Prepositions and Possessive Pronoun

Example 21 illustrates a message in which some prepositions and a possessive pronoun were omitted. This type of message illustrates Döring's (2002) category "Telegram style".

Example 21
[I] Stan [i] Jönköping i så fall. [Min] Lillasyster fyller 20, kalas...
([In] Town [in] Jönköping in that case. [My] Baby sister turns 20, party...)

21.5.4 Graphical (Non-Alphabetic) Means

21.5.4.1 Emoticons

As mentioned in the background section, some phone models (especially targeted for young users) include several preformatted emoticons to ease text input. However, the results show that users took the time and effort to insert emoticons that were not preformatted on their phones. It seems to be worth the effort because it saves all the keystrokes that would have been necessary to communicate the same content implicitly in words. Presumably it also helps disambiguate messages by indicating the mood.

Example 22
Yepp, den har jag ! Fixar en kopia på den :-)
(Yepp, I've got it! Will make a copy of it :-))

21.5.4.2 Asterisks

Asterisks to frame words or phrases often serve the same purpose as emoticons. Compared with the use of emoticons, by including the words explicitly (the content of which would have been conveyed implicitly by an emoticon), the message may be even easier to disambiguate. Actions are described in such a way explicitly in words, such as the one illustrated in example 23. Unfortunately, in completing the web questionnaire, the informant did not indicate whether that particular message was composed on a mobile or at a computer. It seems likely, however, that it was typed on a computer keyboard, as it includes both emoticons and asterisks. However, it is also possible that the message was constructed on a mobile keypad, since situational variables such as relationship between communicators and the goal that the sender wishes to reach may exert more powerful influence than economy of effort.

329

Example 23
pet i sidan Hur vågar du ha upptaget när goa pågen ringer..!? ***peta lite till*** ;o)
Hör av dej när linjenär öppen! ***Cjamiz*** /Förnamn.
(*nudge* How dare you keep the line busy when the nice boy calls..!? ***nudge
again*** ;o) Call me when the lineis open again! ***Hugz***/First name.)

Example 24 illustrates a feature that is most often associated with spoken interaction. Extralinguistic cues, such as laughter and gestures, are normally not spelled out in writing. By adding asterisks around the typed version of laughter, the text is marked explicitly as indicating an action and most likely also as an indicator of how the message was supposed to be interpreted.

Example 24
ello babe.. va görs?? spanar du på kalle eller..? :p ***haha*** [skratt] keep rockin' //x.
(ello babe.. what are you doin?? Looking for kalle..? :p ***haha*** [laughter] keep
rockin' //x.)

21.5.4.3 Symbols Replacing Words

By replacing a word with a symbol that stands for the word, several keystrokes can be saved. Example 25 illustrates use of the maximum 160 character limit. This message makes use of a number of symbolic replacements for word, e.g. 1 (en = a, one) and & (och = and). By using symbols instead of words seven keystrokes were saved in this particular message. Additionally, abbreviations were used to save space [**gbg** = Göteborg, **ngt** = något (something)]. Notice also that the last word in the message ("kväl") is misspelled, since the final letter was omitted owing to lack of space. (Had the word been properly spelled, it would have been "kväll".) This misspelling does not seem to pose any difficulties for interpretation. People add and subtract information, and use their experience and pragmatic knowledge to get to the most likely and relevant interpretation of linguistic information (cf. Sperber and Wilson, 1986).

Example 25
Jo, är i gbg sen 1 (en = a, one) vecka. men har gjort ngt idiotiskt. skaffat ett **2** (två
= two) veckors städjobb i hamnen. måste upp **5** (fem = five) på morron **&** (och =
and) är DÖD när jag kommer hem.. Ringer i kväl
(Well, been in gbg **1** week. but have done something stupid. got myself a **2** week
cleaning job in the harbor. have to get up at **5** in the morning **&** am DEAD when
I get home.. will call tonigh)

In example 26, an emoticon was employed to stand not just for an emotion, but actually also for the word "glad" ("happy"). The message is a fixed expression common in Sweden for happy wishes around midsummer.

Example 26
:-) [glad] midsommar!
(:-) [happy] midsummer!)

21.6 Conclusions

As noted above, there are popular concerns about the way in which standard varieties and conventional linguistic and communicative practices are affected by the use of new communication technologies, such as SMS (cf. Kasesniemi, 2003; Thurlow, 2003). However, the present study found that language use in text messaging is to be regarded as a variant of language use, creatively and effectively suited to the conditions of SMS and the aims for which it is used. Language use in mobile text messaging was found to be adapted to the constraints of production and perception conditions owing to the technical constraints and affordances of the medium, in addition to situational parameters. The study revealed a number of interesting strategies to adapt language use to the conditions of mobile text messaging.

As we have seen, there are both technical and pragmatic factors behind language use in SMS. Mobile text messaging is an asynchronous mode of communication, and messages were mostly created either with multi-tap technique or predictive text entry on the tiny keypad of a mobile phone; entry takes physical effort and is time consuming. There is also a limit to the size of each message, and the text has to be composed with this in mind. The full message size was seldom used to its full advantage. The mean message length was 14.8 words, or 64 characters, well below the upper limit of 160 characters. The users also seemed to employ various devices to color and tone their messages. Emoticons and asterisks were used to convey meaning in a creative way, save time, space and effort and also to help disambiguate text.

Different types of syntactical and lexical reductions were found; in many cases reasons were probably to save time, effort and space. Syntactical reductions by omitting subject pronoun or even whole verb phrases were common. Creative lexical reductions in the form of unconventional and not yet established abbreviations were frequently used. Many of these usages were found to be based on Swedish words, but some were also transferred from norms in other modes of computer-mediated communication, e.g. web chat, and most of the time based on English words. Results from the other studies (e-mail, web chat, instant messaging) in the researcher's larger investigation of computer-mediated communication (Hård af Segerstad, 2002) all showed the use of unconventional abbreviations.

Language use in SMS showed many features associated with spoken language, omission of subject pronoun, verbalization of features of "communication management" (cf. Allwood, 2000) such as hesitation sounds and laughter. Spelling reminiscent of spoken interaction served to save time, effort and space, and to render an informal touch and serve as in-group markers.

Often, messages seem to serve several purposes at once, and economizing is certainly not the only driving force in influencing language usage. Although saving effort in some cases, users seemed also to be expending effort to communicate with a special set of phrases, words and constructions in other cases. The balance of saving effort and writing concisely versus composing a special message for their interlocutor is one that probably has to do with the activity in which they are involved i.e. the context around the message itself: the relationship between users, their shared background knowledge what they want to achieve with the message, etc. As we have seen, most messages are sent between friends who already know each other well. This relationship between users, and the accompanying, allows them to be brief, inexplicit, use community-specific slang and abbreviations – or to labor carefully over the text in a flirtatious message, if the situation and wishes so demand.

Predictive input technologies have the capacity to reduce significantly the effort required to enter text – if the prediction is good. As MacKenzie and Soukoreff (2002) point out, there are a few caveats to consider in basing a predictive language model on a standard (traditional) written corpus. The corpus may not be representative of the particular SMS user's language; the corpus does not reflect user's editing process when composing messages and the corpus does not reflect differences in input modalities. The problems of relying on predictive models are illustrated with the case of Swedish compounds. In contemporary spoken Swedish, there seems to be a tendency to split Swedish compound words. Theories as to the reasons for this vary. (One theory implicates influence from norms of written English.) The difficulty that predictive text entry software has in accepting or establishing elements in the compound makes it easier to split rather than to enter the word manually in the memory of the phone. Whether all split compounds are due to software mispredictions or the senders' own tendency to split compounds is not always evident.

At the Department of Linguistics at Göteborg University, a study of whether, and if so how, the diligent use of writing in new media is a threat to the standard written language among school children in Sweden is in progress. The project is called "Learning to Write in the Information Society". Further research more specifically focused on text messaging that would be interesting to investigate would be a careful comparison of text messages created on the keypad of mobile phones versus messages created with full keyboards on web-based SMS services. Another study would be to compare language use in IM and SMS. All of these studies would make very interesting comparisons across languages.

21.7 References

Allwood, J. (2000) An activity based approach to pragmatics. In Bunt, H. and Black, B. (eds.), *Abduction, Belief and Context in Dialogue: Studies in Computational Pragmatics*. John Benjamins, Amsterdam, pp. 47–80.

Androutsopoulos, J. and Schmidt, G. (2001) SMS-Kommunikation: etnografische Gattungsanalyse am Beispeil einer Kleingruppe. *Zeitschrift für Angewandte Linguistik*.

Clark, H. and Wasow, T. (1998) Repeating words in spontaneous speech. *Cognitive Psychology*, **37**, 201–242.

Döring, N. (2002) "Kurzm. wird gesendet" – Abkürzungen und Akronyme in der SMS-Kommunikation. Muttersprache. *Vierteljahresschrift für Deutsche Sprache 2*.

Eldridge, M. and Grinter, R. (2001) Studying text messaging in teenagers. Presented at the CHI 2001 Workshop No. 1: Mobile Communications: Understanding User, Adoption and Design, Philadelphia, PA.

Grinter, R.E. and Eldridge, M.A. (2001) y do tngrs luv 2 txt msg. Presented at ECSCW2001, Bonn.

Grönqvist, L. (2000) *The TraSA v0.8 User's Manual. A User Friendly Graphical Tool for Automatic Transcription Statistics*. Department of Linguistics Göteborg University, Göteborg.

Hård af Segerstad, Y. (2002) *Use and Adaptation of Written Language to the Conditions of Computer-mediated Communication*. Department of Linguistics, Göteborg University, Göteborg.

Kasesniemi, E.-L. (2003) *Mobile Messages. Young People and a New Communication Culture*. Tampere University Press, Tampere.

Kasesniemi, E.-L. and Rautiainen, P. (2002) Mobile culture of children and teenagers in Finland. In Katz, J.E. and Aakus, M. (eds), *Perceptual Contact – Mobile Communication, Private Talk, Public Performance*. Cambridge University Press, Cambridge, pp. 170–192.

MacKenzie, I.S. and Soukoreff, R.W. (2002) Text entry for mobile computing: models and methods, theory and practice. *Human–Computer Interaction*, **17**, 147–198.

Sperber, D. and Wilson, D. (1986) *Relevance: Communication and Cognition*. Blackwell, Oxford.

Schlobinski, P., Fortmann, N., Gross, O., Hogg, F., Horstmann, F. and Theel, R. (2001) *Simsen. Eine Pilotstudie zu sprachlichen undkommunikativen Aspekten in der SMS-Kommunikation*. Networx Nr. 22, Hannover.

Telia/Temo (2002) *SMS-undersökning* (a survey of SMS use, made by Temo by order of Telia Mobile, Sweden).

Thurlow, C. (2003) Generation Txt? Exposing the sociolinguistics of young people's text messaging. *Discourse Analysis Online*, **1**(1).

Turkle, S. (1995) *Life on the Screen: Identity in the Age of Internet*. Simon and Schuster, New York.

The Sociolinguistics of SMS: An Analysis of SMS Use by a Random Sample of Norwegians

Rich Ling

22.1 Introduction and Method

Since the late 1990s, the use of the short messaging system (SMS, also known as "texting") available on mobile phones has seen phenomenal growth. Statistics show that on average there are more than 280 000 SMS messages sent every hour in Norway, which means more than 6.7 million per day – and this in a country with only 4 million inhabitants (Sandvin et al., 2002). Among teens, SMS is the preferred form of mediated interaction, surpassing instant messaging, e-mail, voice mobile telephony and even traditional fixed-line telephone calls. SMS messages have several characteristics that make it useful for teens and increasingly for other groups. First, they are relatively cheap and they are personal, conveying a message directly from one person to another. Since mobile telephones are now ubiquitous among Norwegian teens, one knows that if one sends an SMS to a certain telephone number then it will come to that person, and not to another individual. SMS is asynchronous, meaning it does not necessarily require the immediate attention of the receiver. In addition, it is relatively unobtrusive. If, for example, a person turns off the ringing sounds on his or her mobile telephone, nobody is the wiser that a teen is sending and receiving communications. Indeed, research shows that teens send and receive SMS messages in class at school and through the night. Around 20% of teens say that they send and receive SMS messages after midnight on a weekly basis (Ling, 2004)

In some ways SMS is an odd duck. It is difficult to write messages on mobile phones since there is no traditional keyboard or writing instrument. Moreover, message length is limited to only 160 characters, the displays for reading the messages are small and transmission relies on terminals limited by poor batteries. When SMS was originally designed, the system was not even primarily intended as a form for personal interaction. Rather, it was created as a way to alert users to voice mail messages and perhaps as a system to broadcast weather or stock information.

Nonetheless, SMS is now a vibrant medium. Text messages are used to coordinate everyday events, to maintain social networks and to help entertain oneself in the open moments of one's day. In the words of one 17-year-old boy, "Often when you are sitting on the bus and subway it is boring and so you can write messages and that entertains you in those boring moments." To be sure, the culture of SMS is centered among teens and in particular among female users. In spite of the fact that adults males were early adopters of mobile telephones (Ling, 2000), it is among females that one currently finds the great motor of SMS culture. In the words of a focus group informant, "Most of the messages I get from boys are pretty short because they don't think it is so fun to sit there and punch in on the phone. That is more of a girl thing" (Erin, 17). In contrast, female teens write longer more complex messages. They include aspects of standard written language such as capitalization and punctuation. Moreover, they are more likely to include emotional elements in their communications (such as emoticons and items such as "xxx"), and they are more inclined to include in their SMS messages such refined formalities of traditional written letters as salutations and closings.

A number of studies are beginning to appear that examine the linguistic properties of e-mail and other computer mediated communication (Werry, 1996; Baron, 1998, 2000, 2003; see also Herring, 1996). A central question in this context is the nature of the communication and the effect of the medium on the formation of the language. Is e-mail more like speech in that it is spontaneous or does it have the more rehearsed and contemplative qualities of writing? With a few exceptions (Doering, 2002; Hård af Segerstaad, 2003a), SMS messages have not received the same linguistic attention. It is the intention of this chapter to explore the linguistic aspects of this form of communication, as practiced in Norway. The chapter will also consider the ways in which the urge to communication drives users to overcome the technological limitations of the system.

In this chapter I will examine a corpus of SMS messages gathered from a random sample of 2003 Norwegians. The data were collected in May of 2002 by telephone. Along with demographic, behavioral and attitudinal questions associated with voice mobile and SMS use, we asked the respondents to read (and, where necessary, to spell out) the content of the last three messages they had sent. This technique resulted in a body of 867 SMS messages, drawn from 463 (23%) of the 2002 respondents.[1] The

---

[1] The data shows that 64.2% of the sample reported sending an SMS at least once per week. Hence, a significant number of SMS users did not provide any messages. This is often because they had not stored the messages on their terminals or were not willing to share the content. There are no statistically significant age- or gender-based differences between those who did or did not provide messages.

sample population was 50% male and 50% female; males reported sending 388 (45%) of the SMS messages in the analysis and females sent 479 (55%).

These messages were categorized (and the categorization was checked by a second judge). In addition, the use of various forms of speech, the breadth of vocabulary, word and message length, use of abbreviations, capitalization, punctuation and the formality of the writing were examined. Finally, the data collection allowed for an examination of the material by socio-demographic variables such as age and gender.

This data-gathering approach entailed a number of limitations. First, it was only possible to gather messages that the survey participants had sent themselves, not messages they had received. There are both ethical and methodological reasons why we asked for the last messages sent as opposed to those received. Ethically, it is not possible for the researcher to ask for messages a respondent has received, since to do so would be to include data from people who have not given their consent to participate in the study. Moreover, there is the methodological problem that the researcher would not know the demographic characteristics (age, gender, background, etc.) of such senders. Without such information, it would not be possible to analyse the material in a meaningful way.

Second, the data-gathering approach was potentially problematic in that it necessitated that the messages had to be read by the respondent to the interviewer and transcribed into the database. There may have been selective filtering of content since the respondents may not have wished to read particularly revealing or piquant messages to the interviewer. Moreover, as noted, interviewees do not always save the messages they send, making it impossible for the interviewees to contribute SMSs to the database. In addition, given the possibility of using both intended and unintended abbreviations and misspellings (Hård af Segerstaad, 2003a) in the messages, it is plausible that the transcription process resulted in some errors. A limited point of control is offered via a study of 67 SMS texts generated by 82 teens from Grimstad, a small town in southern Norway (Ling and Sollund, 2002). In the Grimstad study, the respondents filled out a paper questionnaire as opposed to being interviewed on the telephone. That is, subjects were asked to transcribe their SMS messages themselves. Although it is difficult to quantify the differences given the smaller size of the Grimstad sample, the reader is left with the impression that this more direct form of data collection resulted in a slightly "rougher" corpus where profanity, racism and unguarded remarks are more in evidence.

A further weakness with the survey technique used in the present study is that the messages are often taken out of context from a sequence of messages sent to another person. It is difficult to estimate the degree to which this is an issue given the stricture against examining incoming messages. Obviously, this decontextualization can make interpretation difficult; it also precludes the possibility of doing discourse analysis.

Those are the weaknesses. The strength of the corpus it that it is far more representative of the general Norwegian population's use of SMS than are the more usual convenience samples. The results of this analysis provide a far better picture of Norwegians' use of the medium that do studies where informants are recruited on a non-random basis.

22.2 Who Uses SMS and How Often?

There are an enormous number of SMS messages being sent and received in Norway on a daily basis, and some groups turn out to be more prolific users than others. Data from the study shows that females (as a group)[2] and teens and young adults (again, as a group)[3] of are the most active users of SMS. When considering the frequency of use, more than 85% of teens and young adults report sending SMS messages on a daily basis.[4] By contrast, only 2.5% of those over 67 years old reported using SMS with this frequency. When considering gendered differences, the data show that whereas 36% of the males reported daily use, more than 40% of the females said that they send SMS messages on a daily basis. In the sections that follow, I will look into the themes of the messages, the choice and variety of words used, the length of the messages, the use of abbreviations and dialects and finally the use of capitalization, punctuation, openings and closings.

The data also indicates that the same groups were the most intense users. Females[5] and teens/young adults[6] report sending significantly more messages than their counterparts. As can be seen in Table 22.1, 16–19-year-old females who were SMS users – and only about 2% of this group does not use SMS – reported sending a mean of slightly more than nine messages per day. If one compares this with the rate for their mothers, chances are good that daily use will lag behind the use of the younger generation and, if they do use SMS, their usage rates are less than half that of their offspring.[7]

---

[2] χ^2 (4) = 15.85, sig. = 0.003.

[3] χ^2 (28) = 793.72, sig. <0.001.

[4] According to Hashimoto, this picture is similar to that in Japan. He reports that 96% of 20–24 year-old women are mobile telephone users. "Only" 81% of the men in the same age group report being mobile phone users (Hashimoto, 2002). A somewhat similar use patter is also found in the work of Roessler and Hoeflich in Germany (Roessler and Hoeflich, 2002).

[5] $f(1,1775)$ = 9.58, sig. = 0.001.

[6] $f(7,1769)$ = 72.89, sig. <0.001.

[7] Data from later studies indicates that middle-aged persons are starting to use SMS more often, but that they still lag behind teen users.

Table 22.1 Mean number of SMS messages sent per day by age group and gender for Norway, 2002

Age (years)	Males	Females
13–15	6.28	6.87
16–19*	5.29	9.03
20–24	5.97	6.32
25–34**	2.15	2.95
35–44	0.88	1.21
45–54	0.55	0.56
55–66*	0.17	0.52
>67	0.35	0.08

*$p < 0.10$; **$p < 0.05$.

22.3 Linguistic Analysis of SMS Texts

In order to help understand the nature of SMS, I looked into several characteristics of the messages. This included the themes in the messages, the most frequently used words, length of the messages, the use of abbreviations, capitalization, punctuation, salutations and closings.

22.3.1 Themes in the Messages

Beyond the sheer numbers of messages, there are reasons why all of these communications are being sent out. SMS has found a niche in our communication needs. It is used to coordinate, to send and receive endearments, to pose and answer questions, to make requests for information, to provide personal news, etc. In the words of Gro, an 18-year-old female SMS user, "I use it if I am just going to send a short message. For example if I am just going to ask if they are going to go out. It goes a lot faster." She continues by saying, "I send messages if I am planning something, if I am bored or if it is something that is important." Her reported use spans several of the categories found in the data.

The SMS messages were coded according to the apparent themes in the messages. This is admittedly somewhat slippery analysis. The messages were often only a single utterance in an ongoing dialogue. Given this caveat, about 75% of all messages fell into the categories of (1) coordination, (2) what I call "grooming" – those messages that did not contain any hint of planning, coordination or responses to questions but rather were in essence small "gifts" from one person to another (Johnsen, 2000), (3) answers and (4) questions. The remaining 25% of the messages included categories such as information, commands and requests, personal news and diverse other comments such as location requests. Interestingly, some types of answers were quite common. For example,

Table 22.2   Examples of various types of SMS messages

Theme	%	Examples
Coordination	33	is anything happening tonight? (M, 23)
		When does school start today? (F, 16)
Grooming	17	Good night and sleep with a picture of a sleeping bear (F, 16)
Answers	14	i don't have the car this evening (M, 25)
		yes, no, ok (2.3% of all messages)
Questions	11	have you colored your hair? (W, 15)
		Which operating system do you use? (M, 15)
		are you awake (M, 31)
Information	6	odd (a soccer team) is ahead 2 0 over of viking (M, 27)
		have installed a new software program (M, 44)
Commands or requests	6	Call me (1.5% of all messages)
		remember to buy cola (K, 19)
Personal news	5	i didn't pass the exam (W, 19)
		We are enjoying ourselves in the sun and good weather (M, 58)
Diverse other categories	9	Where are you? (1.4% of all messages)
		Damn you, Rune (M, 13)
		I am at home alone on Friday, things are going to happen then (W, 16)

the answers "yes", "no" and "ok" made up about 2.3% of all the messages in the corpus (22 messages). The message "call me" made up 1.5% and the message "where are you" made up 1.4% of all messages (12 and 11 messages, respectively). The major categories that arose from the data are shown in Table 22.2.

A socio-demographic analysis of the material shows that when it comes to using SMS messages to plan activities, males are more likely to use them for planning activities in the middle future[8] as are older teens and young adults. Females, however, are more likely to use SMS to make plans for the immediate future.[9] Females[10] and to a less significant degree teens and young adults[11] were more likely to send "grooming" SMS messages. Females were also more likely to send emotionally based "grooming" messages.[12]

---

[8] χ^2 (1) = 4.76, sig. = 0.029. Planning in the middle range future was defined as making agreements for activities that had not already started and were to take place within the next few days. These data are subcategories of the theme "coordination". The categories near, middle and long-range coordination are summed into the category "coordination" cited in Table 22.2.

[9] χ^2 (1) = 4.77, sig. = 0.029.

[10] χ^2 (1) = 8.77, sig. = 0.003.

[11] χ^2 (7) = 13.28, sig. = 0.066.

[12] χ^2 (1) = 9.634, sig. = 0.002. Emotionally based grooming messages were typically greetings that included declarations of love.

Table 22.3 Percentage distribution of the most often used words in the SMS corpus by gender

	Females			Males		
Rank	%	Norwegian	English	%	Norwegian	English
1	4.7	*du*	you[a]	4.8	*du*	you[a]
2	2.7	*på*	on/in/at/to	2.9	*i*	in/at
3	2.6	*i*	in/at	2.6	*på*	on/in/at/to
4	2.2	*er*	are	2.1	*er*	are
5	2.1	*det*	it	1.8	*skal*	shall
6	2.0	*jeg*	I	1.7	*vi*	we
7	1.7	*har*	have	1.7	*jeg*	I
8	1.5	*vi*	we	1.7	*til*	to
9	1.4	*til*	to	1.5	*det*	it
10	1.3	*kommer*	come	1.3	*kan*	can
11	1.2	*og*	and	1.3	*ja*	yes
12	1.2	*kan*	can	1.3	*når*	when
13	1.2	*skal*	shall	1.2	*har*	have
14	1.1	*ikke*	not	1.2	*kommer*	come
15	1.1	*med*	with	1.2	*og*	and
16	1.1	*når*	when	1.1	*ikke*	not
17	0.9	*meg*	me	1.0	*hjem*	home
18	0.9	*å*	to	0.9	*å*	to
19	0.9	*hjem*	home	0.8	*ok*	ok
20	0.8	*deg*	you[b]	0.8	*meg*	me

[a] Subject form.
[b] Object form.

22.3.2 Most Frequently Used Words

The SMS messages were analyzed in order to find the most frequently used words in the corpus, by gender. This analysis provides some further insight into the way messages are used and potentially insight into the gendering of messages in that some suggest that females are more prone to use pronouns in general and first person pronouns (Baron, to be published).

Du (you) is the most commonly used word for both males and females. The 20 most common words used for each gender are shown in Table 22.3. These words represent about 33% of all words used in the corpus for each gender.

The words *på* (on/in/at/to), *i* (in/at) and *er* (are) are among the most frequently use words for males and for females. In addition, although it is not obvious in the table, *jeg* (I), including its various alternative spellings,

---

[13] There are several alternative spellings of jeg that include *eg, ei, e, j, je* and *æ*. Several of these spellings are based on the dialect pronunciation of the words.

is in a somewhat weak second place.[13] In a similar way, the word *det* (it) becomes one of the top two to three words if its alternative spelling of "d" is included. In general one finds many prepositions *på* (on/in/at/to) and *i* (in/at) in the SMS messages and no adjectives or adverbs in the most often used words.

The analysis of word frequency in SMS also points to the role of SMS as a coordination tool. One uses SMS to coordinate and to make agreements with others. This necessarily involves a vocabulary describing where and when events will take place. This communication function may well account for the predominance of words such as *på* (on/in/at/to), *i* (in/at), *kommer* (come), *hjem* (home) and *du* (you) in the SMS sample.[14] There were adjectives and adverbs in the sample but they made up a smaller portion of the entire corpus. The material here does not support the idea that women are more likely to use first person pronouns more often than males. However, women do seem to use dialect spellings of "I" somewhat more frequently.

22.3.3 Word Length, Message Length and Message Complexity

Another dimension for examining the material is the mean length of the words and messages and the complexity of the message structure. One of the characteristics of SMS is that it is limited to 160 letters. This is sometimes seen as a limitation in that it does not allow for the development of extended communications. The evidence shows, however, that the average message used only about 20% of the available space (about 32 letters). There is a significant gender-based difference in the number of words per SMS message. When counting the number of words, females write longer SMS messages than males (5.54 words per message for males vs 6.95 for women).[15]

In addition, I considered the complexity of the messages. Complexity is defined here as the number of separate clauses or sentences in an SMS. The material was separated into two groups, simple and complex messages. Simple messages included only a single sentence or clause. An example of a simple message is one sent by a 15-year-old boy: "*Ka du gjer på?*" (What are you doing?). The message is short, direct and shorn of all unnecessary grammatical niceties and formal courtesies such as a salutation, closing, etc. This type of message made up about 66% of all the messages in the sample. By contrast, about one-third of the messages were more complex in their construction. An example of a complex SMS is "*Ja da, det går fremover. Har lest i hele dag, så jeg er ganske stolt. Jeg har*

---

[14] Words such as ring (call) and kl (the abbreviation of o'clock) are also among the more often used words though not among the top 20.

[15] $f(1,478) = 10.445$, sig. $= 0.001$.

følt at jeg har har konsentrasjons problemer. :).men det er min tur å span-dere..." (Yeah, things are progressing. I have studied all day, so I am pretty proud. I have felt that I have a problem concentrating. :). But it is my turn to buy ...). There is a lot happening here. The message tells about the writer's activities, a bit about her sense of herself and perhaps her openness for socializing. It is not a simple one-off message.

Examination of the material from this perspective showed that females write more complex messages than males. More than 74% of the messages sent by males were simply one-sentence or one-clause messages. Only 60% of the messages sent by females fell into this category.[16] Looking at partic-ular age groups, it seems that 16–19-year-old girls are particularly prolific at writing complex SMS messages (52% of all their messages are complex). At the same time, boys in this age group are particularly oriented toward simple "one-phrase" messages (85% are simple and only 15% are com-plex).[17] When looking across the various categories, one finds socio-demographically based differences. Males, for example, are slightly more prone to using short one-word answers in their SMS messages.[18]

22.3.4 Use of Abbreviations

A popular issue in the press is teen's use of abbreviations in e-mail and SMS. One can find directories listing acronyms, abbreviated spellings and emoticons.[19] This has also been a topic of academic interest (Werry; 1996, Hård af Segerstaad, 2003b). One clarification is that the limited space on the screen and the demand for speedy interaction result in pruning words and creating abbreviated forms of interaction. The material here points to the notion that it is teens, and in particular teen girls, who use these inno-vative forms of writing in the case of SMS. Although this group may feel a need to produce text quickly, it also seems that the use of these forms of interaction also contribute to a sense of the group. That is, the coining and use of the various forms of abbreviation are seen as ways of identify-ing group membership. Fine, for example, has observed this in the devel-opment and use of verbal repartee among pre-teen boys (Fine, 1987).

In spite of the prominence in the popular press, only about 6% of the SMS messages in this corpus had abbreviations, acronyms or emoticons.[20]

---

[16] χ^2 (1) = 9.87, sig. = 0.001.

[17] χ^2 (1) = 9.64, sig. = 0.001.

[18] χ^2 (1) = 3.35, sig. = 0.067.

[19] A simple search on Google, for example, turns up hundreds of examples of books on SMS abbreviations in English and dozens in Norwegian.

[20] In the case of the study carried out in Grimstad, the number of abbreviations was about 10% higher. This leads one to suspect that some of the abbreviations did not survive the data collection process. In terms of the demographic analysis presented here, it is hoped that the bias imposed by the transcription process was similar across all the messages.

In a study done in Germany, Doering (2002) points to the same general finding. In support of the popular stereotype, however, teens and young adults SMS seem to be biggest users of abbreviations.[21] The data also show that females use abbreviations and emoticons significantly more than males.[22] About 20% of 13–15-year-old females used abbreviations whereas only 3.5% of 35–44-year-old females used them. Females in the 35–44 age group and those who were 13–15 years old[23] used more abbreviations than like-aged males.

22.3.5 Capitalization and Punctuation

A part of the discussion surrounding SMS is that this form of communication plays into the development (or degradation) of the language. On the one hand, there are those who say that SMS degrades the language in that conventions are ignored and that writing in other media and in other situations suffers from the practices learned in the world of SMS. Others simply cheer the fact that people – and in particular teens – are taking the time to write. On the one hand, it is difficult really to know how one should interpret the material here. When thinking of formal writing, and in particular writing of longer pieces, the expectation is that capitalization and punctuation should be in place.[24] However, when considering a quickly dashed off note, a shopping list or a Post-it reminder, the rules seem to be looser. This is also the case with PC-based chat in various forms (Werry, 1996; Baron and Ling, 2003). Although the material here does not allow direct comparisons to writing in other contexts, an examination of capitalization and punctuation practices helps us to understand how different socio-demographic groups are using SMS.

The data here indicate that SMS messages from the younger users were more likely to have more prescriptively correct capitalization and punctuation than other groups. Thinking first about capitalization, there are three levels of use, i.e. SMS messages with (1) no capitalization, (2) first letter only capitalization[25] and (3) "multiple" capitalization wherein the writer manually capitalized names, proper nouns and at the beginning of secondary sentences, etc.

Examination of the corpus shows that 82% of the messages had no capitalization. It also shows that 11% had only "first letter" capitalization, often the default setting for mobile telephones. The final 7% had "multi-

---

[21] $\chi^2 (1) = 35.19$, sig. <0.001.

[22] $\chi^2 (1) = 9.30$, sig. $= 0.002$.

[23] $\chi^2 (1) = 4.17$, sig. $= 0.002$ and $\chi^2 (1) = 3.41$ sig. $= 0.06$.

[24] For a historical analysis of the development of both punctuation and capitalization, see Baron (2000, pp. 167–196).

[25] In some cases, the first letter of an SMS is, by default, capitalized.

ple" capitalization. This seems to place SMS into the same general realm as shopping lists, Post-it notes and the like. By a light but insignificant margin, more males than females used first letter capitalization (12.4% vs 9.9%).[26] The SMS messages written by females were significantly more likely to have complex capitalization (4.9% for males vs 8.5% for females).[27] Interestingly, it is young adults aged 20–24 years who are most likely to use capitalization in any form[28] and also most likely to use first letter capitalization.[29]

The use of punctuation gives us a sense of SMS vis-à-vis other forms of writing. When compared with writing with a pen and paper or on a full keyboard, punctuation in SMS is more difficult to produce. As with capitalization, punctuation varies in terms of the writing situation.

The analysis of punctuation compared those persons who used no punctuation in their message with those who used punctuation. The corpus was examined for the various types of punctuation used in each message (Table 22.4). The messages were coded in terms of those who used or did not use punctuation. The data shows that 57% (491) of the messages had some form of punctuation. The period was used in 32% of the messages and the question mark in 27% of the messages. These were the most frequently used forms of punctuation.

Young adults aged 20–24 years are also the most likely to use punctuation in their SMS messages.[30] Females use punctuation slightly more than males but the relationship does not appear to be significant. There is a sense in the use of punctuation that it can be a way for the individual to

Table 22.4 Use of punctuation in the SMS corpus for the 491 messages containing punctuation (these could be several punctuation marks in a single message)

Punctuation	No.	%
Quote	6	0.85
Colon	11	1.56
Hyphen	28	3.96
Decimal	27	3.82
Ellipsis	38	5.37
Exclamation point	42	5.94
Comma	43	6.08
Question mark	231	32.67
Period	281	39.75

[26] χ^2 (1) = 2.22, sig. = 0.136.
[27] χ^2 (1) = 7.35, sig. = 0.007.
[28] χ^2 (7) = 21.33, sig. = 0.003.
[29] χ^2 (7) = 14.99, sig. = 0.036.
[30] χ^2 (7) = 25.87, sig. < 0.001.

add extra-textual emphasis to the messages. The strategic use of ellipses to indicate a dramatic pause or perhaps intrigue "*Traff vidar på bussen ... skal vi hilse*" (Met vidar on the bus ... shall we say hi) (Female, 23). Exclamation points might indicate surprise or excitement: "*hei! eg ska var hima heile sommeren-ska t bergen neste veke! Må gjedna komma på besøk!!*" (hi! im at home all summer-going to bergen next week! [You] must certainly come and visit!!) (Female, 19), or several question marks may indicate advanced confusion: "*Når ska æ kom te dæ??*" (When shall I come to you??) (Female, 14). Interestingly, an analysis of texting in the Philippines shows the use of extremely stylistic punctuation (Ellwood-Clayton, 2003).

22.3.6 Openings and Closings

Another measure of SMS messages is the degree to which the writers included salutations and closings of the sort familiar from traditional letter writing. In coding the SMS data, we distinguished between what we called "simple" or "informal" openings and closings and more "advanced" or "formal" versions. Informal openings could be a casual hei (hi). In a few messages we found more formal openings including both a greeting and the name of the person being addressed. Informal closings included the sender's name or initial and perhaps a simple closing such as *Koz* (a stylized spelling of hug) or an emoticon. Formal closings used the formulation of, for example, *Hilsen Jens* (greetings Jens) following a period.

On the whole, there are relatively few messages that had either of these formulations. The data show that ~10% of the messages had either an opening or a closing. The most common were the simple forms with about 3.5% of the messages having a simple opening and 4.5% having a simple closing. The remaining 2% were distributed between messages with formal openings, formal closings and those that contained both an opening and a closing. Thus, when considering only the roughly 90 messages with these features, simple closings were most common. Amongst these, about half were the name or the initial of the sender and the other half were endearments, emoticons or both.

In general, one finds salutations and/or closings in the SMS messages written by females more often that those written by males.[31] Both males and females under age 19 years were also more likely to include these formulations in their messages than older users.[32] As with the other dimensions examined here, teens seem to use the most flourishes in their use of openings and closings.

---

[31] χ^2 (1) = 4.98, sig. = 0.025.
[32] χ^2 (7) = 17.48, sig. 0 0.014.

22.4 Conclusion

22.4.1 Written vs Spoken Language

SMS seems to be a trans-modal phenomenon with features of both spoken and written culture. Some aspects cause one to think that SMS is more like speaking than writing. We see, for example, that there is an immediacy to communications such as the first person present tense message sent by a 15-year-old girl: *"Eg kjeder meg"* (I am bored).

As with many uses of speech, SMS often makes the assumption of informality. Ironically, this lack of ceremony varies by age: female teenagers use salutations and closings more frequently than other SMS users, giving their messages a tone of being a formally constructed letter. At the same time, SMS messages are more short-lived than letters. There is a certain capacity to save (or even transcribe) SMS messages, but they are probably not a form of communication that will survive the generations as do letters (Krogh, 1990).

There is a lot of familiar interaction being played out in the messages. In this respect, SMS seems to be an extension of verbal interaction. The messages in this corpus often underscore the fact that the sender and receiver are carrying on an ongoing discourse with a common collection of familiar reference points such as in the in a sequence of three messages sent by a 35-year-old man: *(1) jeg er på veg hjem, (2) lag middag, (3) hva skal jeg handle inn?* [(1) I am coming home, (2) make dinner, (3) what should I buy?]. The reader can imagine the interlocutor sending the messages "Where are you?" "Shall I pick up the children or make dinner?" "Buy fish and 2 litres of milk." interlaced into the three messages in the corpus. In other words, we see the machinations of everyday life being worked out. Just as with a telephone conversation, the messages point to the user developing a strategy for dinner with his interlocutor.

At the same time, SMS messaging is like writing in that it does not assume that the interlocutors are physically proximate. Hence it is a message that goes across time and space, just as a letter, e-mail or telegram. Therefore, it cannot rely on intonation, proximics or forms of gesture. Rather, these elements need to be coded on to the text if they are included at all.

In addition to the characteristics that are more like writing or speaking, SMS seems to have characteristics that are ambiguous. For example, SMS is an asynchronous form of communication. I send a message with the assumption that the addressee will eventually read it and respond when he or she gets around to it. It is assumed that one cannot necessarily command the attention of one's counterpart in the same way that one does in spoken interaction. SMS, as with e-mail and traditional letter writing, is not like an active conversation where pauses in turn taking are interpreted as being impolite. This said, among teens SMS dialogues can take on the characteristics of a conversation with the development of topics, the inclusion of farewell sequences and the interpretation of pauses in turn taking. Indeed, teens in focus groups we have studied speak of how carefully they think through the timing of SMS responses and how they interpret others' timing (if you answer too quickly you are overeager).

There are also unique formulations in SMS that have a slight foundation in written and/or spoken language, but seem to be distinctive. One example is the word *Koz* or *Kozz* (hug). This is a stylized version of the Norwegian word *kos*. The traditional spelling (and pronunciation) is used in a collection of settings. However, the "z" spelling seems to be found only in SMS (and perhaps among graffiti users).

Finally, SMS is a spontaneous form of interaction. Like other types of writing, it can be edited before sending, something that is often done in order to increase clarity and avoid embarrassment. However, the fact that one can simply send a message from wherever to wherever literally at the drop of a hat means that ill-advised and embarrassingly formulated messages can be sent. Indeed, informants describe so called "drunken messages" where one perhaps says too much or is unguarded in the evaluations that are offered. Unlike speech, these messages do not simply disappear into the wind. Rather, they can be saved and can even be re-sent to others.

22.4.2 Sociolinguistic Analysis

At the broader social level, the results here indicate that teen and young adult females are more adroit users. Simply in terms of raw numbers, teen girls send more SMS than other groups. In addition, their messages are often more complex, and they use more advanced techniques in the production of the text (e.g. capitalization). In spite of the fact that teen males were early adopters of mobile telephones (Ling, 2000), females have since become the most active users. Females, and in particular younger females, seem to have a broader register when using SMS. They use them for immediate practical coordination issues and also for the more emotional side of mobile communication. As with other writing, females seem "to show more command of the standard" than males (Baron, to be published).

22.5 References

Baron, N. (1998) *Language and Communication*, 18, 133–170.
Baron, N. (2000) *Alphabet to Email: How Written English Evolved and Where it's Heading*, Routledge, London.
Baron, N. (2003) In Aitchison, J. and Lewis, D. (eds), *New Language Media*. Routledge, London, pp. 102–113.

Baron, N. (to be published) *Journal of Language and Social Psychology*.

Baron, N. and Ling, R. (2003) In *AOIRAOIR*, Toronto.

Döering, N. (2002) "Have you finished work yet? :)" Communicative functions of text messages. *Receiver* No. 6, Vol. 2002, Vodafone. http://www.receiver.vodafone.com/06/articles/inner05-2.html

Ellwood-Clayton, B. (2003) In Nyiri, K. (ed.), *Mobile Democracy: Essays on Society, Self and Politics*. Passagen Verlag, Vienna, pp. 35–45.

Fine, G.A. (1987) *With the Boys: Little League Baseball and Preadolescent Culture*. University of Chicago Press, Chicago.

Hård af Segerstaad, Y. (2003a) Department. of Linguistics, Göteborg University, Göteborg.

Hård af Segerstaad, Y. (2003b) In Ling, R. and Pedersen, P. (eds), *Front Stage/Back Stage: Mobile Communication and the Renegotiation of the Social Sphere*. Grimstad, Norway.

Hashimoto, Y. (2002) The spread of cellular phones and their influence on young people in Japan. In Kim, S.D. (ed.), *The Social and Cultural Impact/Meaning of Mobile Communication*. School of Communication, Hallym University, Chunchon, pp. 101–112.

Herring, S.E. (1996) *Computer-mediated Communication: Linguistic, Social and Cross-cultural Perspectives*. John Benjamins Publishing, Amsterdam.

Johnsen, T.E. (2000) *Department of Ethnology*, University of Oslo, Oslo.

Krogh, H. (1990) In *Sociology*, University of Colorado, Boulder, CO.

Ling, R. (2000) *Information technology and people*, 13, 102–120.

Ling, R. (2004) *The Mobile Connection: the Cell Phone's Impact on Society*. Morgan Kaufmann, San Francisco.

Ling, R. and Sollund, A. (2002) Final Report for Youngster Project, EU IST Program.

Roessler, P. and Hoeflich, J. (2002) Mobile written communication, or e-mail on your cellular phone. In Kim, S.D. (ed.), *The Social and Cultural Impact/Meaning of Mobile Communication*. School of Communication, Hallym University, Chunchon, pp. 133–157.

Sandvin, H.C., Dagfinrud, A. and Sæther, J.P. (2002) Norwegian Post and Telecommunications Authority, Oslo.

Werry, C.C. (1996) In Herring, S. (ed.), *Computer-mediated Communication: Linguistic, Social and Cross-cultural Perspectives*. John Benjamins, Amsterdam, pp. 47–63.

The Construction of Symbolic Values of the Mobile Phone in the Hong Kong Chinese Print Media

Vicki Yung

23.1 Introduction

The impact of the mobile telephony on everyday life is demonstrated in recent research (see, e.g., Brown et al., 2001; Plant, 2001; Katz and Aakhus, 2002; Ling, 2004). The social change brought about by a technology is not limited to the consequences of its technical functions. All artifacts acquire ideological meanings in society through time (Dant, 1999). These meanings may not be directly generated by the technical or material properties of a technology or a material object such as making calls and sending text messages with the mobile. Symbolic values of artifacts are symbolic meanings and psychological links, which are often constructed, shared and reinforced through social practice and discourse in society. Not all mobile phone users share the same symbolic values of a technology, to a certain extent, because they do not share the same "everyday life", and therefore social practices as well as everyday discourses such as ordinary conversations and popular media, which often shape our assumptions of the world. With popular media operating as a major social force in society, they may play an important role in generating and reflecting symbolic values for consumer technologies through various forms of representation.

This chapter explores how mobile phones are linked to symbolic values particularly through the use of language in popular Chinese print media in Hong Kong as the mainstream media may mirror and shape the ways in which users relate to this class of technology. The media texts on which this chapter is based come from a set of data collected between 2000 and 2001 for a research project studying mobile phones in Hong Kong and

Beijing (Yung, 2003). The main goal of the chapter is to identify key discourse strategies in the construction of symbolic values for mobile phones in the print media. I argue that the media construct certain symbolic values for local users using familiar discourse strategies including the language style, renaming practices and metaphors, with which the local people can identify themselves.

23.2 Everyday Discourses and Symbolic Values

Language is a powerful tool for constructing reality (Spender, 1984). Ordinary use of language in everyday life reflects social and cultural values (Gumperz, 1982; Schiffrin, Tannen, 1984; 1994); hence the discourse of mobile phones in everyday life reveals our sociocultural knowledge, assumptions, perceptions and attitudes toward technology. For instance, attention to the issue regarding the use of mobile phones as a discourse topic may provide clues to "the immediate concern" (Keehan and Schieffelin, 1976, p. 380) of the speaker/user, as users of mobile phones in different cities do not share the same concern about the topic. How mobile phone users relate to their phones varies from community to community. Based on a general observation, three different topics, ownership, brand models and needs, usually emerged in conversation about mobile phones in three different cities, Beijing, Hong Kong and Washington, DC (Yung, 2003). These observations indicate that divergent social and cultural values are associated with this class of technology in different communities. This chapter looks into the media representation in Hong Kong for an understanding of a broader sociocultural context, which may give rise to such values.

Material objects and technologies are multifaceted and their functions depend on the ways in which we use them at a particular moment. However, not all functions and consequences of technologies are represented in discourse. For example, of all the Chinese media texts I collected in Hong Kong, not one article has taken up the issue of mobile phone etiquettes – a topic I often found in the US media (Cohen, 1999; Frank, 1999). The media may play a role in upholding certain values while downplaying other functions or consequences that are directly generated in our interactions with technology.

According to Dant (1999, p. 2), material things are "never distinct from language or interaction; things are often the topic of talk or the focus of action and they often facilitate interaction or mediate by providing a means of interaction rather like language." In a study of mobile phones in China (Yung, 2003), I developed a taxonomy of functional categories to capture the multiplicity of mobile phone functions and consequences observed in Hong Kong and Beijing during 2000–2001. These categories are based on the three types of functions, technomic, sociotechnic and symbolic, developed in archeology and material culture (Binford, 1962; Deetz, 1973; Ames, 1984). During the period when fieldwork was conducted, a set of authentic conversations and media texts relevant to mobile phones were collected. The goal was to understand how everyday discourses of the mobile phone correspond to the actual physical interactions with technology.

To clarify the kind of social change that the mobile brings, it is important to identify the multiple functions of the technology to discern the consequential network including the material, social and psychological consequences to which they lead.

According to Deetz (1973), an object such as a candle can serve all three functions in different contexts. Its technomic function is the lighting effect, "serving the practical function of solving a problem directly imposed by the environment" (Deetz, 1973 p. 19). When it is used at a dinner party, a candle serves a sociotechnic function. The use of candles in the Catholic Church is an example of symbolic function (Deetz, 1973). Modifying the term, I define "symbolic" values as psychological attachments or symbolic values enacted with an artifact through time and interactions in social activities (Yung, 2003).

As Dant (1999, p. 53) suggests, material things sometimes acquire "fetish properties" in many different ways, including "the object may provide a surface for linguistic or quasi-linguistic texts to play across." Symbolic meanings and psychological links are often added to consumer products originally designed for their technical functions; for instance, watches and automobiles are often seen as a status symbol rather than for their technical functions. Unlike technomic values of an object, which are largely determined by its pragmatic functions, symbolic values are recognized and interpreted by social and cultural conventions. A national flag, for example, functioning mainly as an symbolic tool, materializes psychological and emotional links between a group of people and a country. The flag is a semiotic means to those who recognize the meanings when it is waved or raised for symbolic purposes at occasions such as the Olympic Games. The way in which a flag is handled in public is governed by law and the use of flags is socialized through institutional discourse (e.g. Code of Etiquette for flag use in the USA). The symbolic meanings of national flags are not constructed from the material qualities of the object. Anthropological studies of material culture often look into the role of objects in ritualistic practices as ways of denoting cultural values and beliefs.

23.3 Mobile Phones and the Print Media in Hong Kong

The density of mobile phone users in Hong Kong has been among the highest in the world since the 1990s. According to statistics available in

2003 from the International Telecommunication Union (ITU), there are 105 subscribers per 100 inhabitants in Hong Kong. The mobile phone craze in Hong Kong is evident when one enters the MTR subway, the train, shopping malls and restaurants. It is not simply about the frequency of phone use and the density of users in the city. One can see the importance of the mobile phone in other ways. People are alert users who usually answer calls after two to three rings. Users are sensitive about where the reception is not satisfactory. Hong Kong phone users willingly wait in a queue for hours and pay an extra cost for a new model. Young phone users change their phones frequently for their new style rather than new functions. Most phones are embellished with a variety of accessories such as stickers, straps, phone covers and bags.

Unlike other ordinary consumer products and technologies, mobile phones attract a great deal of interest in everyday discourses in Hong Kong. In addition to generic references for the mobile in Cantonese such as sau gei or sau tai, meaning "hand machine" or "hand carry", some common ways of referencing mobile phones in everyday talk are noted in the following examples:

> Where is your Nokia? (a brand).
> Look at my 8210. (Nokia 8210 – a brand model).
> They are the third generation. (3G – the third generation).
> He has a little V. (a nickname for the Motorola V6 series).

In this way, mobile phone brand models and features are regarded as common knowledge in the community. The strength of reception is a common topic when users enter a restaurant or a different reception zone. Special terms such as "blind spots" and "dead corners" are used to describe these zones. Using the mobile as a topic or a metaphor is common in everyday conversation and popular media in the city. For example, an energetic person can be described as a phone with a fully charged battery. During a radio phone-in program on RTHK2 (20 August 2001) about "things to do" to revitalize a romantic relationship, a young male host said to a female caller, "A steady girlfriend is like your mobile phone, you don't have to pay attention to it until it makes noise." The caller did not respond to the analogy and turned the focus on her relationship problem. Using the mobile phone as an analogy, the host illustrated how invisible a steady girlfriend is to a man. This metaphorical use in everyday talk reflects the habitual state of mobile phones as part of everyday life in Hong Kong.

The Chinese print media in the community reflect the intensity of the local mobile phone craze. There is almost daily coverage in most popular publications. The texts selected in the analysis come from the most popular publications with a wide range of readership from high school students to professionals in the Cantonese community. The selected texts

discussed in this chapter include three popular publications in Hong Kong:

- *Next Magazine*: a popular weekly infotainment magazine written in both Standard Chinese and colloquial Cantonese.
- *East Magazine*: an infotainment magazine, similar to *Next Magazine*, written in both Standard Chinese and colloquial Cantonese.
- *Apple Daily*: a popular daily newspaper, produced by the same publication group as *Next Magazine*.

Mobile phones appear in various sections in the print media, including headline news technology sections, business sections, entertainment news and advertisements. This study only looks at non-promotional articles from the technology section in the selected publications. Forms of representation for the mobile phone which construct symbolic values, thus not directly representing the technical aspects of the device, in these texts are elicited for the analysis. It is worth noting that the English translation of these texts may sound similar to advertisements for readers who are not familiar with Asian or Chinese infotainment. They are not advertisements or promotional texts in the sense that they are not directly sponsored by advertisers. Having said that, we should also bear in mind that mobile phone advertisers could be these publications' major clients.

23.4 The Language Issue

Instead of a journalistic style of writing, the technology reports selected for the study from the popular Chinese publications use a new style of writing by widely mixing the local vernacular – Cantonese in Standard Chinese, with an occasional mix of English and Japanese vocabulary items (Yung, 2005). The writing style, proven to be a selling point, closely resembles the spoken form of Cantonese with frequent use of local slang, utterance particles and emotional expressions. The choice of such language style in reporting mobile technologies, I suggest, is a discourse strategy to create a sense of familiarity for local readers.

The language issue in Hong Kong may contribute to a certain extent to an understanding of localization in the media. The sovereignty of Hong Kong was returned to China in 1997 as a Special Administrative Region after British colonial rule for 150 years. With a British colonial history, the government of Hong Kong uses both English and Chinese as official languages but English is regarded as the language of the elite and the professionals. The term "Chinese" may refer to both the local dialect 'Cantonese' and the national language "Putonghua" or Mandarin in China. Note that

Cantonese and Putonghua are two languages that are mutually unintelligible, similar to English and German.

Although Cantonese is the mother tongue of about 95% of the population in Hong Kong, it is not a prestigious language. English has long been accepted among the local people as the language of government, law and school, and Putonghua has gained in status in recent years as the symbol of another political power. Although the status of Cantonese has gradually been elevated especially in commercial and entertainment arenas along with the economic growth during the 1980s, it is not accepted as a standard language in the city. The tension among the three languages, English, Putonghua and Cantonese, was intensified with the political change in 1997. Cantonese is sometimes regarded as a tool for in-group solidarity among the local people, especially when the unprecedented political transition was a major concern in the community.

23.5 The Construction of the Mobile World

In this section, I will discuss how the print media construct symbolic values for the mobile phone through discourse strategies including the language style, naming practices and metaphorical constructions that are familiar to the local community.

23.5.1 The Mobile Identity

One main discourse feature which creates symbolic values for mobile phones in the media texts is the naming of products in the market. As Rymes (1999) states, personal names "are not simply arbitrary labels" and they are associated with "their function in different societies" which can be "powerful markers of social identity" (Rymes, 1999, p. 163). Van Leeuwen (1996) illustrates an array of ways to represent social actors in media texts by expanding the idea of newsmakers labelling and naming practices in media discourse.

Reference terms for everyday concrete objects do not seem to be a subject of potential controversy as they appear to be unarguable and less complex than abstract nouns for emotions (Goddard, 1998). However, linguistic research has shown that references and naming practices for material objects are complex as they relate to human categorization, functional contexts and cultural perception (Labov, 1973; Lakoff, 1986). Dant (1999, p. 146) states that "The serial object is one which is regarded as just another one, the same as the one next to it ... The model is in contrast an object that claims a value of distinctiveness and singularity even though it might be manufactured on a production line." Different ways of naming mobile phones individualize and add symbolic values to these products.

On the one hand, global forces such as international marketing strategies and branding efforts have tremendous impacts on creating images for consumer technologies (Yung, 2005). On the other hand, the local forces may produce values that are not cultivated by the marketers.

Each mobile phone generally embodies at least six layers of identity: generic, generation, retail status, national origin, brand and model. These features can be highlighted, inhibited or reconstructed in media texts through various ways of representation. In the selected media texts, mobile phones are often localized with Cantonese nicknames, which characterize particular features of each model for local users. With these features, mobile phones take on different social identities in these texts and carry sociocultural characters as they are localized through linguistic strategies. With these naming practices, these texts highlight layers of mobile identity and draw attention to particular symbolic values of the products including fashion sense, retail status, national identity and individuality.

23.5.1.1 Generation

Mobile phones are represented in terms of generation as a way to group mobile phones based on their level of technology and the period of time in the market. The following terms are often directly borrowed from the English official references:

- 1G   First generation: analog
- 2G   Second generation: GSM, PCS and CDMA (9 kbps)
    [2.5G   2.5 generation: GPRS (144 kbps)]
- 3G   3rd Generation: WCDM or CDMA 2000 (2 Mbps)

These three generations refer to various technological systems and speed of information transfer.

Generating symbolic values, the media texts in the data often use their local nicknames for big, bulky, first generation mobile phones introduced in the late 1980s. Large-sized phones, considered antique items today, are called dai go dai in Cantonese, meaning "bigger than big brother" literally, because they were expensive and only bigwigs or gang leaders (often referred as dai go "big brother" in Cantonese) could afford them in the late 1980s and early 1990s. They were also labelled sui woo, "water bottle", because of the size and shape of these phones. Sometimes, the two labels are combined as dai go dai sui woo, "big brother's water bottle". At the time, they were considered desirable objects to be carried around for showing off one's wealth. These nicknames mark this kind of ownership, which was particularly for display. The classic nickname dai go dai, "big brother", was later diffused to other Chinese-speaking communities, including those in Taiwan and China. It was also used as a form of ridicule as these bigwigs carried big bulky phones around the way in which little kids carry water bottles to school.

Dant (1999, p. 150) argues that "objects are 'of their time' in a number of ways, as products of an era, as artefacts that age and as objects whose value changes". He also states that, "Social relations with objects change in time in two ways. First, our response to the object changes with time. Second, the form and the meaning of objects change as society changes" (Dant, 1999, p. 131). As everyday technologies mark time, the social meanings of these nicknames also change. Today dai go dai sui woo, "big brother's water bottle", means outdated phones or old-fashioned, bulky phones that nobody wants to carry any more. These nicknames were once used to elevate the status of the technology and its users. Now they have become denigrating terms.

23.5.1.2 Retail Status

Electronic products are often evaluated based on their retail status in two terms, hong fo and sui fo, in relation to price and value in the texts I analyze. These terms, however, are not officially defined as it is regarded as basic knowledge of technological products for the local consumers. Electronic products in the Hong Kong retail market can be divided into hong fo/zeng fo (licensed merchandise) and sui fo (parallel imports or gray market copies). Licensed merchandise is imported through authorized distribution channels with original licenses and manuals, whereas parallel imports are imported through various private channels without distribution rights, and therefore usually do not include product warranties or manuals from the manufacturer. However, they can be legally sold and owned in Hong Kong (note that parallel imports are not the same as pirated or fake products).

Licensed copies versus parallel imports of mobile phones are always in competition. The value of sui fo (parallel imports) to consumers is that they are often cheaper, come out in the market earlier and have different features from the hong fo (licensed stock). The following text translated from *Next Magazine* is a typical example, which shows how the media represent products by highlighting their retail status:

Samsung High-end Gold and Silver Versions

(Since) Super expensive "dual-face king" Samsung A288 came out, hong fo (licensed stock) has been checked out but not sold. (If you) wanna buy, of course (you) should pick sui fo (parallel imports), which are half the price and currently in stock (about HK$3280). Recently, hong fo (licensed stock) still 'plays no. 1 brother' [slang meaning "stuck up"]. Two new colors came out. They are gold and silver. And the price has been lowered a bit to HK$6380.

Tips: The look (of the model) is pretty but the price is still too high.

[Source: Samsung high-end gold and silver versions, nextdigital,
Next Magazine, 24 May 2001, p. 24.]

These reports often provide information and recommendations based on retail status and price. Hence the economic values added on these products have little relevance of their technical properties.

The status of the mobile in terms of hong fo and sui fo shows the relationship between retail status and market values of these products. The labels also symbolically manifest the hierarchical grades of legal residency in human society, such as by birth, by immigration and other means. These grades are marked on identification cards of Hong Kong residents, and they designate the kind of social status and benefits individuals receive.

23.5.1.3 National Origin

Mobile phones are often characterized with their national origin in the print media. In the mobile world, phones metaphorically carry their "national identity", usually based on the country, where the headquarters of the manufacturer are located or where they are manufactured. For example, Telital GM220E is labelled as fat zik sau gei, "French National". The word "zik", referring to "national" or "regional" origin, is a formal usage for immigration forms or other legal documents for one's birthplace or national identity. Another example is Trium Astral, identified as yat yue fat zik, "Japanese born French national", which indicates dual national identities. The character 'yue' denotes a person's cultural or racial heritage, for example, Chinese, Japanese or French, whereas "zik" refers to an individual's political national identity or birthplace.

Identifying products with national origin which carry symbolic meanings, the media attach a layer of national identity to mobile phones. The values or meanings constructed for the technology through these identities are not linked to the technical functions but they may provoke perceptions of particular countries prevailed in the local community.

23.5.1.4 Brands

In the data, brands are divided into mainstream brands and alternatives. Mainstream brands include Nokia, Motorola and Ericsson, while all other brands are considered alternative brands; for example, Sony, Samsung, Siemens and Panasonic are a few popular alternative brands.

Mobile brand names are often translated in to Standard Chinese in promotional discourse but they are not always used in the media. Even when a brand name has an official Chinese name, the Chinese media in Hong Kong still use the name in both Chinese and English interchangeably, especially with the three popular brands. The official Cantonese names of the three mainstream brands are lok kei ah for Nokia, mo tok lo lai for Motorola and oi lap shun for Ericsson.

Localization achieved through naming practices is prevalent in Hong Kong. Although international brand names are translated into Standard Chinese or Cantonese names, they are always localized as neighborhood family stores in the following form:

First syllable of the name + gei (Store)

The word gei (store) is taken from a common usage for traditional family stores in small communities or villages, for example, ling gei, in which ling is a person or a family name, ling, and gei means the store of ling. For example, Motorola is usually referred as mo gei, Nokia as lok gei and Ericsson as oi gei. Keeping the first syllable of the name mo and adding the Cantonese word gei for 'store', this linguistic strategy in many cases completely erases the foreignness of the name. This form of localization is also frequently used in the print media for other franchised corporations such as McDonalds, which is called mak gei in Cantonese. The strategy not only tailors to the preference for an informal and colloquial style of language in the local popular print media but also downplays the foreign identity and cultivates a local identity for foreign brand names.

23.5.1.5 Brand Models

Brand models are also localized through renaming practices as official brand models are labelled with local nicknames based on size, shape and special features of the phone. These nicknames are borrowed from personal nicknames zai for boys and wong (king) or seung min (dual faces) from comic or entertainment characters. These brand models are also designated as entertainment stars when they are ranked in terms of popularity. Some examples of nicknames that are based on size and shape are given in Table 23.1:

Motorola V6 (and the V6 series) has a nickname V zai, meaning "Little V" or "V boy", because it is the first model that is of palm size. This nickname is associated mainly with its size. The 1997 model from Nokia (8110), with a half-moon shape, acquires the nickname ziu zai, meaning "little banana", because of its shape. Table 23.2 gives examples of nicknames that characterize special features of phone models.

Here, Samsung A288 is called "dual-face hand phone king" because it has two screens, one on the front cover and the other inside, whereas Sagem's nickname "Prince of radio broadcast" refers to the model's superb sound quality.

Table 23.1 Nicknames based on size and shape

Brand model	Cantonese nickname	English translation
Motorola V66	V zai	V boy or little V
Motorola Startac	gwai	Turtle
Nokia 8110	ziu zai	Little banana
Nokia 8850	so pau	Shaver
Trium Mars	fo sing zai	Mars boy or little mars

Table 23.2 Nicknames that characterize special features

Brand model	Cantonese Nickname	English Translation
Samsung A288	seung min wong sau din dan ngan zai	Dual-face hand phone King Single-eye boy
Samsung A399	sau sun seung min A399	Slim body dual-face A399
Sagem	bo yum wong zi	Prince of radio broadcast

It is a cultural practice for Cantonese speakers to use nicknames for close friends, family members and co-workers in everyday social activities. In public discourse, the media also use nicknames for politicians and entertainers. Blending them with brand models certainly attaches socio-cultural characteristics to mobile phones. Linking these local practices to mobile phones, these print media recreate layers of social identities for the products.

23.5.2 Social Relationship

Another strategy that the media uses to construct a world of mobile phones resembling a human society is to create social relationships for brand models. Mobile phones are constructed as romantic couples, siblings and schoolfellows using labels and kinship terms.

The label ching lui gei (couple phones) is commonly used in the texts to refer to identical phone models owned by couples. This usage is also applied for a phone model – DNET688 – to construct a sense of romantic relationship for a pair of identical phones with two different drawings on their covers designed by a Taiwanese artist [Source: Taiwanese hand phones carry their special features. Everyday Technology, *Apple Daily*, 25 April 2001]. The label of this set of phones imposes an interpretation of the drawings of a man and a woman on the corresponding covers. Hence consumers who own these products would inherit such added symbolic value.

Family relationships are also constructed in the mobile world with the use of kinship terms. Kinship relationships are often created for a series of phone models from the same brand. For instance, the term T hai hing dai, meaning "T brothers", is used to refer to the three Ericsson models T65, T66 and T68 [Source: Nokia looks alike, Ericsson T65. Technology Platform, *East Magazine*, 17 October 2001, p. 111]. Another example comes from an article about two of Samsung's models, R200 and R220 [Source: CommunicAsia 2001: 2.5G goes on, bluetooth disappears. nextdigital, *Next Magazine*, 28 June 2001, p. 11]. Since their functions and their bodies are almost identical, they are labelled sun lung fung toi ("the new dragon–phoenix twins"). The traditional Chinese phrase lung fung ("dragon–phoenix") is used to symbolize a "male" and "female" pair/couple/twins. The only difference between the two models is that the R220 has a cover for the keypad.

Mobile phones from the same brand are also referred as tong mon ("school fellows"). This term is usually used for two phones that share similar features but produced at different times – that is, they belong to different generations of the same model. For example, this term is used in a headline of an article for Sony models, saying tong mon leung doi peng ("school fellows in competition"). These "school fellows" refer to the two Sony models Z28 and Z18 [Source: Sony Dial King upgrades. nextdigital, *Next Magazine*, 26 July 2001, p.106].

Constructing social relationships for brand models, the media metaphorically create a social world for mobile phones resembling the local community, in which kinship terms are widely used among friends and colleagues as a way to cultivate in-group solidarity. These social relationships also give rise to an imagined social world in which mobile phones take up their membership in the imagined community.

23.5.3 The Mobile in Action

Mobile phones in this representational world have their physical bodies. They are referred as sun or tai, borrowing the same Chinese characters for "human body", for example, bok **sun** WAP gei (a thin body WAP phone) and PDA sau gei wen hap **tai** (a mixed body of PDA and hand phone). In this way, in the world of mobile phones, they are also represented as social actors, and thus they take actions similar to human beings. This kind of representation is constructed through linguistic and visual strategies. In terms of linguistic strategies, mobile phones take action verbs as they compete in the market, fight for popularity and battle in the mobile world.

The arrival of a new model can be represented as entertainment stars appearing on the stage, or fighters attacking the market. For example, an article from *Apple Daily* presents the arrival of a new model of Samsung using three action verbs for the agent, Samsung SGH-N300: dunk cheung (**rise** on stage), chut yin (**appear**) and sheung cheung (**ascend** the stage) [Source: Changing face of Samsung SGH N300. Everyday Technology, *Apple Daily*, 20 August 2001]. Two of these verbs, "rise" (on stage) and "ascend" (the stage), shape the arrival of this new model as an elegant entertainment star making an appearance on stage. The next example demonstrates the use of a different metaphor for the arrival of a new type of phones, PDA phones. The headline of the article is "PDA Phone go tsok zap gong" (PDA phones attack Hong Kong at high speed) [Source: PDA phones attack Hong Kong at high speed. nextdigital, *Next Magazine*, 26 July 2001, p. 102]. The storm metaphor is used to illustrate the impact of these new products in the market through the use of both linguistic and visual strategies. Linguistically, the headline says these PDA phones attack Hong Kong at high speed. Visually, the phones are presented with an icon for storms used in weather reports.

In the mobile phone world, brand models constantly fight for popularity and they are represented as entertainers or fighters. In a headline of an article from *East Magazine*, "Samsung wins all Chinese hearts", Samsung, the brand name, is represented as an agent, which "wins" all Chinese hearts like an entertainment star [Source: Samsung wins all Chinese hearts. E Frontline, *East Magazine*, 17 October 2001, p. 108]. In another headline of a report from *Next Magazine*, "Siemens' new phones snatch the spotlight", the new brand models are constructed as actors aggressively "snatch" the spotlight [Source: Siemens' new phones snatch the spotlight. nextdigital, *Next Magazine*, 2 August 2001, p. 108]. With the same verb "snatch", brand models are depicted as aggressive competitors in the headline of an article from *Next Magazine*, "Ericsson's new phones snatch customers" [Source: Ericsson's new phones snatch guests/customers. nextdigital, *Next Magazine*, 2 August 2001, p. 108]. In a competitive mobile world, phone models are even fighters in a battle with other phones when they are represented as agents in the texts. For example, a headline from *East Magazine* says, "Silver version remote control king Sony J70: Low price attacks". Labelling the Sony J70 with a nickname "Silver version remote control king", this headline constructs the model as a social actor using low price as a weapon to attack [Source: Silver version remote control king Sony J70. *East Magazine*, 24 October 2001, p. 110].

Another report taken from *Next Magazine* illustrates a competition between GSM phones [Source: colour screen dual band phones arrive. nextdigital, *Next Magazine*, 25 October 2001, p. 134]. The translation of the report is given below:

GSM hand phones finally farewell black and white,
(and) join the generation of color screen.
Because the first GSM hand phone with 256 color screen display –
OKWAP i108 has already fought its way to Hong Kong.

363

> Through a rocky path, the first one is not Ericsson T68 as it has been said in the rumor.
> But it turns out that the Taiwanese made *ying zi dat* OKWAP i108 gets out of the race first.

In this text, GSM phones are positioned as actors again in the Hong Kong retail market. They have to say farewell to black-and-white screen display because the first GSM phone with a colored screen display has arrived in the market. The proverb fung wui lou zhun (through a rocky path), referring to a sudden change in a battle, functions as a prelude to the anticipation in the market that Ericsson T68 would be the first one to come out with a colored screen as it has been said in the rumour. But it is a Taiwanese product that gets the first place. The last phrase, cheung zap pau zhut, "get out of the race first", is often used in horse racing. It refers to the one who gets the first place coming out of the gate at the start. Here, it is used metaphorically with phones with new features appearing in the market first.

Granting agency to mobile phones in the texts, the print media construct new identities and memberships resembling human society for these technological products for their readers and consumers. In this way, mobile phones are constructed as entertainment stars or comic characters in the local context. The use of metaphors by the print media reinforces the imagined social world they create for the technology.

23.6 Conclusion

This chapter identifies ways with which symbolic values of mobile phones are constructed in the print media in Hong Kong. These are symbolic meanings and psychological attachments that are not direct consequences of the technical use of the phone. These values linking to the social and cultural environment in which the technology is used are often constructed and shared through everyday discourses in society. The analysis in the chapter demonstrate how the local Chinese print media use a particular language style and linguistic strategies to reconstitute a social world in which mobile phones acquire social identities and social relationships resembling the local community. These symbolic values constructed in the press seem to overtake the technical values from the technology.

Mixing the local vernacular in Standard Chinese as a discourse strategy, the print media localize this class of products and create a sense of familiarity for local users. The linguistic strategies including the naming practices and metaphors reconstitute social images and placements for these inanimate objects in an imagined social world as camaraderie for the local community. Linking social identities for mobile phone characteristics, the media texts construct a social reality for the technology, in which phone models are also associated with each other similarly to social relationships in human society. In this representational world, mobile phones are ever changing with fierce competition. These products take up their own social identities, social relationships and even actions for competition in their own social world. This type of representation cultivates sociocultural characteristics for individual phones. These characterizations carrying symbolic meanings for display, ownership and membership may supersede their technological values in discourse. They also create psychological attachments for the local users as local images of mobile phones are created.

This chapter raises two issues regarding the process of commodification (Silverstone and Haddon, 1996): (1) the role of the media, especially those not directly sponsored by advertisers or marketers, in generating and reflecting social meanings of the mobile technology in a community and (2) the social meanings generated by the media in relation to users' sociocultural knowledge and perceptions of technology. This analysis shows that the Chinese print media in Hong Kong play a role in constructing local images linking the features of technology and the local social systems for the local consumers in the community.

As mentioned earlier, the mobile phone is a frequent topic in everyday discourses in Hong Kong. They are often referred by their brand models and nicknames in daily conversation and serve as metaphors, for example, using the mobile phone as an analogy for a steady girlfriend and the battery as human energy. The symbolic values created with various discourse strategies, particularly mixing the spoken form of Cantonese in the print media, seem to correspond to the way in which mobile phones are represented in everyday talk in the community. What I suggest here is that the media may have contributed to such an environment for local mobile users.

23.7 References

Ames, K.L. (1984) Material culture as non-verbal communication: a historical case study. In Mayo, E. (ed.) *American Material Culture: the Shape of Things Around Us*. Bowling Green State University Popular Press, Bowling Green, OH, pp. 25–47.

Binford, L.R. (1962) Archaeology as anthropology. *American Antiquity*, 28: 217–225.

Brown, B., Green, N. and Harper, R. (eds) (2001) *Wireless World: Social and Interactional Aspects of the Mobile Age*. Springer, London.

Cohen, S. (1999) Sound and fury – portable phones in public places. *The Washington Post*, 23 April, p. C4

Dant, T. (1999) *Material Culture in the Social World: Values, Activities, Lifestyles*. Open University Press, Milton Keynes.

Deetz, J.J.F. (1973) Ceramics from Plymouth, 1620–1835: the archaeological evidence. In Quimby, I.M.G. (ed.), *Ceramics in America*. University Press of Virginia , Charlottesville, VA, pp. 19–20.

Frank, S. (1999) Cell phone users: are they clueless or just plain rude? *Chicago Tribune*, 6 May.

Goddard, C. (1998) *Semantic Analysis: a Practical Introduction.* Oxford University Press, New York.

Gumperz, J.J. (1982) *Discourse Strategies.* Cambridge University Press, Cambridge.

Katz, J.E. and Aakhus, M. (eds) (2002) *Perceptual Contact: Mobile Communication, Private Talk, and Public Performance.* Cambridge University Press, Cambridge.

Keehan, E.O. and Schieffelin, B. (1976) Topic as a discourse notion. In Li, C.N. (ed.), *Subject and Topic.* Academic Press, New York.

Labov, W. (1973) The boundaries of words and their meanings. In Bailey, C.N., Shuy, R.W. (eds), *New Ways of Analyzing Variation in English.* Georgetown University Press, Washington, DC, pp. 340–373.

Lakoff, G. (1986) *Women, Fire, and Dangerous Things: What Categories Reveal About the Mind.* Chicago University Press, Chicago.

Ling, R. (2004) *The Mobile Connection: the Cell Phone's Impact on Society.* Morgan Kaufmann, San Francisco.

Plant, S. (2001) *On the Mobile: the Effects of Mobile Telephones on Social and Individual Life.* Motorola.

Rymes, B. (1999) Names. *Journal of Linguistic Anthropology,* **9**(1–2), 163–166.

Schiffrin, D. (1994) *Approaches to Discourse.* Blackwell, Oxford.

Silverstone, R. and Hadden, L. (1996) Design and the domestication of information and communication technologies: technical change and everyday life. In Robin, M. and Silverstone, R. (eds), *Communication by Design: the Politics of Information and Communication Technologies,* Oxford University Press, New York, pp. 44–74.

Spender, D. (1984) Defining reality: a powerful tool. In Kramarae, C., Schulz, M. and O'Barr, W.M. (eds), *Language and Power,* Sage, London, pp. 194–205.

Tannen, D. (1984) *Conversational Style: Analyzing Talk Among Friends.* Norwood.

Van Leeuwen, T. (1996) The representation of social actors. In Caldas-Coulthard, C.R. and Coulthard, M. (eds), *Texts and Practices: Readings in Critical Discourse Analysis.* Routledge, London, pp. 32–70.

Yung, V. (2003) Mobile phones in interaction: discourse analysis and the material world. PhD Dissertation, Department of Linguistics, Georgetown University, Washington, DC.

Yung, V. (2005) Brand names, nicknames and numbers: reconstructing layers of identity for mobile phones in Hong Kong. In Krause, K.L. and Gibbs, D. (eds), *Cyberlines: Languages and Cultures 2.* James Nicholas Publishers.

Part 5

Innovative Adoption and Commercialization of New Mobile Services

Introduction

Per E. Pedersen

24

There is a considerable discrepancy between the attention paid to new mobile data services in the professional and academic telecommunication press and among mobile operators. When reading academic telecommunication press articles, one gets the impression that mobile data services represent large proportions of operators' revenue, but the truth is different. Today, mobile data services represent a rather small proportion of mobile operator revenues, and revenues from peer-to-peer SMS and similar messaging services represent the main part of this revenue (Netsize, 2004). While expectations and ambitions have been adjusted and reduced with respect to the revenues to be generated by mobile data services in both 2.5G and 3G networks, considerable growth in the revenues generated by these services is expected in the next few years. One of the most important preconditions for converting these expectations into operator and third-party profits is the innovative adoption of these services by mobile phone users. We apply the term innovative adoption to illustrate that the success of new mobile data services seems to depend much less on the adoption of innovative services than on innovation in service use by end-users themselves. Examples of such innovative adoption are the creation of new forms of language and genres of photography to utilize SMS and MMS services, the redefinition of individuals' roles in working and everyday life to utilize accessibility potentials in mobile e-mail and instant messaging services and the generation of new forms of contact initialization norms to utilize the matching potentiality of mobile contact and chatting services. Thus, innovative adoption of mobile services and technologies is associated with behavioral change, for example in the way we communicate, collect and share information or consume products and services. Innovative adoption is also one of the reasons why new mobile data services have been adopted mostly by younger users. It seems that routine behavior and social norms that are challenged by the use of these services are not too fixed among these users, making innovative adoption much easier. In this part of the book, the authors investigate some of the uses and application areas of these new mobile data services, they explore the requirements for end-users' innovative adoption of such services, some of the effects of using these services and the pattern of cross-media use into which new mobile data service use seems to fit.

Both in information systems and in uses and gratifications research on mobile services, it is often hypothesized that the adoption process is a family simple process where end-users' utilitarian motives explain the adoption of instrumental services whereas hedonic motives explain the adoption of entertaining services (e.g. Leung and Wei, 2000; Kleijnen, et al., 2004). To challenge this hypothesis, Pedersen investigates a wide set of *motives for adopting* an instrumental mobile parking service. The motives are organized within the framework of a mobile data services adoption model. Trial users of the mobile parking service are studied, and the results clearly show that the motives of personal and social identity expressiveness are important determinants of these trial users' intention to continue using the mobile parking service. End-users who have innovated new parking habits utilizing the mobile parking service offerings enjoy how the service fits into their new parking behavior and find this innovative behavior an expression of their own identity as innovative individuals.

Nysveen and Thorbjørnsen go beyond the descriptive analysis of adoption and user behavior to investigate the *relationship effects* of using mobile data services. The empirical context is broadcast media and the services investigated are mobile data services in the form of SMS and MMS services supporting the broadcast of regular programs over the traditional TV medium. Taking a relationship investment perspective (Rusbult, 1987), Nysveen and Thorbjørnsen show that the relationship between the broadcast company and the consumer using mobile services supporting the programs develops differently from that among those not using these supplementary services or accessing them using PCs. This finding is used to argue that the mobile phone seem to have an important set of attributes related to its instrumentality as a relationship building and relationship maintenance tool. These attributes make the mobile phone a valuable instrument in the hand of innovative service providers using the mobile phone to strengthen the relationship between the offered brand and the end user both as a consumer and as a social individual.

Nielsen et al. present the first of two chapters investigating single and *cross media use* of new mobile data services and related data services distributed through traditional media such as the TV and the PC. For some categories of services, mobile phone access to these data services may substitute the services offered through traditional media channels, whereas for other services, mobile phone access is more or less just a supplement to these services. However, for most services it is believed that access to these services using mobile phones will complement the traditional way of accessing these services. While complementarity of media use for media such as TVs, VCRs and PCs have been much investigated (see, e.g., Neuendorf, Jeffres og Atkin, 2000; Morrison and Krugman, 2001; Lin, 2002), we know very little of how the complementarity of mobile phones and PCs or TVs will be. We see some examples of cross media complementarity in the simple use of SMS for voting and chatting on the TV, but new cross-media services are likely to develop in the next few years. One category of such services is alert services on mobile phones. In Nielsen et al.'s chapter, alert services related to the content offered through broadband programs (Web-TV) are focused. Research questions of substitution and complementarity are raised and 39 subjects participating in a trial of new broadband content services and mobile alerts related to these content services are investigated using qualitative interviews. The authors conclude that the mobile alerts complement the broadband content services in a set of accepted situations and for a set of accepted or preferred content categories. These findings suggest that context and content relevance are important antecedents of successful cross-media use.

Miyata et al. investigate the issue of cross-media use resulting from mobile phones having access to data services traditionally accessed using PCs. A survey of 1002 mobile phone and PC users in the Yamanashi Prefecture in Japan is reported, including data on their single- and cross-media user behavior. Miyata et al. investigate both "one-medium" users primarily using a single medium for surfing the web or sending e-mail and cross-media users accessing the web and communication by e-mail using both media. Their study reports the characteristics of "one-medium" and cross-media users and also how user contexts, communication patterns and social network attributes may be used to describe and explain the complementary use of mobile phones and PCs among cross-media users.

These four contributions of new mobile data services illustrate some of the examples of future research to be expected in social science research on mobile services. Traditional models explaining the appropriation, adoption, usage and effects of ICT usage are challenged by the observations made when investigating end users' innovative adoption of new mobile services. Also, traditional approaches investigating single-medium user behavior are likely to be insufficient when trying to understand and explain the behavior of innovative cross-media users combining a variety of ICT-based media. The approaches and variety of background disciplines of the authors of this part illustrate the multitude of research currently occupied with new mobile services as an empirical research arena for investigating innovative consumer behavior.

24.1 References

Kleijnen, M., Wetzels, M. and de Ruyter, K. (2004) Consumer acceptance of wireless finance. *Journal of Financial Services Marketing*, **8**, 206–217.

Leung, L. and Wei, R. (2000) More than just talk on the move: uses and gratifications of the cellular phone. *Journalism and Mass Communication Quarterly*, **77**, 308–320.

Lin, C.A. (2002) Perceived gratifications of online media service use among potential users. *Telematics and Informatics*, **19**, 3–19.

Morrison, M. and Krugman, D.M. (2001) A look at mass and computer mediated technologies: understanding the roles of television and computers in the home. *Journal of Broadcasting and Electronic Media*, **45**, 135–161.

Netsize (2004) *The Netsize Guide 2004: Developing the Multimedia Mobile Market.* Netsize, Paris.

Neuendorf, K., Jeffres, L.W. and Atkin, D. (2000) The television of abundance arrives: cable choices and interest maximization. *Telematics and Informatics*, **17**, 169–197.

Rusbult, C.E. (1987) Commitment in close relationships: the investment model. In Peplau, L.A. et al. (eds), *Readings in Social Psychology: Classic and Contemporary Contributions.* Prentice Hall, Englewood Cliffs, NJ.

25

Instrumentality Challenged: The Adoption of a Mobile Parking Service

Per E. Pedersen

25.1 Introduction

Previous studies in uses and gratifications and domestication research have investigated the adoption and uses of current mobile services users. Naturally, the findings in these traditions do not necessarily generalize to current non-users. On the other hand, findings in these research traditions consistently emphasize the importance of non-instrumental motivational factors in mobile service adoption and use (Leung and Wei, 2000; Taylor and Harper, 2001a,b; Kaseniemi and Rautiainen, 2002; Skog, 2002). In a recent series of surveys in adoption research, we have also investigated the adoption of mobile data services (Pedersen et al., 2002). Our findings so far indicate that even current non-users are influenced by non-instrumental gratifications in their potential adoption and use of mobile services.

To investigate further the influence of motivational, attitudinal, social and resource-based determinants of mobile service adoption in new user groups, a study of new users exposed to instrumental services may provide a kind of "crucial test" of the importance of different motivational and social processes in mobile service adoption. To investigate this issue, mobile parking services were chosen as an appropriate instrumental mobile service. An empirical study of the adoption requirements of new users of mobile parking services was conducted.

Mobile parking services are used to pay for car parking at selected parking sites. These services are typically used by calling or texting to a central server the starting and stopping of parking time. In the car window, a bar-code identifying the customer is placed so that parking site personnel can scan the bar code to check if parking has been paid for. Alert services are also typically provided so that customers may extend their parking time without having to return to their car. Except for alert services, very few other value-added services are currently provided. Compared with other mobile data services, mobile parking services of this kind provide no communication or coordination support typical of many successful mobile data services. The service investigated here also provides very limited informational content both directly and in the form of value-added services. As such it is an instrumental transactional service often believed to be adopted for instrumental reasons of ease of use, usefulness, relative advantage, availability and flexibility only.

25.2 Model

The theory of planned behavior (TPB) has its origin in social psychology but has long been applied to the adoption of ICTs. It includes instrumental, social and resource-based antecedents of technology adoption (Ajzen, 1991). We have previously applied this theory to the adoption of mobile services with considerable success (Pedersen, 2001, 2002; Pedersen et al., 2002). The model suggests that adoption of ICTs and ICT services is the result of motivational, attitudinal, social and resource based factors. While motivational factors are often seen as instrumental variables, our model includes motivational, attitudinal and social influences that play on the individual's expressive dimensions. Thus, we ask not only what the ICT or ICT service can do, but also how the adopted object or service plays into the individual's sense of self. In this study, we applied a re-specified and extended version of the model to explain the adoption of mobile parking services. In Figure 25.1, this modified TPB model is illustrated. We use this illustration as a basis for the discussion of how the general TPB model has been extended and modified.

The model includes four primary influences of adopters' intention to use mobile services. The motivational influences include intrinsic, extrinsic and derived motivations for using mobile services. The attitudinal influence stems from motivational determinants and social norms. The last two influences are social, represented by subjective norm, and resource-based, represented by behavioral control.

Two issues are of relevance with respect to *ease of use* in the model. Because many early adopters of mobile services are expected to be younger, more skilled and more innovative, the higher competence of these users and their more exploratory and advanced use of service functionality suggest that ease of use should have less influence in adoption models of new mobile data services. However, studies also report a more playful use of mobile phones among younger and innovative users and consequently they are more focused on exploring the functionality of a service (Oksman and Rautiainen, 2001). Studies have also indicated a relationship between digital capital and symbolic capital, suggesting that

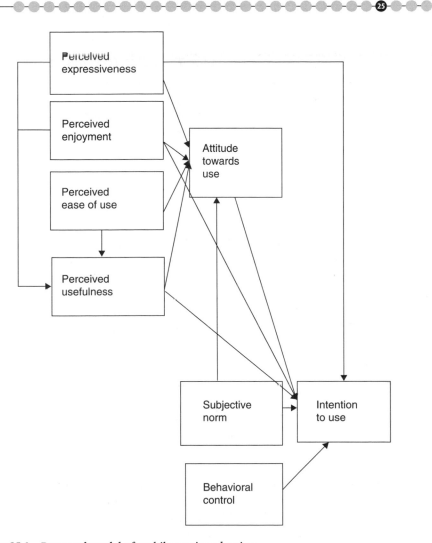

Figure 25.1 Proposed model of mobile service adoption.

services designed for young users should not be too easy to use (Taylor and Harper, 2001b). If they are too easy to use, no status would stem from being able to handle the device, application or service. This explanation may also generalize to innovative users. The other issue is that of service differences in the importance of ease of use. For example, studies applying the perspective of "flow" and "telepresence" have shown that to provide intrinsic motivation, some services must represent a certain challenge to the user. Challenge positively influences flow through increased telepresence (Novak et al., 2000; Hunter and Kalafatis, 2001). This, in turn, implies that we might expect a negative effect of ease of use (challenge inversed) on perceived enjoyment for highly involved users and for services that are used for reasons of intrinsic motives.

Perceived *usefulness* was originally seen as a fairly simple concept including components of effectiveness and efficiency that are mainly related to extrinsic motivation in work contexts. As seen from uses and gratifications studies, the extrinsic motivations of mobile services are not limited to effectiveness and efficiency. Motivations of accessibility, flexibility, sociability and security typical of communication (as opposed to information) services have all been mentioned in these studies. In addition, motivations of enjoyment, fashion and status, and expressiveness have been mentioned (Leung and Wei, 1999a,b, 2000; Höflich and Rössler, 2001; Fortunati et al., 2003). Some of these motivations are intrinsic, but others may perhaps best be characterized as derived, meaning that they provide an instrumentality or gratification that was not intended by or anticipated during design, and that perhaps also was not considered or anticipated by the user at the time of the adoption (Anderson et al., 2002; Pedersen, 2002; Pedersen et al., 2002).

Hence studies suggest that the usefulness concept should be extended and supplemented to cover the issues of intrinsic and derived motivations. For example, *enjoyment and entertainment* go beyond ease of use and usefulness, and are perceived as instrumental of services primarily designed for entertainment (mobile games, mobile video and audio streaming, chat and flirt services) (Leung and Wei, 1999b, 2000). The instrumentality of these services is enjoyment and entertainment in itself, not the efficiency or effectiveness of being able to access mobile entertainment services ubiquitously. This indicates that enjoyment should be included in adoption models developed for users of mobile services as a separate concept contributing to perceptions of usefulness, ease of use and attitudes towards use. However, we expect that the effects of enjoyment on these concepts are not significant for instrumental mobile data services like mobile parking services.

In CMC research, *expressiveness* is compared with instrumentality as two styles of communication (Boneva et al., 2001). Expressiveness is used for communication in relationships of emotional intimacy and sharing, whereas instrumentality is used for communication in relationships based on common activities. For example, Boneva et al. (2001) believes female communication to be more expressive, and male communication more instrumental. Based on these assumptions, services that communicate expressiveness in this form are more likely to be appreciated by female users. In social psychology, recent contributions have suggested replacing the well-known concept of self-identity as a determinant of intended behavior with self-expression (Mannetti et al., 2002). Research on the influence of self-identity on intended behavior is, however, still relevant. Typically, the relationship between behavior and self-identity is given a social interpretation based on Mead's and Goffman's theories of the social construction of the self (Mead, 1934; Goffman, 1959), a structuration interpretation based on Giddens' theories (Giddens, 1991), or a role-oriented personality interpretation. In the first case, self-identity is the result of social identification, in the second case, it is the results of the interaction of social identity and repeated actions maintaining a "personal biography", and in the final case, it is a more rational frame of reference for behavioral decisions.

When applying the term expressiveness here, we focus on the importance of behavior as being interpreted by others in the social construction of identity and by oneself in the repeated self-construction of identity. Thus, expressiveness is a more operational concept applied to the use of technologies or services or the consumption of products and services that are important both to social identity and to role-oriented self-identity. In other words, we adopt ICTs and ICT services, at least to some degree, based on the identity we wish to project and to develop or maintain a particular sense of self. Indeed, this notion is also reflected, at a social level, in the work of the so-called domestication theorists (Haddon, 2001).

Consistent with this conception of expressiveness, consumer psychology characterizes value-expressive products as expressing the consumer's identity both in social networks and to oneself. We suggest that expressiveness in terms of both the social expression of identity and self-identification are important elements in the adoption and use of mobile services. Expressiveness is an instrumental attribute of a communication service partly influencing usefulness and partly influencing attitudes directly. For information services, expressiveness is an unanticipated service characteristic. Thus, we expect that expressiveness is more relevant when explaining the adoption of communication services than transaction oriented information services like mobile parking services.

Attitudes are generally believed to be the results of personal and social influences. However, in the original technology acceptance model (TAM), attitudes towards use are determined by personal influences only. When including subjective norm in the model, it is possible to create a relationship between norms and attitudes that may be particularly relevant to users' adoption of mobile services. However, it is also important to conceptually discriminate norms and attitudes in adoption models. Hence we suggest accepting an influence of subjective norm on attitudes, but reject including effects of external and interpersonal influence on attitudes directly. We also suggest extending the determinants of attitudes towards use from purely instrumental determinants to more derived determinants such as enjoyment and expressiveness. However, the attitude formation process is believed to be similar for usefulness, ease of use, enjoyment and expressiveness in that the individual sees a service as instrumental in fulfilling intrinsic, extrinsic and derived gratifications, and consequently develops a positive attitude towards using it. The relationship between attitudes and intentions may be different for different service categories. For example, for services that are widespread and well known, it is easy to obtain information on other users' experience and also to gain experience from actually using the service oneself. This indicates that for established services, instrumental and experiential motives are the most important explanations of user intentions. On the other hand, if services are new and unknown as they are for mobile parking services, intentions may be based on general attitudes and less on experientially derived motives.

Subjective norms are developed through external and interpersonal influence. In general, Webster and Trevino (1995) suggest social influences, and hence subjective norms, to be more influential in explaining the adoption and use of new media. The question, however, is which services should be considered new media in the Scandinavian market for mobile services. In an international setting, most mobile services may be considered new media, but in Scandinavia, mobile data services – specifically in the form of text messaging – are now well integrated in the everyday lives at least of young users. Consequently, even though social motivations for adoption may be important, these motivations may by now be more instrumental than norm based, and should be identified through instrumental determinants of attitude toward use rather than through subjective norm. To give an example, young users may find text messaging instrumental in social coordination because all other members of their social network use it, but still feel little social pressure towards using text messaging services as a norm.

The inclusion of *behavioral control* in TPB has been an important contributor to its explanatory power. In general, we argue that the determinants of behavioral control are believed to be less important to young and innovative users than others. This is because of their experience and skill in using mobile services and the providers' facilitation of mobile services such as text messaging services to the young user segment. Financial resources and pricing, however, are indirectly believed to be important determinants of behavioral control due to both limited resources among young users and recent findings that these users are more price sensitive than previously assumed (Karlsen et al., 2001). Behavioral control is a general term composed of elements of individual traits and perceptions of operators' and providers' facilitation. It is also likely that the influence of behavioral control will vary across mobile services. In general, we suggest that the influence of self-efficacy on behavioral control is greatest for complex, new, integrated, expensive and terminal demanding services. The same may thus be argued of the influence of behavioral control on intention to use these services. For example, the adoption of technically complex services, services requiring advanced terminals, services with hidden costs and generally expensive services will be more influenced by behavioral control than simple and cost-efficient services. On the other

hand, behavioral control will be more relevant to the less skilled and price-sensitive user than to the innovative and price-insensitive users. Given that innovative users are the first adopt new mobile data services, the adoption of mobile parking services should *currently* be less influenced by behavioral control.

25.3 Method

To investigate the motivational, attitudinal, social and resource-based influences on the intention to use mobile parking services, a survey of new mobile parking services users was designed. The survey was designed as a simple one-group post-test design. A quasi-experimental setting was applied by selecting respondents in the population who had recently signed up for a free test trial of the service or who had recently signed up for full membership. Of the two categories, most respondents were in the first category. The trial service was announced using large posters at major parking areas including individual folders explaining how users could phone or SMS the provider to obtain a free 1-hour parking service. A total of 2550 respondents were identified in the population, and a list of these users was used as a sample frame.

Subjects were given the opportunity to visit a web site to answer the questionnaire online or use a pre-paid postal version attached to the introductory letter and procedure material. Forty-seven subjects chose to answer the questionnaire online and 418 offline using the postal alternative. Thus, a total of 465 questionnaires were returned. Six of the questionnaires were excluded from the analysis owing to late arrival. The final response rate obtained was 18.2%. Sample demographics of the mobile parking service subjects are shown in Table 25.1.

Table 25.1   Sample demographics – parking study

Age (years)	N = 452	Income (Norwegian Krone)	N = 454
0–19	2.4	<200,000	13.7
20–29	24.1	200,000–399,000	44.9
30–39	33.8	400,000–600,000	24.4
40–49	23.0	>600,000	17.0
50–59	11.3		
≥60	5.3		

Education	N = 457	Sex	N = 456
Primary	2.2	Male	72.1
Secondary	23.6	Female	27.9
University <3 years	37.0		
University ≥4 years	37.2		

The sample included a larger proportion of men than women, a larger proportion of subjects with university education and a larger proportion of subjects with higher level income when compared with the general Norwegian population. However, these differences are not very large and the differences in the distributions are according to what one might expect of new users of a mobile parking service. Hence we assume that the sample demographics correspond well to the population demographics of new mobile parking service adopters.

The model suggested in Section 25.2 includes eight concepts: ease of use, usefulness, expressiveness, enjoyment, attitudes towards use, subjective norm, behavioral control and intention to use. Most of these concepts are well founded in adoption, uses and gratification, or domestication research literature. Consequently, the construct validity of these concepts is in general considered acceptable. To measure the concepts, a questionnaire was designed containing multiple measures of each of the eight concepts. In general, the concepts were measured by the subjects indicating their agreement with a set of statements using a seven-point scale ranging from "strongly disagree" to "strongly agree". Some concepts were measured using seven-point scales of bipolar adjectives. For each measure, the items were adapted to the mobile parking service context of the study. The measures were investigated for reliability with the lowest values of Cronbach's α of 0.71 and 0.77 and an average value of 0.86.

Ease of use was measured using four items developed from adapting the original items of Davis et al. (1989) to our setting. Similar operations can also be found in Taylor and Todd (1995) and Battacherjee (2000). Usefulness was measured using three items covering the original dimensions of time saving, improvement, usefulness and quality suggested by Davis (1989). Attitude towards use was measured using four bipolar adjectives indicating different aspects of the subjects' attitude towards use. The items were very similar to those used by Davis (1989), Taylor and Todd (1995) and Battacherjee (2000). The enjoyment concept was defined as incorporating a group of gratifications identified in studies of the Internet as "enjoyment" (Papacharissi and Rubin, 2000), of ICQ as "entertainment" (Leung, 2001), of mobile phones as "relaxation" (Leung and Wei, 2000), of pagers as "fun-seeking" (Leung and Wei, 1999b) and of text messaging as "nutz-spaz" (Höfflich and Rössler, 2001). To cover these elements of enjoyment, a four-item scale was developed collecting items from uses and gratification scales including the "entertainment", "relaxation", "excitement" and "fun-seeking" conceptions of enjoyment.

The term expressiveness has been used in social psychology of individuals' general ability to express their emotions or identity (Cassidy et al., 1992). In research on identity formation and personality, it is used as a measure of the relationship between what a person believes about his- or herself (what his or her potentialities are; see Schwartz et al., 2000, p. 507) and how he or she expresses him- herself, using the concept of "personality expressiveness" (Waterman, 1993). In this line of research, a person expresses herself through activities, and expressiveness is measured by subjects indicating how important particular activities are in expressing their identity. In social psychological research on the prediction of behavior, the term is closely related to self-identity, which has been found to be a significant predictor of intention to perform specific behaviors (Sparks and Guthrie, 1998). In this literature, self-identity is typically measured using statements challenging the relationship between behavior and the subjects' perceptions of their own personality. In consumer research, the expressiveness concept has been extended from individuals to products indicating how well a product expresses values beyond instrumental utility (Mittal, 1994). Hence value-expressive products are seen as expressing the consumer's identity. This is typically measured by subjects indicating how products are used to "express my personality" and are "compatible with how I like to think of myself" (Mittal, 1994, p. 258). Therefore, items measuring these conceptions of expressiveness have been included. In addition, expressive gratifications have been identified in uses and gratifications research. For example, Arnett (1995) included "identity formation" as a particular gratification of young users, Leung (2001) included "express affection", "fashion" and "inclusion" as gratifications of ICQ use and Leung and Wei (1999b, 2000) included "fashion and status" as a gratification of both pager and mobile phone use. From these studies, a status-related expressiveness item was suggested.

Studies of text messaging use have shown how one of the most important ways of expressing one's service use is to discuss the service with others and to share it with others (Larsson, 2000; Grinter and Eldridge, 2001; Kaseniemi and Rautiainen, 2002). Therefore, items referring to this particular form of expressiveness were included. Similar items, measuring the gratification of sharing technology use with others – social interaction – have been included in studies of video games (Sherry et al., 2001) and TV (Lee and Lee, 1995). This element in expressiveness is also consistent with social perspectives of self-identity and items covering the social element of expressiveness are also included in our measure. Hence, of the four items used, two were oriented towards self-expressiveness and two were oriented towards social expressiveness.

Subjective norm was measured using three items almost identical with the items used by Mathieson (1991) and Battacherjee (2000). In addition, a general norm item was included, inspired by sociological research on mobile service use (Skog, 2002). The measure of behavioral control was almost identical with the measure applied by Taylor and Todd (1995) and Battacherjee (2000). Finally, intention to use was measured with a two-item scale adapted from Mathieson (1991) and Battacherjee (2000).

All our traditional measures are based upon previously validated measures (Venkatesh and Morris, 2000), and their reliabilities were considered acceptable. To test the discriminant and convergence validity of the independent variables in our model, the items of all six independent variables were included in a confirmatory factor analysis. The analysis showed minimum crossloadings below 0.34 for all variables, indicating acceptable convergence and discriminant validity.

25.4 Results

Using the data from the parking services study, the adoption model of Section 25.2 was estimated. The results of this estimation are shown in the adoption model for the mobile parking services illustrated in Figure 25.2.

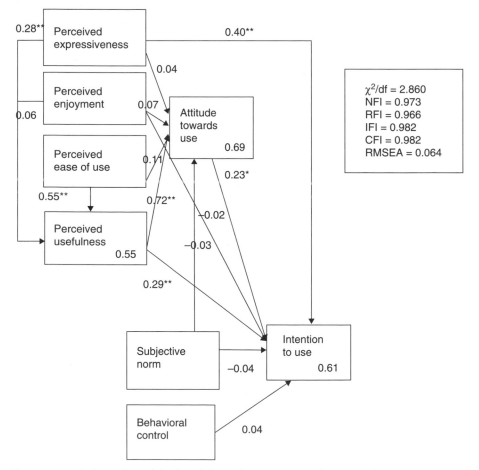

Figure 25.2 Estimated model of mobile parking services adoption. Values on arrows are standardized regression coefficients, * and ** indicate significance at $p < 0.05$ and $p < 0.01$, respectively. Values in variable squares indicate explained variances (R-squared).

From Figure 25.2, we see that model fit is very good when evaluated by all fit indexes.[1] The model explains 61% of the variance in intention to use the parking service. This is generally considered a large proportion of the variance, so the explanatory power of the model is very good.

When investigating model relationships, we first see that intention to use mobile parking is explained by (1) direct instrumentality of usefulness and (2) expressiveness and by (3) attitudes towards use. The effects of subjective norm and behavioral control are not significant. This gives a fairly simple model for explaining intention to use mobile parking. The services are used because they are instrumentally useful and are considered as a way of expressing oneself that is consistent with the users' idea of themselves. One should be careful in comparing the influences by standardized coefficients, but expressiveness is a very important determinant of intention to use mobile parking services. This is rather surprising given the utilitarian instrumentality of mobile parking services. In addition, the effect of attitudes towards use is significant at the 5% level, and attitudes are influenced by usefulness and ease of use only. Thus, attitudes seem to be mainly influenced by traditional utilitarian instrumentality. Usefulness, on the other hand, is significantly influenced by ease of use and expressiveness. Thus, enjoyment plays no direct or indirect role in the adoption of mobile parking services. On the other hand, the derived motivations of expressiveness play both an indirect and an important direct role.

From these observations, we conclude that users' intentions are simply explained by the direct motivational influence of usefulness and expressiveness and by attitudes towards use. Even though the processes influencing the social and resource-based variables may be fairly complex, these variables do not themselves seem to play any significant role in explaining behavioral intentions to use mobile parking services.

25.5 Discussion

In Section 25.2, a model of mobile data services adoption was suggested based on the theory of planned behavior and modified and re-specified using findings of mobile service end-user behavior in uses and gratifications and domestication research. The model included four primary influences of adopters' intention to use mobile services. The motivational influence included intrinsic, extrinsic and derived motivations for using

---

[1] We generally employ parsimony-adjusted measures of fit only. According to Browne and Cudeck, cited in Arbuckle and Wothke (1999), an RMSEA of <0.08 is acceptable. According to Bentler, cited in Battacherjee (2000), χ^2/df should be <5, preferably <2, and all other indexes should be close to 1 (Taylor and Todd, 1995). In general, we apply the rules of $\chi^2/df \approx 2$ or better, RMSEA < 0.08 and all other indexes ≈1.

mobile services. The attitudinal influence stemmed from motivational determinants and social norms. The last two influences were social, represented by subjective norm, and resource-based, represented by behavioral control.

This model was tested with empirical data on users' adoption and intention to adopt mobile parking services. These services were selected for their focus on instrumental gratifications primarily believed to meet users' extrinsic motivations for using mobile data services. Data from 459 users who had tried mobile parking services were used to estimate the model. The general results showed that usefulness, expressiveness and attitudes towards the use explained the trial users' intentions to use mobile parking services, that attitudes towards use of these services were determined by ease of use and usefulness and that usefulness was determined by ease of use and expressiveness. Thus, expressiveness showed a far more important role in influencing the adoption of mobile parking services than what was expected of an instrumentally oriented mobile data service that we believed, at the point of departure, would be adopted for utilitarian reasons.

When looking at the social and resource-based influences, neither subjective norms nor behavioral control were found to influence significantly intention to use mobile parking services. As proposed in Section 25.2, lack of influence from subjective norm may partly be explained by the instrumentality of the service and partly by the innovativeness of the subjects. The lack of influence from behavioral control was somewhat surprising given that mobile parking services are generally considered to be pricey and complex services relying on a well-functioning underlying infrastructure. However, the simplicity of the service interface may have been revealed through trial user experience, and these innovative users may also be less price sensitive than average mobile data services users.

The study was based on the developed procedures, measures and results of five previous studies of mobile service adoption. Therefore, we propose that the theoretical and methodological foundations for acceptable internal and external validity are sufficient. In general, all theoretical concepts have been discussed in Section 25.2 and 25.3 or in previous work (Pedersen, 2001, 2002; Pedersen et al., 2002), and are well founded in adoption, uses and gratifications and domestication research. Thus, construct validity is considered acceptable. Furthermore, analyses of measurement items showed that measures were reliable and that constructs had acceptable convergence and discriminant validity.

With respect to internal validity, the procedure used to recruit subjects in this study may have resulted in subjects with a more positive attitude towards the service than the population of trial users. Even though we may have recruited subjects with positive attitudes, many of the comments found in the survey questionnaire were also negative, and an equally important motivation for participating in the survey may have been to express negative experience from using the trial services. Therefore, we assume that the recruitment and experimental procedures did not make the sample systematically different from the sampling frame representing the population of trial users.

With respect to external validity, consumers with no experience in trying this kind of service may perceive service characteristics as different and their intention to use mobile parking services may be based on different judgments. As indicated above, it is not unlikely that these subjects will be more influenced by social pressure, and that perceived behavioral control may be more relevant to their intention to adopt these services. Still, users are not likely to adopt these kinds of services without some initial trial, making our findings externally valid anyway. In a recent survey of mobile users versus shoppers in Finland, Germany and Greece, Vrechopoulos et al. (2002) found few demographic differences between the two user categories in the three countries. Hence these findings may indicate that the early adopters of mobile data services and adopters of traditional mobile services are not that different.

Another issue with respect to the subjects recruited is the skewed distributions of demographic variables such as age, gender and income. However, we have investigated model differences and perception differences by age, gender and income. Some differences were revealed between different demographic segments, but the model proved remarkably stable across segments. For example, the pattern of significant and not significant influences on intention to use mobile parking services was exactly the same for male and female users and for highly and less educated users. This shows that even though the number of female users participating in the study was low, the motivational processes of male and female users were very similar. For example, we found no indications that male users were more instrumentally influenced by usefulness and attitudes towards use whereas female users were more influenced by social pressure or expressiveness, often proposed in other studies of gender differences in ICT adoption (Venkatesh and Morris, 2000). Further documentation of model stability can be found in Pedersen and Nysveen (2002).

We argue that the selection of mobile parking services represents a "crucial test" of the external validity of our TPB model extensions (intrinsic and derived motivations of enjoyment and expressiveness). We propose that when finding expressiveness to be influential in the adoption of this service, we may generalize this finding to most other instrumental data services. Hence we conclude that our finding of expressiveness as influential in the adoption of mobile parking services makes it very likely that expressiveness is a unique gratification of most mobile data services, and that developers should take this adoption determinant into consideration when designing and marketing their services. Still, our findings should be interpreted with care because more attention was given to internal than to external validity in the design of the study.

For researchers, the results of this study provide a "crucial test" of non-instrumental influences on the adoption of instrumental mobile services. Given this setting, it is surprising to see such a consistent influence of expressiveness. This indicates that the motivational process of adoption is more complex than previously assumed and suggests that this process should be further elaborated on. For example, the relationship among intrinsic, extrinsic and derived motivations requires further analysis. For mobile parking services in particular, the influences of the self-identity elements of expressiveness are particularly interesting. In applied social psychology and consumer psychology, the element of self-identity in consumption has been given some attention (Belk, 1988; Sparks and Guthrie, 1998; Mannetti et al., 2002). Self-identity in many of these contributions is seen somewhat different from the socially constructed self-identity of Mead and Goffman (Mead, 1934; Goffman, 1959) and of the structuration theory of Giddens (1991). This line of research has mainly been applied to the consumption of value expressive products (Belk, 1988; Mittal, 1994) such as objects of display or style and products related to personal lifestyles, such as environmentally relevant products (Cook et al., 2002).

In IS research, these concepts have been given little attention. Instead, symbolic elements of media choice and use have been investigated in this tradition, focusing more on the symbolic effects of using specific technologies and services. For example, Trevino et al.'s (2000) operation of symbolism in the use of a particular medium was done by asking the subjects if they considered the use of a particular medium as symbolizing "low priority", "formality" or "urgency". This conceptualization of symbolism is much more instrumental and rather different from the conceptualization of symbolic media use as an instrument in the continuous expression of users' self-identity and social identity. As mobile services are introduced in work contexts, the influences of social identity and self-identity in the process of adopting these services should be given more attention in IS research also. Our development of the expressiveness concept, the evaluation of the validity and reliability of the concept and the demonstration of its influence on the intention to adopt these instrumental mobile parking services represent significant contributions to this research.

25.6 References

Ajzen, I. (1991) The theory of planned behavior. *Organization Behavior and Human Decision Processes*, 50, 179–211.

Anderson, B., Gale, C., Gower, A.P., France, E.F., Jones, M.L.R., Lacohee, H.V., McWilliam, A., Tracey, K. and Trimby, M. (2002). Digital living – people centered innovation and strategy. *BT Technology Journal*, 20, 11–29.

Arbuckle, J.L. and Wothke, W. (1999). *Amos 4.0 User's Guide*. SmallWaters, Chicago.

Arnett, J. (1995) Adolescents' uses of media for self-socialization. *Journal of Youth and Adolescence*, 24, 519–531.

Battacherjee, A. (2000) Acceptance of e-commerce services: the case of electronic brokerages. *IEEE Transactions on Systems, Man and Cybernetics*, **30**, 411–420.

Belk, R.W., (1988). Possessions and the extended self. *Journal of Consumer Research*, **15**, 139–168.

Boneva, B., Kraut, R., and Frohlich, D., (2001) Using e-mail for personal relationships: the difference gender makes. *American Behavioral Scientist*, **45**, 530–549.

Cassidy, J., Parke, R.D., Butkovsky, L., and Braungart, J.M. (1992) Family–peer connections – the roles of emotional expressiveness within the family and childrens understanding of emotions. *Child Development*, **63**, 603–618.

Cook, A.J., Kerr, G.N. and Moore, K. (2002) Attitudes and intentions towards purchasing GM food. *Journal of Economic Psychology*, **23**, 557–572.

Davis, F.D. (1989) Perceived usefulness, perceived ease of use, and user acceptance of information technology. *MIS Quarterly*, **13**, 319–340.

Davis, F.D., Bagozzi, R.P. and Warshaw, P.R. (1989) User acceptance of computer technology: a comparison of two theoretical models. *Management Science*, **35**, 982–1002.

Fortunati, L., Katz, J.E. and Riccini, R. (eds), (2003) *Mediating the Human Body: Technology, Communication and Fashion*. Lawrence Erlbaum, London.

Giddens, A. (1991) *Modernity and Self-identity: Self and Society in the Late Modern Age*. Basic Books, New York.

Goffman, E. (1959). *The Presentation of Self in Everyday Life*. Doubleday, New York.

Grinter, R.E. and Eldridge, M. (2001) Y do tngrs luv 2 txt msg? In Prinz, M., Jarke, Y., Schmidt, K. and Wilf, V. (eds), *Proceedings of the Seventh European Conference on Computer-Supported Cooperative Work ECSCW'01*. Kluwer, Dordrecht, pp. 219–238.

Haddon, L. (2001) Domestication and mobile telephony. In Katz, J.E. (ed.), *Machines That Become Us*. Transaction, New Brunswick, pp. 43–56.

Höflich, J.R. and Rössler, P. (2001) Mobile schriftliche Kommunikation oder: E-Mail für das Handy. *Medien & Kommunikationswissenschaft*, **49**, 437–461.

Hunter, L. and Kalafatis, S.P. (2001) *Nirvana on the Web – Do Business Users Experience Flow When Using the Internet?* Presented at EMAC, Rotterdam.

Karlsen, M.A., Helgemo, I. and Gripsrud, M. (2001) *Useful, Cheap and Fun: a Survey of Teenagers' Demands for Mobile Telephony*. Research Report, Telenor R&D, Grimstad, Norway.

Kaseniemi, E. and Rautiainen, P. (2002) Mobile culture of children and teenagers in Finland. In Katz, J.E. and Aakhus, M. (eds), *Perpetual Contact*. Cambridge University Press, Cambridge.

Larsson, C. (2000) En mobiltelefon är inte bara en mobil telefon. Magisteruppsats, Göteborgs Universitet, Gotenburg.

Lee, B. and Lee, R.S. (1995) How and why people watch TV: implications for the future of interactive television. *Journal of Advertising Research*, November/December, 9–18.

Leung, L. (2001) College student motives for chatting on the ICQ. *New Media and Society*, **3**, 483–500.

Leung, L. and Wei, R. (1999a) Seeking news via the pager: an expectancy–value study. *Journal of Broadcasting and Electronic Media*, **43**, 299–315.

Leung, L. and Wei, R. (1999b) The gratifications of pager use: sociability, information-seeking, entertainment, utility, and fashion and status. *Telematics and Informatics*, **15**, 253–264.

Leung, L. and Wei, R. (2000) More than just talk on the move: uses and gratifications of the cellular phone. *J&MC Quarterly*, **77**, 308–320.

Mannetti, L., Pierro, A., and Livi, S. (2002) Explaining consumer conduct: from planned to self-expressive behavior. *Journal of Applied Social Psychology*, **32**, 1431–1451.

Mathieson, K. (1991) Predicting user intentions: comparing the technology acceptance model with the theory of planned behavior. *Information Systems Research*, **2**, 173–191.

Mead, G.H. (1934) *Mind, Self and Society: from the Standpoint of a Social Behaviorist*. University of Chicago Press, Chicago.

Mittal, B. (1994) A study of the concept of affective choice mode for consumer decisions. *Advances in Consumer Research*, **21**, 256–263.

Novak, T.P., Hoffman, D. and Yung, Y.F. (2000) Measuring the customer experience in online environments: a structural modeling approach. *Marketing Science*, **19**, 22–42.

Oksman, V. and Raitiainen, P. (2001) "Perhaps it is a body part". How the mobile phone became an organic part of the everyday lifes of children and teenagers. Presented at the 15th Nordiska konferensen för medie- ock kommunikationsforskning, Reykjavik, 11–13 August.

Papacharissi, Z. and Rubin, A.M. (2000) Predictors of Internet use. *Journal of Broadcasting and Electronic Media*, **44**, 175–196.

387

Pedersen, P.E. (2001) *Adoption of Mobile Commerce: an Exploratory Analysis*. SNF-Report 51/01, Institute for Research in Economics and Business Administration, Bergen.

Pedersen, P.E. (2002) *The Adoption of Messaging Services Among Norwegian Teens: Development and Test of an Extended Adoption Model*. SNF-Report 23/02, Institute for Research in Economics and Business Administration, Bergen.

Pedersen, P.E. and Nysveen, H. (2002) *The Adoption of a Mobile Parking Service: Instrumentality and Expressiveness*. SNF-Working Paper 76/02, Institute for Research in Economics and Business Administration, Bergen.

Pedersen, P.E., Nysveen, H. and Thorbjørnsen, H. (2002) *The Adoption of Mobile Services: a Cross Service Study*. SNF-Report 31/02, Institute for Research in Economics and Business Administration, Bergen.

Schwartz, S.J., Mullis, R.L., Waterman, A.S. and Dunham, R.M. (2000) Ego identity status, identity style, and personal expressiveness. An empirical investigation of three divergent constructs. *Journal of Adolescent Research*, 15, 504–521.

Sherry, J., Lucas, K., Rechtsteiner, S., Brooks, C. and Wilson, B. (2001) Video game use and gratifications as predictors of use and game preference. Presented at the 51th convention of the International Communication Association, Washington, DC, 24–28 May.

Skog, B. (2002) Mobiles and the Norwegian teen: identity, gender and class. In Katz, J.E. and Aakhus, M. (eds), *Perpetual Contact*. Cambridge University Press, Cambridge.

Sparks, P. and Guthrie, C.A. (1998) Self-identity and the theory of planned behavior: a useful addition or an unhelpful artifice? *Journal of Applied Social Psychology* 28, 1393–1410.

Taylor, A.S. and Harper, R.H.R. (2001a) The gift of the gab?: a design oriented sociology of young people's use of 'mobilZe! Working Paper, Digital World Research Centre, University of Surrey.

Taylor, A.S. and Harper, R.H.R. (2001b) Talking activity: young people and mobile phones. Presented at the CHI 2001 Workshop on Mobile Communications 30 March–5 April, Seattle, WA.

Taylor, S. and Todd, P.A. (1995) Understanding information technology usage: a test of competing models. *Information Systems Research*, 6, 144–176.

Trevino, L.K., Webster, J. and Stein, E.W. (2000) Making connections: complementary influences on communication media choices, attitudes and use. *Organization Science*, 11, 163–182.

Venkatesh, V. and Morris, M.G. (2000) Why don't men ever stop to ask for directions? Gender, social influence, and their role in technology acceptance and usage behavior. *MIS Quarterly*, 24, 115–139.

Vrechopoulos, A.P., Constantiou, I.D. and Sideris, I. (2002) Strategic marketing planning for mobile commerce diffusion and consumer adoption. Presented at M-Business 2002, Athens, 8–9 July.

Waterman, A.S. (1993) Psychometric properties of the Personally Expressive Activities Questionnaire. Unpublished manuscript, College of New Jersey, Ewing, NJ.

Webster, J. and Trevino, L.K. (1995) Rational and social theories as complementary explanations of communication media choices: two policy-capturing studies. *Academy of Management Journal*, 38, 1544–1572.

26

Relationship Deepening Through Mobile and Interactive Services

Helge Thorbjørnsen and Herbjørn Nysveen

26.1 Introduction

Mobile and interactive media are proposed as powerful channels for both distribution and marketing communication. A vast stream of academic literature focuses on the potential positive effects of utilizing Internet- and mobile-based services for strengthening consumer–brand relationship ties and maximizing firm revenues. This paper deals with the topic of "channel additions", that is, adding new channels of distribution and marketing communication to existing ones. Specifically, we investigate the effects of such channel additions on consumer–brand relationship ties and brand usage. We focus on the Norwegian broadcaster TV2, and the positive effects on consumer response of adding SMS services and WebTV services to this brands' main channel (i.e. TV broadcast). SMS and WebTV channel additions were selected on the basis of their widespread usage and strong growth, the fact that they are both interactive media and because the characteristics of the two media are sufficiently different to make us able to pinpoint fundamental differences in channel characteristics.

Implementing channel additions is a challenging task, as it is very difficult to anticipate whether new channels will serve as substitutes, supplements or complements to existing ones, and whether the channel additions will produce the necessary marketing synergies with traditional channels (Dutta-Bergman, 2004). When adding new channels, the risk of channel conflict is always present, that is, one may experience that usage of the new channel substitutes sales or usage of a brand's main channel – something that often is referred to as channel cannibalism. Consequently, we argue that it is essential to examine carefully the properties and characteristics of potential channel additions for identifying possible synergistic effects on consumer usage and gratifications sought by consumers.

The purpose of the Chapter is twofold: first, to contribute to the development of a general theoretical framework for examining effects of channel additions. And second, to investigate empirically the differences in effects between SMS and WebTV channel additions on consumer–brand relationship ties and main channel usage. Concurrently, our endeavor should be of interest to both industry players and scholars of mobile telephony for several reasons: we provide initial insight into how mobile channels interact with other essential vehicles of distribution and communication, and we offer a conceptual mindset for evaluating potential effects of mobile channel additions on consumer–brand relationship ties. Moreover, we pinpoint how mobile services directly and indirectly may influence the usage of other channels from the same brand/vendor.

26.2 Channel Additions

According to Geysken et al. (2002), there are three different ways in which a channel addition may increase sales: through market expansion, brand switching and relationship deepening. A market expansion occurs when new segments of consumers are reached who previously did not buy in the category. An example could be if the webTV channel of the Norwegian broadcaster TV2 attracted a new and global web audience for, let us say, a new season of Seinfeld episodes. Brand switching is a second way to expand demand, and this simply refers to winning new customers from competing brands, that is, gaining new groups of customers who previously bought in the category, but from a different vendor. The final way to expand demand is through relationship deepening, which merely pertains to selling more to existing customers. We shall concentrate on such relationship deepening in this chapter, and focus on both distribution and marketing communication effects of adding new channels to existing customers.

In marketing journals, a particular focus has recently been directed towards the potential positive effects of adding new interactive communication channels for building strong consumer–brand relationship ties (Alba et al., 1997; Holland and Baker, 2001; Thorbjørnsen et al., 2002b, Aaker et al. in press). Strong consumer–brand relationships are highly correlated with consumers' (re-)purchase intentions, positive word-of-mouth, tolerance for price increases and other positive consequences sought by brands (Fournier, 1994; Thorbjørnsen et al., 2002a). Consequently, identifying potential antecedents of strong consumer–brand relationships has been given considerable attention within the science of marketing. Interactive marketing media, such as the Internet and cellular phones, are proposed as effective relationship-building tools as they allow for a two-way (reactive) multi-platform dialogue between the consumer and the brand. According to most relationship researchers, iterative and reactive exchanges of information are necessary for relationships to develop. Such "intimate" information exchanges between the brand vendor and the consumer are easily facilitated through both machine-interactive and person-interactive channels of marketing communication. Whereas machine-interactive communication refers to communication with a "machine" (such as PC software or a mobile application), person-interactive communication pertains to mediated communication with another human (such as e-mailing or SMS messaging) (Hoffman and Novak, 1996). Past research has revealed that both machine-interactive communication (e.g. with personalized web sites) and person-interactive communication (e.g. user groups or brand communities) on the Internet strengthen consumer–brand relationship ties (Mathwick, 2002; Thorbjørnsen et al., 2002b). However, virtually all studies of interactive marketing communication effects investigate only the effects of single channels, not combined effects of multiple channels or effects of adding a new channel to a brand's existing ones.

Although the majority of academic contributions emphasize the potentially positive effects of channel additions on consumer response, several authors also claim that adding new marketing channels may be directly dysfunctional. Although additional channels can significantly strengthen a brand's position, it may also compete with the brand's main channel for customer attention and customer usage (Chan-Olmsted and Ha, 2003). A traditional perspective is that the consumption of traditional channels will be reduced when new channels are introduced on the market (Lazarsfeld, 1940; McCombs, 1972). Such a competitive perspective is based on the notion that the total time used on channel consumption remains constant, and that an increase in the consumption of a newly introduced channel therefore reduces the use of traditional channels (Dutta-Bergman, 2004). In this perspective, new channels *substitute*, displace or cannibalize traditional channels. An alternative perspective is that new channels serve as a *supplement* to traditional channels (Anderson and Narus, 1995; Riel et al., 2001). This means that the new channels can add value to the core service presented in traditional channels without influencing the use of the traditional channels. This may be additional services offered in new channels that add value to the core service without influencing the use of the core service in the traditional channel. According to Chan-Olmsted and Ha (2003, p. 612), an important business model for broadcast companies is to "utilize the Internet as a supplemental medium for developing a relationship with the audience of an offline core product". The *complementary* perspective introduced by Dutta-Bergman (2004) argues for synergies between traditional and new channels. Dutta-Bergman found that users of online news are more likely to use news services distributed in traditional channels, pointing to a

Table 26.1 Channel substitutes, supplements and complements

New channel use	Traditional channel use		
	Increase	Constant	Decrease
Increase	Complementary	Supplementary	Substitute
Constant	Supplementary	Supplementary	Supplementary
Decrease	Substitute	Supplementary	Complementary

situation where the use of new channels also increases the use of the traditional channels. The three perspectives are summarized in Table 26.1 in relation to how the introduction of a new channel may influence the use of traditional channels.

As can be seen from Table 26.1, complementary channels are defined as a situation where consumption in new channels also increases consumption in existing channels. When the channels are a supplement to each other, the use of new channels does not influence the use of traditional channels. Substitution points to a situation where new channels displace traditional channels. Consequently, we argue that it is of vital importance to consider both the direct effect of the channel addition in itself on new channel usage and consumer–brand relationship strength and also the indirect effect the channel addition has on *main channel usage*.

26.3 Relationship Deepening

As suggested by Geysken et al. (2002), Alba et al. (1997) and others, implementing interactive channel additions is an important strategy for strengthening consumer–brand relationships and leveraging total sales volumes. Hence studying the effects of channel additions on customer–brand relationships and main channel usage in a complementary perspective (Dutta-Bergman, 2004) seems relevant and important. Two theoretical models from social psychology are among the most influential perspectives for studying relationships, namely the Interdependency Model (Thibaut and Kelley, 1959) and the Investment Model (Rusbult, 1980). The latter model builds upon and extends the former model. Although frameworks for measuring the strength of consumer–brand relationship ties already exist in marketing (e.g. the BRQ framework of Fournier, 1994, 1998), we argue that the Investment Model of Rusbult (1980) is more applicable as it specifies structural relationships between the relationship dimensions and contains fewer predictor variables [see Thorbjørnsen et al. (2002a) for a comparison and test of the different consumer–brand relationship models].

The Investment Model postulates four main antecedents of relationship stability, namely satisfaction with the partner, quality of alternative partners and direct and indirect relationship investments. In a marketing context, the two first variables can be translated to brand satisfaction and perceived quality of alternative brands. *Brand satisfaction* is defined as "the consumer's response to the evaluation of the perceived discrepancy between prior expectations (or some other norm of performance) and the actual performance of the product as perceived after its consumption" (Tse and Wilton, 1988, p. 204), and is assumed to have a positive effect on the relationship stability between a brand and its customers. *Quality of alternative brand partners* refers to "the perceived desirability of the best available alternative to a relationship" (Rusbult et al., 1998, p. 359). If the perceived quality of alternative brands increases, this will have a negative effect on the stability of the relationship between a brand and its customers because of the reduction in perceived relative quality towards an existing brand relationship partner. Moreover, the model proposes that *direct relationship investments*, defined as "the magnitude and importance of the resources that are attached to the relationship – resources that would decline in value or be lost if the relationship were to end" (Rusbult et al., 1998, p. 359), will have a positive effect on relationship stability. Such resources may be time, money and other efforts invested in a relationship. The other additional antecedents put forward by Rusbult (1980) in the Investment Model is *indirect relationship investments*. These are investments that "come into existence when originally extraneous resources such as mutual friends, personal identity or shared material possessions become attached to the relationship" (Rusbult et al., 1998, p. 359). Also, indirect investments are proposed to influence relationship stability positively.

26.4 Characteristics of WebTV and SMS Services

In this section, effects of *mobile* channel additions and *webTV* channel additions on consumer–brand relationships are focused upon. Mobile channel additions are limited to the use of *short message services (SMSs)*. SMSs are a facility for sending short text messages between mobile phones (Turban et al., 2002). The text messages can be sent or received by cellphones via network operators or from SMS gateways on the Internet (Lai, 2002). *WebTV* is defined as television shows or programs distributed on the web. Interactive TV (Pramataris et al., 2001) is also used to describe online television services, and interactivity is definitively also an integral part of the webTV construct used in this chapter.

In the following, we discuss the characteristics of webTV and SMS services across four different dimensions; *information accessibility*, *information personalization*, *information dissemination*, and *information richness* (Nysveen, Pedersen and Thorbjørnsen, 2005). These dimensions are all relevant aspects of interactive communication (Doyle, 2001;

Siau et al., 2001; Te'eni, 2001; Balasubramanian et al., 2002; Dogac and Tumer, 2002) and serve as useful concepts for pinpointing the differential properties of webTV versus SMS services. Moreover, the four dimensions also construe important determinants for building consumer–brand relationship ties through two-way communication.

26.4.1 Information Accessibility

Channel additions make it easier for customers to given access to a brand. According to Balasubramanian et al. (2002), time is a resource that is very limited in a modern person's life and, therefore, very costly. Channels giving flexibility in time and space access to information would therefore be highly valued by customers. Watson et al. (2002) discuss what they label the "u-commerce" construct. Three characteristics of u-commerce are relevant for *information accessibility*: ubiquitous access (access everywhere), universal access (the possibility to stay connected wherever the customers are) and unison access (the integration of various communication systems that permit a single interface or connection point). WebTV makes a brand more flexible relative to location and time because a customer can access the brand both from a television and from a PC (location) and both when the customer is using a television and when the customer is using a PC (time). However, the flexibility made possible regarding time and location is much better for mobile services such as SMS. Ubiquity was also mentioned by Siau et al. (2001) as one of the unique features of mobile commerce. Moreover, portability and availability at all times are also pinpointed as important features by other researchers (Kannan et al., 2001; Yunos and Gao, 2002). Doyle (2001) pointed out the potential of everywhere access by taking advantage of location-based services, whereas Andersson and Nilsson (2000) referred to the place and time independence of mobile channels as a key advantage compared wth other channels.

26.4.2 Information Personalization

Another unique characteristic of webTV and SMS is *personalization*. Personalization is defined as the possibility of tailoring information and content to each customer's unique needs and preferences (Bezjian-Avery et al., 1998; Roehm and Haugtvedt, 1999). Often, such personalization is based on user profiles and identification. Interactive webTV increases broadcast companies' competitive advantage because personalized and relevant information for the TV viewers are enabled (Pramataris et al., 2001). Personalization entails that the individual customer – based on his/her preferences – can receive tailor-made broadcast services and gain access to support services that are adapted to their personal profile, and that he/she also has the possibility to subscribe to services that are in accordance with their individual profile. Lot21 (2001) argued that mobile phones are very personal and that only friends, family and co-workers are allowed access to their cellular phone number. This was supported by Siau et al. (2001), who argued that mobile communication can be personalized to represent information or services appropriate for the individual customer. Furthermore, uniqueness (that the information customers receive is adapted to the time of the day, customer location and customer roles and preferences) is one of the dimensions of the "u-commerce" construct presented by Watson et al. (2002), which describes the potential for personalization in mobile commerce. Another dimension enabling personal services through mobile channels is the possibility of sending relevant and time-sensitive information to, say, a loyal card customer (Doyle, 2001). Moreover, Kannan et al. (2001) argue that wireless devices are ideal for maintaining customer relationships. The reason for this, they contended, is the ability to provide truly personalized content and services by tracking personal identity, by the ability to track consumers across media and over time, by the ability to provide content and service at the point of need and by the capability to provide highly engaging content. According to Andersson and Nilsson (2000), personalization enabled by mobile channels improves the possibility for interactive relationships between a brand and its customers.

26.4.3 Information Dissemination

Through mobile channels, information can be sent to all mobile users within a specific geographic region or to consumers with a specific demographic background. Thus, brands have the opportunity to disseminate information to specific customer populations (Siau et al., 2001). In addition, mobile services are typically used to coordinate social networks. Information received by one member of a network is often forwarded to other members of the network (Doyle, 2001). An SMS broadcast by a brand, informing about a new version of a product, may be forwarded to other people not currently members of the brand's customer database. Hence brand information can be distributed on a broader level than the brand's own customer database, thanks to the social interaction among the members of the customer database. Studies within the uses and gratification theory have also focused on the unique gratifications of mobile channels. A study by Leung and Wei (1998) revealed that pagers were viewed as a mark of status and social identity. Pagers were used to express fashion and status and to integrate with peer social networks. A study by Ling (2001) also showed that mobile phones are used to express fashion and for presentation of self. Results from these studies indicated that gratifications for using mobile devices are related to the expression of characteristics of the individual. Ling (2001) also pointed out that mobile phones are often used in public spaces. This makes SMS suitable for customers to express their values and attitudes to other people. Hence this public use makes it possible for customers to express themselves in an open social context. The forwarding mechanism and the coordination of social networks are much more salient for SMS than for webTV, making information dissemination possibilities particularly relevant for SMS compared with webTV.

26.4.4 Information Richness

The literature on media and information richness is vast, especially within the information systems (IS) research tradition (Daft and Lengel, 1986; Evans and Wurster, 1997; Haythornthwaite et al., 1998). The conceptual content of media richness is fairly wide, and often includes traditional dimensions of interactivity, and also synchronicity, compatibility and transparency, multimedia properties, information storing, control over information flow, etc. (Haythornthwaite et al., 1998). According to Evans and Wurster (1997), the primary dimensions of media richness are the level of interactivity and bandwidth of communication. As various properties of interactivity are included in our remaining information dimensions (accessibility, personalization, dissemination), we here focus on richness very narrowly, namely as interface bandwidth. The richness entailed in the interface is thus influenced primarily by the data transfer bandwidth, but also screen size and multimedia features allowed for by the interface. When comparing webTV and SMS services along this dimension, it seems fairly obvious that webTV allows for significantly richer information than does SMS. SMS services are today limited to text messages only, and even the next generation of messaging services – MMS – which allows for pictures, jazzy graphics and audio and video clips, are currently not even near webTV when it comes to information richness. webTV allows for real-time feed of broadband content, delivered through large TV or PC screens, often with hi-fi sound equipment.

26.5 Propositions

26.5.1 Main Channel Use – Direct Effects

As discussed in previous sections, it is of vital importance when evaluating channel additions not only to look at the isolated effects of the channel addition itself, but also to pay close attention to the effects that channel addition have on main channel usage. After all, the brand's main channel is usually the strongest contributor for overall brand revenues.

Table 26.2 Comparison of channel characteristics

Information dimensions	Main channel	WebTV	Mobile (SMS)
Information accessibility	Low	Medium	High
Information personalization	Low	High	High
Information dissemination	Low	Low	High
Information richness	High	High	Low

Consequently, to be able to evaluate whether the channel addition will substitute, supplement or complement a brand's main channel, one need to examine carefully the characteristics of both the potential channel addition and the characteristics of the main channel. If, for instance, the same gratifications are being met in both channels (that is, the channel characteristics are very similar), then we expect the channels to supplement or even substitute each other, instead of complementing each other. In Table 26.2, we compare a brand's main channel (TV2 broadcast channel) with the two channel additions (SMS services and webTV) across the four dimensions discussed above.

The relative advantages of the two channel additions (WebTV and SMS services) were discussed in detail above. When comparing the TV2 main channel across the four information dimensions, this one-way, non-interactive medium receives a high score only on information richness. The dimensions of information accessibility, personalization and dissemination – as defined in this chapter – all pertain to and require some level of interactivity or connectedness to interactive media. An analog, non-interactive TV channel is rich in the sense of bandwidth, screen size and multi-media features, but will not permit personalization, consumer-controlled accessibility to information dissemination of information content.

Table 26.2 clearly indicates that the level of complementarity between channel additions and the main channel will depend on the gratifications sought by the consumer in the different channels. However, as we do not have any real insight into the configuration (relative weight) of which gratifications consumers will seek from each channel addition, an evaluation of potential channel complementarity will have to be based on a subjective inspection of channel characteristics. Even at first glance one can easily argue that webTV channel additions have most features and gratifications in common with the main channel, and we would therefore expect this channel to be a supplement rather than a complement to the main channel. We know that information richness is very important to consumers when watching TV – whether offline or online – and both the webTV channel and main channel do a good job in this dimension. Moreover, when it comes to information accessibility and dissemination, the webTV channel is not very different from the main channel, at least not when compared with mobile services.

By the same token, we observe that mobile SMS services perform totally different from regular TV (main channel) in all information dimensions listed. In the dimensions in which SMS services have a strong position, the main channel has not, and vice versa. Consequently, we propose that webTV primarily will serve as a *supplement* to a brand's main channel and that mobile SMS services will serve as a *complement* to a brand's main channel. This, in turn, gives us the following initial propositions:

- *Proposition 1a:* Consumers' use of webTV channel additions will have no direct effect on main channel use.
- *Proposition 1b:* Consumers' use of SMS channel additions will have a positive direct effect on main channel use.

By direct effect we mean effects that are *not* being mediated via strengthened consumer–brand relationship ties, that is, an added effect that may be attributed to, e.g., the degree to which the channels are complementary versus supplementary.

26.5.2 Main Channel Use – Indirect Effects

As discussed at the beginning of this chapter, channel additions may be implemented for both marketing communication and distribution purposes. The distinction between the two is often blurred and difficult to identify – especially for informational/digital products and services – where consumers often pay for the informational content being exchanged. For the brand case utilized in this paper, TV2, this is certainly the case. WebTV and SMS channel additions are here implemented both for the purpose of strengthening consumer–brand relationships through interactive communication and for increasing revenues through distributing subscription-based services through new channels of distribution. The final and dual purpose of both marketing communication and distribution is still the same: increasing revenues through the brand's main channel and channel additions. The direct effects of webTV and SMS channel additions as distribution channels on main channel usage were discussed above. In the following, we shall discuss the effects that the two channel additions have *on* and *via* consumer–brand relationship ties.

The investment model consists of four relationship dimensions: satisfaction, quality of alternative partners, direct relationship investments and indirect relationship investments. In short, we believe that webTV and SMS channel additions will have a significant (positive) effect on all these dimensions. First, for satisfaction and quality of alternative partners, we believe that all the properties of information listed in Table 26.2 to some extent will have a significant and positive influence on the overall satisfaction and quality perceptions of consumers.[1] There is sufficient evidence in the literature on the positive effects of information accessibility, personalization and richness on brand satisfaction and quality perceptions (Luedi, 1997; Novak, et al., 2000; Moon, 2000; Balasubramanian et al., 2002; Watson et al., 2002; Thorbjørnsen et al., 2002b) for proposing a positive relationship between webTV and SMS channel additions and these two relationship dimensions. However, it should be noted that an increase in perceived brand partner quality will decrease the perceived relative quality of alternative brand partners (other brands) and the directional relationship between the channel additions and quality of alternative partners will therefore be negative. Turning to the effects on direct relationship investments, we also expect a positive effect of webTV and SMS channel additions. Through using these channel additions, i.e. learning new features, spending time, money and cognitive efforts on the channel, the brand continuously becomes a more intertwined part of consumers' everyday life. We also postulate a positive relationship between the two channel additions and indirect relationship investments. Although SMS channel additions have a particular high standing on the property of information dissemination (which is very central for connecting external resources such as friends and family to the brand relationship), we still expect webTV channel additions to influence positively indirect relationship investments through word-of-mouth, networked resources, etc.

Consequently, as the consumer–brand relationship ties also to a large extent drives behavior (usage of the main channel and channel addition), we need to consider relationship strength when estimating the effects of the channel additions on main channel usage. In proposition 1a, we stated that webTV channel additions will not have any effects on main channel use, as the channels to a large extent are perceived as supplements or even substitutes. Of course, as we argue that webTV channel additions will leverage consumer–brand relationship ties, one can easily extend this line of argument by saying that these relationship ties, in turn, will translate into increased use of both the main channel and the channel addition. However, as stated in proposition 1a, this would not be a "direct effect", but rather an effect being fully mediated by increased relationship strength. For SMS services, we do *not* expect the effect to be fully mediated by relationship dimensions, because the two channels would also be complementary in usage motives and gratifications. That is, for SMS services we would expect a *direct*, unmediated, effect on main channel

---

[1] Increased perceived quality of the brand partner will, according to the Investment Model, deflate the relative quality of alternative partners. A negative relationship between channel addition usage and quality of alternative partners therefore represents a positive contribution to consumer–brand relationship ties.

usage (proposition 1b), in addition to an *indirect*, mediated, effect via increased consumer–brand relationship ties. Consequently, we can now extend the effects proposed in propositions 1a and 1b by also taking into consideration the indirect effects on main channel use via relationship dimensions:

- *Proposition 2a:* Consumers' use of webTV channel additions will have an *indirect* positive effect on main channel use, mediated through strengthened consumer–brand relationship dimensions.
- *Proposition 2b:* Consumers' use of SMS channel additions will have an *indirect* positive effect on main channel use, mediated through strengthened consumer–brand relationship dimensions.

26.6 Methodology

The study reported here was conducted for TV2, which is one of several companies offering broadcast services in Norway. The TV2 brand was used because it offers both webTV and SMS channel additions. The webTV channel addition offers services such as news, entertainment, sport, weather and current event programs. The programs can be downloaded and watched when it suits the customers, or they can be viewed in real-time. It is also possible for the customers to edit the programs. For example, they can choose to watch only part of news or sport programs. Some of the services are free, but customers have to pay for most of the services. Payment services are typically organized as subscription services, and subscription periods range from 1 week to 1 year. Links to relevant additional information are also available on the webTV. When a customer is watching a concert by an artist, links are made available e.g. to the artist's website, reviews of the artist's last album and shops where one can by CDs by the artist. There is also a "My webTV" menu where customers can personalize their services. It is possible for customers to view their purchase history on the webTV, get an overview of the services to which they are actually subscribing, initiate or quit subscription services and personalize the quality of the program presentation to their hardware and bandwidth capacity. It is also possible, of course, for the customers to have a dialogue with the brand through e-mail on the webTV site. The SMS channel addition includes the possibility to subscribe to services related to weather forecasts, news, sport and entertainment on SMS. This means that the customers specify their preferences for what kind of information they want to subscribe to and how often they want to be notified about the subjects and topics they have specified. The TV2 SMS alert and -subscription services are all available through the brand's website, www.tv2.no.

26.6.1 Design, Sample and Procedure

To test the propositions set forth, a quasi-experimental design, including a pre-test and a post-test study, was utilized. The pre-test study was announced on the two sections of the TV2 website relating to webTV services and SMS services, respectively. Recruitment was based on a self-selection procedure, and respondents participated in the study through clicking on an interactive announcement text on the webTV or SMS services web page. After having clicked on the announcement text, they gained immediate access to an online questionnaire (pre-test). Respondents were informed that the survey included another questionnaire, of which they would be reminded of by e-mail or regular mail 2 weeks later. The respondent decided themselves whether they preferred to receive an electronic- or paper version of the post-test. Sample characteristics of the two groups are listed in Table 26.3.

26.6.2 Measures

The Investment Model, presented earlier, includes four basic constructs; satisfaction, quality of alternatives and direct and indirect relationship investments. In addition to these relationship concepts, channel addition usage and main channel usage were also measured. It should be noted that the variable "Use of main channel" was measured in the post-test, whereas the remaining variables were measured in the pre-test. By measuring main channel usage at a later point of time, we therefore have a

Table 26.3 Sample characteristics

	SMS ($n = 226$)	WebTV ($n = 230$)
Age		
0–19	22.2	19.2
20–29	43.6	48.0
30–39	20.4	16.6
40–49	7.6	10.9
50–59	4.9	4.8
≥60	1.3	0.4
Education		
Primary	9.4	12.7
Secondary	50.9	39.7
University ≤3 years	21.9	28.8
University >3 years	17.9	18.8
Sex		
Male	75.7	73.3
Female	24.3	26.7

more reliable proxy of how channel addition usage and the consequent increase in consumer–brand relationship ties influence main channel usage over time. Measuring both the proposed antecedent (channel addition usage) and consequence (main channel usage) simultaneously would threaten the internal validity of the study. Moreover, consumer–brand relationships evolve over time, and the effects of such relationship will therefore materialize over time – not instantly. The measures of investment model constructs were based on Rusbult (1980), Rusbult et al. (1998) and Thorbjørnsen et al. (2002a), although some items were slightly revised to fit the present setting better. Convergent and discriminant validity, and also reliability, were investigated and found acceptable for both the webTV and SMS subsamples, according to the procedures proposed by Hair et al. (1998) and Anderson and Gerbing (1988).

26.7 Results

Our research model consists of four propositions to be tested, that is, the direct and indirect effects of SMS and webTV channel additions on main channel usage. This model may be estimated using structural equation modeling and treated as a "supermodel" consisting of three nested models. The first model merely predicts the effects of relationship dimensions on main channel usage. The second model focuses on the indirect effects of the channel addition on main channel usage via relationship dimensions (propositions 2a and 2b). The third model pertains to the added direct effects of channel additions on main channel use when controlling for indirect effects (propositions 1a and 1b). In order to avoid invalid conclusions of the main effects of channel addition usage on main channel usage, we investigate the propositions by using the "supermodel" as a frame of reference and gradually remove the direct and indirect effects of channel addition usage. Through utilizing this strategy, the nested models can be compared using comparative indexes and measures in structural equations modelling (Rust, Lee et al., 1995).

In Figure 26.1, the two supermodels are displayed, showing the direct effects of brand relationship dimensions on main channel use, the indirect effects of channel additions and the direct (unmediated) effects of channel addition usage on main channel usage. Both the SMS model (CFI = 0.099, RMSEA = 0.051) and the webTV model (CFI = 0.099, RMSEA = 0.062) show a good overall fit.

From Figure 26.1, we see that the explained variance in main channel use is 42.8% of webTV and 49% of SMS channel additions, suggesting that the models have considerable explanatory power. The regression coefficient between channel addition use and main channel use reflects the added effect of channel additions on main channel use, that is, effects that are not mediated through relationship dimensions. Consequently,

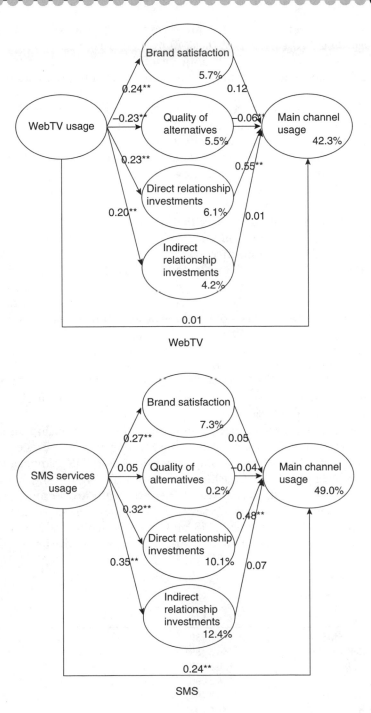

Figure 26.1 Structural models (standardized coefficients), ** and * indicate significance at $p < 0.01$ and $p < 0.05$, respectively.

this coefficient can be used for investigating propositions 1a and 1b regarding direct effects. For the SMS study, this coefficient is significant at the 1% level, whereas no significant effect is observed for the webTV study. Accordingly, we find general support for both propositions 1a and 1b.

When turning to the test of propositions 2a and 2b, we need to investigate the impact on model fit and explanatory power when gradually removing direct and indirect effects of channel additions. In Table 26.4, the χ^2 values for the supermodel and the two nested models excluding direct and indirect effects are shown. First, we remove direct effects (see $\Delta\chi^2$ values for "Supermodel") and then we remove indirect effects (see $\Delta\chi^2$ values for "Indirect").

From Table 26.4, we observe that on removing the direct and indirect effects of SMS channel addition, the χ^2 values increases by 10.5 (d.f. = 1) and 35.5 (d.f. = 4), respectively. This observation has two primary implications. First, the SMS channel additions have both a direct and an indirect effect on main channel usage. Consequently, both propositions 1b and 2b are supported. Second, this finding also suggests that the supermodel is the best model – of the three nested models tested – for explaining main channel use. For the webTV channel additions, removing the indirect effect increases the χ^2 values by 16.5 (d.f = 1), whereas no increase is observed on removing direct effects. Consequently, propositions 1a and 2a are supported. The lack of direct effect of webTV channel additions – which also can be observed by inspecting Figure 26.1 – is in line with our predictions.

Now, an additional aspect to look at is *which* relationship dimensions the effect on main channel is mediated through. WebTV usage has a significant effect on all relationship dimensions. For SMS channel additions, the effects on all relationship dimensions are significant, besides the effect on quality of alternatives. As there currently are far more providers of SMS services than webTV services in Norway, we believe that the difference in effect between SMS and webTV channel additions in this dimension pertains more to the *availability* of alternative brand partners,

Table 26.4 Nested model analysis

Model	χ^2 (d.f.)	$\Delta\chi^2$ (d.f.)	R^2	ΔR^2
"Supermodel" SMS	140.2 (89)	10.5 (1)**	0.490	0.035
"Supermodel" webTV	166.5 (89)	0.0 (1)	0.428	0.000
"Indirect" SMS	150.7 (90)	35.5 (4)**	0.455	0.012
"Indirect" webTV	166.5 (90)	16.5 (4)**	0.428	0.001
"Relationship" SMS	186.2 (94)		0.443	
"Relationship" webTV	183.0 (94)		0.427	

**Significance at $p < 0.01$.

and not so much to the relative quality. Consequently, we do not pay much attention to this differential effect. However, on inspecting the repression coefficients from relationship dimensions to main channel usage, more surprising effects are revealed. For both webTV and channel additions, the indirect effect works through direct relationship investments alone. None of the remaining dimensions – including satisfaction – seem to mediate the effects of channel additions on main channel use.

26.8 Discussion and Implications

The results revealed in this study have several implications for both marketing scholars and industry players. First, interactive channels additions such as webTV and SMS services appear as effective tools for building strong consumer–brand relationships. The regression paths from channel addition usage to relationship dimensions are all significant (except for the effect on quality of alternatives for SMS channel additions). This is an important finding, as strong consumer–brand relationship ties have proven to have positive effects on variables such as commitment, tolerance for price premiums and positive word-of-mouth (Thorbjørnsen et al., 2002a). Adding interactive channels to the brands marketing mix may therefore significantly leverage brand equity even in cases where the short-term effects on profits may be slim.

Second, the effects of webTV and SMS channel additions on main channel usage are mediated through direct relationship investments. That is, when consumers perceive they have invested significant efforts into using, learning and paying for the channel addition, this investment also transforms into increased usage of the brand's main channel. Through increased channel addition consumption, the consumers perceive that the brand has become a more intertwined part of their everyday life, and their usage of both brand channels consequently increases. The implications of this finding for mobile service vendors are therefore fairly straightforward, albeit perhaps more difficult to implement in practice. Since main channel usage primarily is mediated through direct relationship investments, service vendors should put serious efforts into attempting to facilitate such investments. Direct relationship investments may be strengthened through increasing the relevance and salience of the service across a larger set of everyday situations, in other words, increasing the breadth of brand awareness for consumers. Consumers should be encouraged to integrate the use of mobile services into their everyday life tasks and to spend time and money learning and using the service. Moreover, advertising campaigns could, for instance, portray the strong self-congruency between "cool users" and their preferred mobile services, or display the share variety of situations and settings in which the service could be utilized.

Third, when channels additions are perceived as complements – rather then as supplements – to the brand main channel, we observe a direct, unmediated effect on main channel usage. We argue that mobile channel additions such as SMS services have very different characteristics and comparative advantages compared with the brand's main channel (TV), and will therefore fulfil different consumer gratifications. Whereas SMS channel additions receive a high score on the properties of information accessibility, personalization and dissemination, but a low score on information richness, the reverse is true for the brand main channel. Concurrently, SMS channel additions complement the services provided through the brand channel – and vice versa – and the two channels will therefore reinforce each other. WebTV services, on the other hand, do not meet sufficiently different consumer gratifications, and will therefore supplement, rather then complement, the brand's main channel. However, owing to the significant effects of webTV channel additions on relationship dimensions, and subsequent indirect effect on main channel usage, the webTV channel additions must also be considered a significant success in this picture.

As the bandwidth of mobile services increases and interfaces become richer, the properties of the TV, webTV and mobile services converge. Increased convergence of channel properties makes the channels more similar on key dimensions, and hence less complementary – something that potentially will increase channel conflicts and competition. However, owing to the unique features of mobile services when it comes to ubiquitous information access and location-specific services, we argue that their comparative advantages as relationship-building tools remain strong. Although more research is warranted on these issues, the basic and general implications for brand vendors seem clear: interactive channel additions such as webTV and SMS services serve as efficient relationship-building tools, and the primary key to increased main channel usage is facilitating consumers' direct relationship investments. Moreover, investments in mobile channel additions appear to be particularly beneficial when the channels serve as complements to other channels of distribution and marketing communication. Vendors and suppliers of mobile services should therefore carefully examine the properties of the remaining channels of distribution and marketing before launching a new service. Concurrently, after having launched the service, considerable efforts should be put into demonstrating how the new channel is complementary to existing ones, and facilitating direct relationship investments.

26.9 References

Aaker, J., Fournier, F. and Brasel, S.A. (2004) When good brands do bad. *Journal of Consumer Research*, 31, 1–16.

Alba, J.W., Lynch, J., Weitz, B., Janiszewski, C., Lutz, R., Sawyer, A. and Wood, S. (1997) Interactive home shopping: consumer, retailer, and manufacturer incentives to participate in electronic marketplaces. *Journal of Marketing*, **61**, 38–53.

Anderson, J.C. and Gerbing, D.W. (1988) Structural equation modelling in practice: a review and recommended two-step approach. *Psychological Bulletin*, **103**, 411–434.

Anderson, J.C. and Narus, J.A. (1995) Capturing the value of supplementary services. *Harvard Business Review*, **73**, 75–83.

Andersson, A. and Nilsson, J. (2000) Wireless advertising effectiveness. Evaluation of an SMS advertising trial. Master Thesis in Marketing, Stockholm School of Economics, Stockholm.

Balasubramanian, S., Peterson, R.A. and Jarvenpaa, S.L. (2002) Exploring the implications of M-commerce for markets and marketing. *Journal of the Academy of Marketing Science*, **30**, 348–361.

Bezjian-Avery, A., Calder, B. and Iacobucci, D. (1998) New media interactive advertising vs. traditional advertising. *Journal of Advertising Research*, **38**, 23–33.

Chan-Olmsted, S.M. and Ha, L.S. (2003) Internet business models for broadcasters: how television stations perceive and integrate the Internet. *Journal of Broadcasting and Electronic Media* **47**, 597–617.

Daft, R.L. and Lengel, R.H. (1986) Organizational information requirements, media richness and structural design. *Management Science*, **32**, 554–570.

Dogac, A. and Tumer, A. (2002) Issues in mobile electronic commerce. *Journal of Database Management*, **13**, 36–43.

Doyle, S. (2001) Software review: using short message services as a marketing tool. *Journal of Database Marketing*, **8**: 273–277.

Dutta-Bergman, M.J. (2004) Complementarity in consumption of news types across traditional and new media. *Journal of Broadcasting and Electronic Media* **48**, 41–60.

Evans, P.B. and Wurster, T.S. (1997) Strategy and the new economics of information. *Harvard Business Review*, Sept.–Oct., 71–82.

Fournier, S. (1994) A consumer–brand relationship framework for strategic brand management. Unpublished PhD Dissertation, University of Florida, UMI.

Fournier, S. (1998) Consumers and their brands. Developing relationship theory in consumer research. *Journal of Consumer Research*, **24**, 343–373.

Geysken, I., Gielens, K. and Dekimpe, M.G. (2002) The market valuation of internet channel addition. *Journal of Marketing*, **66** (April), 102–119.

Hair, J.F., Anderson, R.E., Tatham, R.L. and Black, W.C. (1998) *Multivariate Data Analysis*, 5th edn. Prentice-Hall, Englewood Cliffs, NJ.

Haythornthwaite, C., Wellman, B. and Garton, L. (1998) Work and community via CMC. In Gackenbach, J. (ed.) *Psychology and the Internet*. Academic Press, London.

Hoffman, D.L. and Novak, T.P. (1996) Marketing in hypermedia computer-mediated environments: conceptual foundations. *Journal of Marketing*, **60** (July), 50–68.

Holland, J. and Baker, S.M. (2001) Customer participation in site brand loyalty. *Journal of Interactive Marketing*, **15**, 34–45.

Kannan, P.K., Chang, A.-M. and Whinston, A.B. (2001) Wireless commerce: marketing issues and possibilities. In *Proceedings of the 34th Hawaii International Conference on System Sciences*, pp. 1–6.

Lai, T.L. (2002) Short message service (SMS): the impact on service quality and perceived value on satisfaction, intention and usage. Working Paper, Nanyang Business School, Nanyang Technological University, Singapore.

Lazarsfeld, P.F. (1940) *Radio and the Printed Page*. Dell, Sloan and Pearce, New York.

Leung, L. and Wei, R. (1998) The gratification of pager use: sociability, iInformation-seeking, entertainment, utility, and fashion and status. *Telematics and Informatics*, 253–264.

Ling, R. (2001) "It is in". It doesn't matter if you need it or not, just that you have it. Fashion and the domestication of the mobile telephone among teens in Norway. Working Paper, Telenor R&D, Oslo.

Lot21 (2001) *The Future of Wireless Marketing*. Lot21, San Francisco.

Luedi, A.F. (1997) Personalize or perish. *Electronic Markets*, **7**: 22–25.

Mathwick, C. (2002) Understanding the online consumer: a typology of online relational norms and behavior. *Journal of Interactive Marketing* **16**, 40–55.

McCombs, M. (1972) Mass media in the marketplace. *Journalism Monographs*, **24**, 1–104.

Moon, Y. (2000) Intimate exchanges: using computers to elicit self-disclosure from consumers. *Journal of Consumer Research* **26**: 323–339.

Novak, T.P., Hoffman, D. and Young, Y.F. (2000) Measuring the customer experience in online environments: a structural modeling approach. *Marketing Science*, **19**: 22–42.

Nysveen, H., Pedersen, P.E. and Thorbjørnsen, H. (2005) Mobilizing the brand: the effect of mobile services on brand relationships and main channel use, *Journal of Service Research*, **7**: 257–276.

Pramataris, K.C., Papakyriakopoulos, D.A., Lekakos, G. and Mylonopoulos, N.A. (2001) Personalized interactive TV advertising: the iMEDIA business model. *Electronic Markets*, **11**, 17–25.

Riel, A.C.R., Liljander, V. and Jurriens, P. (2001) Exploring consumer evaluations of e-services: a portal site. *International Journal of Industry Management*, **14**: 359–377.

Roehm, H.A. and Haugtvedt, C.P. (1999) Understanding interactivity in cyberspace advertising. In Schuman, D.W., Thorson, E. (eds), *Advertising and the World Wide Web*, Lawrence Erlbaum, Mahwah, NJ.

Rusbult, C.E. (1980) Satisfaction and commitment in friendships. *Representative Research in Social Psychology*, **11**, 96–105.

Rusbult, C.E., Martz, J.M. and Agnew, C.R. (1998) The investment model scale: measuring commitment level, satisfaction level, quality of alternatives, and investment size. *Personal Relationships*, **5**: 357–390.

Rust, R.T., Lee, C. and Valentine, E., Jr (1995) Comparing covariance structure models: a general methodology. *International Journal of Research in Marketing*, **12**, 279–291.

Siau, K., Lim, E.-P. and Shen, Z. (2001) Mobile commerce: promises, challenges, and research agenda. *Journal of Database Management* **14**, 4–13.

Te'eni, D. (2001) Review: a cognitive-affective model of organizational communication for designing IT. *MIS Quarterly*, **25**, 251–312.

Thibaut, J.W. and Kelley, H.H. (1959) *The Social Psychology of Groups*. Wiley, New York.

Thorbjørnsen, H., Breivik, E. and Supphellen, M. (2002a) Consumer–brand relationships. A test of alternative models. In Evans, K.R., Scheer, L.K. (eds), *Proceedings of the AMA Winter Educators Conference 13*, Austin, RX, pp. 283–285.

Thorbjørnsen, H., Supphellen, M., Nysveen, H. and Pedersen, P.E. (2002b) Building brand relationships online. A comparison of two interactive applications. *Journal of Interactive Marketing*, **16**, 17–34.

Tse, D.K. and Wilton, P.C. (1988) Models of consumer satisfaction formation: an extension. *Journal of Marketing Research* **25**, 204–212.

Turban, E., King, D., Lee, J., Warkentin, M. and Chung, H.M. (2002) *Electronic Commerce. A Managerial Perspective*. Prentice Hall, Pearson Education International, Upper Saddle River, NJ.

Watson, R.T., Pitt, L.F., Berthon, P. and Zinkhan, G.M. (2002) U_Commerce: expanding the universe of marketing. *Journal of the Academy of Marketing Science*, **30**: 333–347.

Yunos, H.M. and Gao, J. (2002) Wireless advertising. Working Paper, Department of Computer Engineering, San Jose State University, San Jose, CA.

27

The Integration of Mobile Alerts into Everyday Life

Marianne Jensen, Kristin Thrane and Siri Johanne Nilsen

27.1 Introduction

The unexpected success of mobile phones and text messages (short messaging system, SMS) made it clear to companies in the telecommunication industry that the boundaries of the product and services development are not only techno-economic, focusing on bandwidth and memory capacity, but in essence, socio-human (Repo, 2003). The creation of a market means creating totally new categories of products and needs, placing major emphasis on strategies focusing on user-friendliness. The perspective of social shaping of technology (Bijker and Law, 1992) stresses the importance of justifying the consumer as an active player in the dialogue between producers and consumers in the design process of the product. One of the obvious reasons for failed product development projects is the ways in which on the one hand the engineers, marketers and designers perceive the product, whereas on the other hand the needs of the consumer are completely different. An even more serious problem is that the image of the consumer is generally one-sided (Latour, 1996). In many cases, the consumer image tends to be more or less drawn from the technical world of the product developers themselves.

Mobile communications devices and services are in a stage of rapid development. The scenarios for 3G mobile phones are really the radio, record player and television of today. It is at the same time a new invention and a combination of old media technology put in a new situation. Huge expectations have been attached to the upcoming services but, as previously, development has been largely technology driven, even though the end user is frequently mentioned in the visions (Repo, 2003; Ling, 2004).

This chapter is a result of two broadband pilots that were tested and evaluated in Norway during Autumn 2002 and Spring 2003, under the names "News and Sport" pilot and "Lifestyle" pilot. Telenor[1] and NRK[2] collaboratively carried out the pilots, which lasted approximately 4 weeks each. The chapter deals with the topic of "mobile alerts", that is, SMS services to notify mobile users of special events, news or programs offered as broadband content, such as webTV services. We do not yet know the purpose for which it would be natural to utilize the mobile alert capability. The invention and development of alert-related services are also still at an early stage. However, instead of focusing here on the swift development of technology, we are primarily interested in identifying the kinds of situations in which users would consider mobile alert-related services as meaningful. This chapter is based on six group interviews with a total of 39 users. Of most importance to us is to make the voice of the critical consumer heard in product development at an early stage.

27.2 Research Approach

Our research approach can be considered in the context of the following research question: in what kinds of situations is it meaningful to use mobile alerts?

Our study looks for ideas about the kinds of situations in which mobile alerts possibly can be of use, and why; we try to identify social factors explaining the adoption or rejection of this type of mobile communication, including routinized habits, practices and feelings. Using qualitative methods, this study describes how a selection of users integrated the technology into the mental and physical context of their everyday life.

In the following section, we describe the trial and methods used in data collection. Next, there is a presentation of theoretical contributions regarding the adoption and use of new technology. This study draws on different research traditions. We shall look into the work of Rogers' diffusion model (Rogers, 1955), before we describe Silverstone, Hirsh and Morley's concept of domestication (Silverstone et al., 1992). More specifically, we outline key elements of the domestication approach, showing where these are shared in some empirical studies of mobile telephony; analyzing consumers' choices and routinized habits, practices and usage situations in their daily life (Haddon, 2003; Ling, 2004). Following this, we return to the analysis of the empirical data organized into discussions of the metaphorical and physical integration of mobile alerts into everyday life. We provide initial insight into how mobile alerts must match different mental and physical contexts to be meaningful to users. Finally, in the conclusion, we present a model that sums up the kinds of situations in which mobile alerts would be of use.

---

[1] A major Norwegian telecommunication provider.
[2] The national Norwegian broadcaster.

27.3 Description of the Trial and Methods Used in the Study

The two pilots can both be considered to be market and research trials. The research focuses on technical development and user interface issues in addition to user behavior. The commercial aim of the trials was to test the attractiveness of and the willingness to pay for new broadband services. Mobile alerts were introduced to engage users in a more interactive way and to increase the use of broadband services.

The "News and Sport" pilot had 3001 registered users whereas the "Lifestyle" pilot had 2500 registered users. Both had periods where the users had to pay to use the services and explore the content. The users were recruited by advertisements on different websites and through e-mails to ADSL customers (from both Telenor and NextGenel[3]). Because of the content and technical solutions, users were required to have an ADSL connection. This is the main reason why the users had a high level of knowledge and expertise about broadband technology and opportunities and Internet services in general. The skills of the users became very evident during the focus group interviews. Many of the users were working within the IT field. There were also disabled pensioners and retired people amongst the participants in the focus groups. The fact that users had an ADSL connection at the time of the trial and used the Internet and the opportunities provided by broadband access on a regular basis shows that the users are innovators and early adopters (Rogers, 1995; see later in the chapter for more about this).

In the "News and Sport" pilot, only 6% of the participants were females. Because of the lack of female interest, one of the main objectives of the "Lifestyle" pilot was to have more women trying out the services. The services were designed to engage more women. To a certain extent the aim to reach more women in the latter pilot was fulfilled, as 15% of the users were female. It is important to bear in mind that the pilot periods were of limited duration, so the trial users did not have enough time to integrate the services into their everyday life and daily Internet routines. The participants in the focus groups also stressed that their use of the pilots had a character of testing the services rather than the frequent use of a service. These are factors that may bias the results. However, the tests were carried out in real user environments with realistic and high-quality content, and for that reason we argue that the reactions of the users provide valuable insight into many aspects of the adoption process. With this background the study is exploratory, and the results are not generalizable to other populations.

---

[3] One of the broadband providers in Norway.

In the "News and Sport" pilot, eight mobile alerts were distributed during the 4-week pilot, giving notice of sports events, political events and special news. The web-traffic report shows an increase in the use of the services after the alerts had been distributed. It is important to stress the fact that the users did not have the opportunity to choose if they wanted to receive mobile alerts or not in the "News and Sport" pilot.

In the "Lifestyle" pilot, the mobile alerts were not exploited or explored as much as a service. The users could voluntarily sign up for different types of mobile alerts in this test. A result of this was that 75% of the trial users did not have any experience with mobile alerts. The material shows that 460 out of 2500 pilot users did sign up for the general news alerts. However, the pilot did not send out news alerts during the pilot period, as no news was regarded as important enough for such distribution. None of the other categories of alerts in the "Lifestyle" pilot were popular. These were "stop smoking course" (16 persons signed up) and "recipe of the day" (17 persons signed up). At the same time, 460 users did sign up to receive alerts in the category "general news".

In both trials we used both qualitative and quantitative methods for data acquisition. These were traffic logs from web-servers, e-mail questionnaires and focus groups. Since this chapter is of an exploratory nature, we have chosen to build our arguments on the qualitative data from the focus groups and to disregard the quantitative data, as these are less relevant for illuminating the contextualization of the mobile alerts. Some qualitative data will be referred to, but for descriptive and illustrative purposes only. We carried out four focus groups for the "News and Sport" pilot and two focus groups for the "Lifestyle" pilot. The number of participants in the focus groups was between five and eight. In the "News and Sport" pilot only two women volunteered for the focus group interviews. In the "Lifestyle" pilot six women attended the interviews.

The focus groups were organized as informal group conversations. We started with an introduction where the respondents introduced themselves, their age, profession and what kind of ICTs they had at home. We then started a discussion around the various elements of the trials. Through the discussions we obtained unique insight into how the pilots were experienced, used and contextualized by the users. The interviews were recorded and transcribed and quotes have been translated into English from Norwegian. Analytical categories have been created on the basis of analysis of the sound records and text (Loafland and Loafland, 1984).

27.4 Technology and Social Change

Mobile telephony is a technology that is quickly finding its niche. Its functionally has grown beyond simple communication to a system that allows for the communication of text, access to the Internet, the capturing and sending of images and the distribution of location sensitive information. Mobile alerts are just another twist to the technical development.

However, both theoretical and empirical research on digital information and communication (ICT) products such as the Internet, mobile phones and digital TV is still in its infancy, being part of a more fundamental discussion in social science about technology and social change.

Borch (2002) has pointed to several theoretical contributors to the understanding of the interaction between technology and society. For a long time social scientists found inspiration in the theories of Saint-Simon, Marx and Weber about the relationship between technology and society. Suggestions were made that the dominant notion of the IT–social relationship was based on the idea that technology was affecting society without being reciprocally influenced. Based on the opposite notion that technology does not live a life of its own, but is socially constructed, is the work of Bijker et al. (1987), Hughes (1986), Russel (1986) and – not least – Latour (1987), who drop the distinction between the technical and the social and understand technology in terms of relationships formed between human and non-human elements of "actor networks".

Even though the social shaping of the technology concept was a fruitful correction to the dominant notions about technology, it nevertheless shared one fundamental concern with technological determinism: they were both focusing on one side of the technology, the side of conception, invention, development and design (Winner, 1985). Media and culture studies assumed that a limited focus on a socially shaped development and production of technology was incomplete because it failed to consider the social forces at work on the other side of technology: the way in which technology came to be appropriated by their users. One of the directions in media and culture studies is focusing on the domestic consumption of ICT products, arguing that the use and meaning of these kinds of products can only be understood within class, gendered, geographical and generational aspects of their consumption context. Silverstone et al.'s (1992) concept of domestication will be further described and compared with two other theoretical contributors in the next section.

27.5 Theoretical Contributions

Two of the most often cited and used theoretical contributions to the understanding of consumers' adoption and use of new ICT products are Rogers' diffusion model and Silverstone, Hirsch and Morley's concept of domestication. However, since we are dealing with the mobile telephone, it seems to us that Beck and Beck-Gersham (2002) would be a theoretical contribution of relevance, indicating the role of the mobile phone as a supporting actor for the rise of a new institution, that of individualism. This means that, contrary to Silverstone et al.'s perspective, pointing to the relationship between the technology and, in a broader sense, the family structures, Beck and Beck-Gersham point out that the family, the political and all other structures will be carried out against the backdrop of the individual and his or her uniqueness and ego.

Rogers' diffusion model is basically a theory about how innovations are spread from one population to another (Rogers, 1995). According to Rogers, most diffusions follow a predetermined curve, reflecting his notion that the world consists of five adoption groups: innovators, early adopters, early majority, late majority and laggards. His well-known theory of "a critical mass" is based on the idea that the value of a communication device increases with the number of persons owning one. For instance, the utility of having access to the Internet or owning a mobile phone increases with more people having access to or owning the same item.

Rogers' diffusion model has some positive elements that have to be considered: first, the view of the diffusion as a two-step process that emphasizes the role of person-to-person social dynamics in the adoption processes, but also the description of the attributes associated with the item that eventually are being adopted in addition to the role of the critical mass in the adoption process.

One of the negative elements is that the vocabulary of the model heavily reflects the fact that people's behavior is caused by the diffusion of facts and machines and not ICT products. Beyond this, several elements are problematic with Rogers' perspective. These include the assumption of rational actors, a simplistic notion of the diffusion system, the analysis stops with the actual adoption of the innovation and finally that the ideology of anti-adoption is not considered in a realistic way (Ling, 2001; Borch, 2002).

27.6 The Domestication of New Services into Everyday Life

Silverstone et al.'s (1992) concept of domestication concentrates on the end users and the end-user context. The concept of domestication is relevant to the analysis of mobile alert and will be thoroughly described. Where Rogers' approach is usually more focused on the situation in the marketing world, the domestication approach is often more academic in its use. In addition, the domestication approach does not focus on the marketing of particular innovations to the degree of Rogers, but rather takes a more neutral stance.

When new technologies are introduced into the home and everyday life they must, in Silverstone et al.'s terms, be domesticated. On this basis, the authors distinguish four elements or phases (Silverstone et al., 1992):

- appropriation, in which acquisition is central;
- objectification, which focuses on the display of the object;
- incorporation, which considers how the object is integrated into the routines of the home;
- conversion, the phase in which the object is harmonized within the broader social context (Silverstone et al., 1992; Silverstone, 1994: Silverstone and Haddon, 1996).

Whereas objectification and incorporation principally take place within the household, Silverstone et al. (1992) suggest that the last approach – conversion – like appropriation, takes place between the household and the outside world. In this setting, ICT products are not only seen as technologies, but also as media, referring to the observation that they are providing – either actively, interactively or passively – links between households, between household members and the public world in very complex, often unexpected and sometimes unsuccessful ways. At stake are the values of the household and its members – values that are articulated and incorporated in the discussions, negotiations, practices and routines of daily life. Different families will draw on different cultural resources, based on religious beliefs, personal biography or the culture of family and friends, and as a result construct a bounded environment – the home (Borch, 2002).

The elements or phases in the domestication process have been criticized (Ling, 2001; Borch, 2002). The issue reflects the fact that the domestication concept was originally used in British studies to provide a framework for thinking about ICT in the home rather than with relation to portable technologies, such as mobile phones (Haddon, 2003). Haddon reflects on how issues raised in mobile phone studies suggest ways to extend the framework of domestication out of the home despite the connotation of the word domestic. Haddon points to the possibility that public space could be analyzed as a counterpart to regulation of ICTs in the home, usually in a more tacit and less formalized way, sometimes ambiguous and more in the form of expectation about appropriate behavior held by those co-present.

27.7 The Mental and Physical Integration of Mobile Alert into Everyday Life

Implementing mobile alerts is a challenging task, as it is very difficult to anticipate how mobile alerts may connect to ideas of mobile phones as a personal medium (Ling, 2002). In our analysis of mobile alerts, we found an active discussion in the trial home as to the role of the service in the users' everyday life. The discussion draws basically on the conversion and incorporation phases of Silverstone et al.'s domestication.

The discussion here seems, however, to go beyond the notion of conversion as outlined by Silverstone et al. in that it focuses on the development of a mental construct or metaphor with which to describe the system. The interesting finding in our study is that the participants could not simply pick up a description of the service in everyday conversation with others, but rather were pressed to develop their own understanding of the service. Mobile alerts were a largely undefined technology in the minds of the users. Its relationship with equipment and services such as broadband PC and the Internet gave it a somewhat superficial understanding. A mental concept of the system was therefore not always immediately obvious to some of the participants during the early stages of the trial. One participant noted, for example:

> I constantly got messages that I could see this and that, but then I did not understand where to watch? And I did not know it was TV on the web, as it said. I think there could be a reminder on the mobile where to go.

Metaphors may encourage or hamper the social spread of technologies. The social inertia provided by suitable metaphors can support the adoption of new technologies by helping it over various thresholds or boosting interest and critical points. However, although a metaphor may aid the development of a device, it might also become the albatross around its neck (Fischer, 1992; Sawheny, 1996; White, 1997). The mobile alert developers gave some impetus to some glosses or metaphors via the online survey in addition to the information on the web site, associating the mobile alert as "news alert for your broadband services" and "news alert for SMS/e-mail", providing rather blurred cues that did not give any indication of the potential market for Telenor (Akrich and Latour, 1992). In that sense, the mobile alert was largely an undefined system for both users and most of those who worked on various portions of the project.

Given its complexity, it is only natural that the users sought glosses or metaphors with which to describe the mobile alert and thereby provide the technology with meaning and significance. In this section we first consider the metaphors and glosses used by pilot users to describe mobile alerts, and then turn to the physical integration of the mobile alert in everyday life and the way in which mobile alerts either complemented or clashed with existing habits and values.

27.7.1 The Metaphor of Mobile Alerts as Advertisement

Several of the respondents considered the alerts as a kind of advertising.

The responses in the focus groups point to advertising directly to mobile phones being negative. Most likely this reflects the fear of becoming bombarded by alerts: too many, and attracting attention in socially inappropriate settings. As one respondent said, "I fear getting 17 messages in one day". The users want to control what to receive and do not want to overfill their inbox. It also explains why the respondents did not want to pay for such services. One even compared the alerts with "trash SMS". One user said:

> I would like to have them, but it must be about something important.

However, not all of them saw the service as carrying advertising. One respondent said:

> Over time I have become so tired of advertising that I cheer for this (service) that comes without advertising.

This user is very happy not to be exposed to advertising. He either did not receive any alerts, or it could be that he did not regard alerts to the mobile phone as advertising.

27.7.2 "The Thing About the Mobile Telephone was Cool"

Mobile phones are used for business, but also to a large extent for personal communication and micro-coordination (Ling, 2002). In the "News and Sport" pilot, users could receive a mobile alert of an upcoming football match, and then even use the mobile phone to pay for access to the match as broadband content (web TV). This is an introduction of the alert in a media mix that is rather new to most users and may feel superficial and scary. However, some users in the focus group reacted positively to the alert service in connection with payment for broadband services. One said:

> That thing about the mobile telephone was cool, I was on the Internet with something I think about as NRK related, and suddenly my mobile telephone is of interest ...

The use of the mobile phone as a wallet is not familiar to most users. The acceptance of the new service may indicate that this user was an innovator reacting positively to the newness of the wallet function of the mobile phone (Rogers, 1995).

27.7.3 SMS is Unfit to Carry Anything but Short Personal Messages

As to the appropriateness of mobile phones being used to carry new forms of content, one respondent said:

I think SMS as a totality is exaggerated as a medium.

This point of view was discussed further, and one mentioned that SMS was unfit to carry anything but short personal messages. This view may represent a conservative mentality towards media in general, claiming that the function of the media is static. Both focus groups on the lifestyle service displayed such views. In one respondent's words:

I believe TV is TV, telephone is telephone, and PC is PC.

I could get an e-mail (alert), not mobile (alert), it (the mobile) is used enough as it is.

Our data show that the respondents did not develop a general metaphor. However, most of them integrated the mobile alerts in the context of their everyday lives, finding a mental concept using glosses such as "advertisement", "the thing about the mobile phone" or just considering the alert service to be "unfit to carry anything but short personal messages". However, as mentioned above, one user did not connect the mobile alerts to the broadband service at all. The old lay considered the alert "to be a reminder on the mobile where to go", indicating a struggle to find a mental concept that never really succeeded.

It seems to us that the blending of media and content will be difficult for some user groups to understand and manage. In Rogers' terms, the complexity of the innovation (in this case mobile alerts in relation to broadband content) will be negatively related to the rate of adoption (Rogers, 1995). We expect users to bring their prior habits and attitudes towards the use of media into the interpretation of new services, which the quote mentioned above might also illustrate. To support this, we saw very clearly that the news service was seen as the news from TV watched on PC, and not as new and unique broadband news content. This is an indication that the respondent could still be in the appropriation phase as to domestication of the technology. The user is aware of the technology and understands its use, but does not seem to have incorporated the mobile alert in his milieu.

27.7.4 Mobile Alert as an Indicator of Competence in the Public Culture

In the online surveys the users were asked to what extent they appreciated the mobile alerts. The "News and Sport" pilot distributed mobile alerts on a more frequent basis; 29% of the respondents in the online survey said that the mobile alerts made them log on and use the pilot more often. There were also 32% who said that they appreciated the alerts (online survey). One might expect the new use of the mobile phone in relation to a broadband service to be yet undefined and that the connection between the media consequently may feel alien to users. On the other hand, the newness of the service could also produce curiosity and positive attitudes in the innovator users (Rogers, 1995). Several of the respondents told us of checking the broadband service after receiving an alert on the mobile phone. This was evident in one of the focus groups. One user said:

> I watched the live football game to see how they [service providers] managed it. It went really well. I had it running in the background to see if they had programming throughout the whole game.

One might expect innovator users to try out this new connection between the mobile alert and the broadband service on a trial basis. On the other hand, it is here that one can see traces of conversion, as described by Silverstone et al. (1992). This is the portion of the domestication process in which one defines the relationship between the household and the outside world. Some households may resist the aspect of new technology in use; others may use it as an indicator of membership and competence in the public culture. It is here that trial users were called upon to develop a vocabulary to summarize the new alert service. One user said:

> I would like to get reminders on issues I have decided in advance I would like reminders on. I am sure many people do not want it, but I have nothing against building a profile at a content distributor and getting a notification to the channel I am sitting on at the moment that something is happening.

This respondent from the "News and Sport" pilot used the alert service as indicator of membership and competence in the public culture; his positive attitude shows that the service reconfirmed his identity and uniqueness as one with a high level of knowledge and expertise about broadband technology and Internet services in general. We may say that the trial user has successfully domesticated the new technology.

But there is just one thing: he does not at all mention the family – either the home, or any particular activity taking place within the home – as part of any process in the establishment of his identity. He is focusing on himself and only himself. This type of response may indicate a trend towards what Beck and Beck-Gersheim (2002) would call "being individual". We are judged not by our achieved status (as part of a family), but rather through the market segment we inhabit at the moment. Not only are we individually available, but we are also becoming situationally available. Information technologies are in many ways the motor of this process. The mobile communication allows us to determine the rough location of our friends, a nearby taxi or item, or alert services of all kinds.

So what is at stake here? Beck's nomadic individualist is a result of the retreat of the classical institutions: state, class, nuclear family and ethnic group. What has happened is that the respondent in our group has become more of a rule finder himself. He identifies the mobile alert service as a question of uncertainty, of risk, but it also leaves the door much more open to him for innovation. He does not care what others may think or do; he places ego before collective. This means that for the educational system, the family, the political structure and everything else, their tasks will be carried against the backdrop of his uniqueness, ego and particularities.

27.7.5 Media Alerts and Media Habits

Beyond the metaphorical placing of mobile alert, another aspect of domestication is the actual use of the service; the physical integration. Silverstone et al. refer to this as incorporation, which considers how the object is integrated into the routines of the home. As we saw in the section above, the users of the news service were far more positive towards the use of mobile alert than the users of the lifestyle service. In the "News and Sport" pilot, the alerts were reminders of programming of live events to take place up to 2 hours after the alerts were distributed, such as football matches, the funeral of the scientist Thor Heyerdahl and the annual convention of one of the main political parties in Norway (Høyre). In the "Lifestyle" pilot the reminders were issued 15 minutes prior to scheduled programs. The fact that the alerts in this pilot were offered to notify of scheduled programs was commented on by several of our users. They typically referred to their general knowledge of scheduled programming:

> I know the program starts at 19.30, so if I need a reminder I can set one up myself on my mobile.

This is an example of common attitudes to alerts connected to television program viewing among the users. Many viewers develop media habits based on the regularity of TV programs. For these types of ritual media behavior, the alerts tend to be irrelevant. On the other hand, the quote also shows how users draw upon existing patterns of behavior. According to Rogers, attributes of the innovation affect the adoption. The abovementioned quote indicates that the attributes of the service may not be compatible with needs, values and past experiences of the user (Rogers, 1995). Further, alerts in the "Lifestyle" pilot were also connected to regular events such as time and subject for chat sessions after TV programs. Another respondent, however, mentioned a need for such a service:

> I watch so little television that I would like such a service, but I would not be interested in paying more than SMS costs for it.

Being without TV viewing regularity, the attitude to receiving alerts is more positive for this user. This is one of the few respondents who hold positive attitudes to program alerts and can connect to his lack of ritual of media behavior. He can see uses of the alerts for himself based upon his own needs and patterns of technology use. In Rogers' terms, the technology has a relative advantage to him (Rogers, 1995). However, there were also users with negative attitudes in this phase. One user said:

> I do not need a beep on the mobile when I am at my cabin and away from the PC.

This user emphasizes the importance of physical contexts as to the alerts. He could see the practical consequences of the alerts and how they may be in conflict with his own pattern of living. It seems that the negative attitude towards the alerts is connected with disturbance of everyday peace:

> It is ok to have some peace and quiet in everyday life: as little stress as possible.

A social consequence of the mobile alert could be an increase in interruptions, which may create stress in some users and raise the general level of noise in society. Another respondent also emphasized this point:

> I am on the web-based newspapers during the day, and in the evenings when I am in social encounters I do not want to be disturbed by beep, beep.

Here we see that the users had a pre-existing understanding of the quality of everyday life. Rather than attempting to integrate the alert service into the media habits, they felt that it fitted better into what they considered to be "noise in society" or a "disruption of social interference".

27.7.6 Mobile Alerts Must be Adjusted to the Context of the User

A recurring argument in the focus groups was the need for the alerts to be adjusted to the context of the user. This is a question of time and space related to media consumption:

> I do not need an alert when I am in my car on my way to Stockholm. But I need it when my PC is on.

This user was positive to alerts on the mobile, but was sceptical to receiving alerts when too remote to access the broadband service. To the users it is not necessarily an advantage to have the contents and news they need available in every situation and in the same format. This indicated what Beck would consider to be of importance, the need for technology that allows us to be nomadic individualists, technology that monitors the process, not only of being individually available, but also situationally available (Beck and Beck-Gersheim, 2002).

Our data show that we had respondents who expressed a wish for PC-based alerts. Especially when working on the PC, e-mail, sound-based or visual signs may be displayed as alerts instead of on the mobile phone. The explanation for this is that it would be convenient and useful to be alerted when you have the service easily accessible. Further, there was a particular need to be notified on the occurrence of major events. An additional point is that the users may set up their own profiles as to which themes to receive alerts on. Ideally, the content or nature of the message would decide whether the alert should be sent to the mobile phone or to the PC, or any other channel for that matter. One respondent said:

> It would be convenient to say that until five o'clock in the evening I will take alerts to e-mail, otherwise to SMS ...

In sum, our data show that for the majority of pilot users in one focus group mobile alerts were regarded as an interesting additional function to the broadband service, but not as a service on its own worth paying for. The mobile alerts and the broadband service are not connected or perceived as mutually dependent media by the users. There has to be a perceived need to fulfil for the alert to become integrated and attractive.

27.8 Conclusions and Discussion

SMS-based news services are not a new media phenomenon in Norway. Most newspapers on the Internet offer subscriptions to such services. In the study discussed in this chapter, the media mix included the mobile phone and a PC connected to a broadband connection. Mobile alerts in this study included the distribution of alerts by short messages (SMS), to make the user aware of content available over the broadband connection. The nature of the alert was general news and alerts on live programming (news and sport).

In this chapter, an attempt has been made to look for ideas about the kinds of situations in which the mobile alert capability of a mobile phone would be of use, and why. To find these "accepted situations" in relation to two different broadband pilots (see Figure 27.1), this chapter has examined how trial users have domesticated the new technology. Specifically, an attempt has been made to draw out some of the major issues associated with the mental and physical integration of mobile alerts. We have noted that this demonstrates an activity related to Silverstone et al.'s (1992) conversion. It goes beyond their notion of conversion, however, in that it focuses on the use of mental construction. The interesting thing in our study is that the participants could not simply pick up a mental construction of the service in everyday conversation with others, but rather were pressed to develop their own understanding of the service. Our data

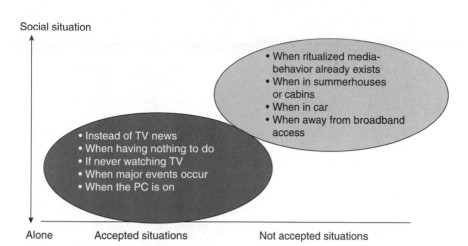

Figure 27.1   Accepted and not accepted situations for use of mobile alerts.

show that the trial participants did not develop a common metaphorical background into which they could place the mobile alert.

The evidence here suggests that in one-third of the cases there was an acceptance, or at least some willingness to change in relation to mobile alerts. Our data show that the trial participants seemed to be more open to being informed about extraordinary events or incidents. Related to this we also know that many of the users in the "News and Sport" pilot consume news from newspapers, the Internet and TV several times a day, and therefore they are in the target group for news alerts; they master the technology and have a wish and need to always be updated on the news headlines. The use of alerts seems to meet their particularly strong need for news consumption, an activity that has been related to Rogers' innovators and early adopters (Rogers, 1995) and also Beck's nomadic individual; we are judged not by our achieved status (as part of a market segment or as part of a family), but rather through the market segment we inhabit at the moment (Beck and Beck-Gersheim 2002).

On the other hand, a negative attitude towards alerts was seen in connection with the disturbance of everyday peace. Some users fear that mobile alerts will contribute to an increase in interruptions, which may create stress in some users and raise the general level of noise in society. Related to this is the concern that alerts are inappropriate in many social situations. Such statements could also indicate a low level of domestication. None of the respondents mentioned, for instance, strategies for avoiding disturbance or minimizing intrusion by the alerts, except for alerts to the PC, while at work or at the home computer.

What can we say about the future use of mobile alerts on the basis of this study? We feel that it says most about initial-stage experience of use.

Although the two pilots have different characteristics, it seems that "accepted situations" generally fit in when people are out of a family or group situation, where their individualized need does not conflict with already routinized habits. Some users answered that this was a preferred service when they had "nothing to do", which indicates that avoiding boredom is a mental situation that fits with this type of mobile service. The users want to control what to receive and do not want to overfill their inbox. It also explains why the respondents did not want to pay for such services. One even compared the alerts with "trash SMS". This indicates the great importance of being cautious with the number and nature of alerts to be distributed. It is clear that developers must take into consideration the voice of the critical consumer early in the product development process.

In addition to the more direct aspects of the alerts, the material here provides us with insight into how new services become domesticated. The material shows how the new services become increasingly integrated into our everyday life, how they metaphorically rearrange the existing furniture and how they become part of our identity vis-à-vis other persons. The development of the mobile alert drew the mobile phone out of its taken-for-granted role as a personal communication medium, but also associated it with the process of a broader meaning vis-à-vis the participants' notion of themselves and their participation in the broader context of Norwegian culture. In some cases this took the form of technological venerations, in which the service occupied a place in the identity of the trial users to show the competence in the public culture. A second attitude was that it clashed with the users' self-image. The material presented here shows how these concerns were made concrete in different placements and situations.

27.9 References

Akrich, M. and Latour, B. (1992) A summary of a convenient vocabulary for the semiotics of human and nonhuman assemblies. In Bijker, W.E. and Law, J. (eds), *Shaping Technology/Building Society: Studies in Sociotechnical Change*. MIT Press, Cambridge, MA.

Beck, U. and Beck-Gersham, E. (2002) *Individualization: Institutionalized Individualism and Its Social and Political Consequences*. Sage, London.

Bijker, W.E. and Law, J. (eds) (1992) *Shaping Technology/Building Society – Studies in Sociotechnical Change*. MIT Press, Cambridge, MA.

Bijker, W.E., Hughes, T.P. and Pinch, T.J. (eds) (1987) *The Social Construction of Technological Systems: New Directions in the Sociology and History of Technology*. MIT Press, Cambridge, MA.

Borch, A. (2002) Taming wild products. Paper presented at the ETE Workshop in Jena, 28 February–2 March. National Institute for Consuming Research, Norway.

Brandth, B. (1999) *Gruppeintervju: Perspektiv, Relasjoner og Kontekst. I: Holter H og Kalleberg R (red) Kvalitative Metoder i Samfunnsforskning*. Universitetsforlagets Metodebibliotek, Oslo.

Fischer, C. (1992) *America Calling: a Social History of Telephony to 1940*. University of California Press, Berkeley, CA.

Fog, J. (1998) *Med Samtalen som Udgangspunkt. Det Kvalitative Forskningsinterview*. Akademisk Forlag, Copenhagen.

Haddon, L. (2003) Domestication and mobile telephony. In Katz, J.E. (ed.), *Machines that Become Us. The Social Context of Personal Communication Technology*. Transactive, New Brunswick.

Hughes, T.P. (1986) The seamless web: technology, science, etcetera, etcetera. *Social Studies of Science*, **16**, 281–292.

Jensen, M. and Thrane, K. (2003) *"Hvor Mye Bruker Vi, og Hva Har Vi Behov For" – Behovserkjennelse ved Anskaffelse av IKT i Hjemmesfæren*. R&D Note N 70/2003. Telenor R&D, Fornebu, Norway.

Kvale, S. (1997) *Det Kvalitative Forskningsintervju*. Ad Notam Gyldendal, Gjøvik, Norway.

Latour, B. (1987) *Science in Action*. Open University Press, Milton Keynes.

Latour, B. (1996) *Aramis or the Love of Technology*. Harvard University Press, Cambridge, MA.

Ling, R. (1999) *"I Am Happiest by Having the Best": the Adoption and Rejection of Mobile Telephony*. R&D Report R 15/99. Telenor R&D, Kjeller, Norway.

Ling, R. (2001) The diffusion of mobile telephony among Norwegian teens: a report after the revolution. Presented at ICUST 2001, Paris, June 2001.

Ling, R. (2002) *The Social and Cultural Consequences of Mobile Telephony as Seen in the Norwegian Context*. R&D Report R 9/2002. Telenor R&D, Fornebu, Norway.

Ling, R. (2004) *The Mobile Connection. The Cell Phone's Impact on Society*. The Morgan Kaufmann Series in Interactive Technologies. Morgan Kaufmann, San Francisco.

Ling, R. et al. (eds) (2001) *Telenor HB@ Project Deliverable D19. Evaluation of Pilot Services – Final Report*. R&D Report R 53/2001. Telenor R&D, Fornebu, Norway.

Loafland, J. and Loafland, L.H. (1984) *Analyzing Social Settings: a Guide to Qualitative Observation and Analysis*. Wadsworth, Belmont, CA.

Nøtnæs, T. (2001) *Innføring i Bruk av Fokusgrupper*. SSB Notat 2001/24. Statistisk Sentralbyrå, Oslo.

NRK and Telenor (2002) *Bredbåndspiloten "Nyheter og Sport". Et Samarbeidsprosjekt Mellom NRK Kringkasting Utviklingsavdelingen og Telenor Plus*.

NRK and Telenor (2003) *Bredbåndspilot for PULS, Barmeny, FBI og STYRK Live. Et Samarbeidsprosjekt Mellom NRK Kringkasting Utviklingsavdelingen og Telenor Plus*.

Repo, P. (2003) *Mobile Video*. Publication 2003:5. National Consumer Research Centre, Helsinki.

Rogers, E. (1995) *Diffusion and Innovations*. Free Press, New York.

Russel, S. (1986) The social construction of artefacts: a response to Pinch and Bijker. *Social Studies of Science*, **16**, 331–346.

Sawheney, H. (1996) Information superhighway: metaphors as midwives. *Media, Culture and Society*, **18**, 291–314.

Silverstone, R. and Haddon, L. (1996) Design and the domestication of information and communication technologies : technical change and everyday life. In *Communication by Design: the Politics of Information and Communication Technologies*. Oxford.

Silverstone, R., Hirsch and Morley (1992) Information and communication technologies and the moral economy of the household. In Silverstone, R. and Hirsch (eds), *Consuming technologies. Media and Information in Domestic Spaces*. Routledge, London.

SSB Internet-målinga 2. kvartal (2002) [online]. URL: www.ssb.no/emner/10/03/inet/main.html.

Thrane, K. and Ling, R. (2000) *IKT-basert Underholdning i Hjemmet*. R&D Note N 83/2000. Telenor R&D, Kjeller, Norway.

Thrane, K. and Ling, R. (2002) Broadband @ home: how will broadband access change our lives over the next ten years and how will our lives change broadband? *Telektronikk*, **98**(2/3), 2–8.

White, P.B. (1997) On line services and transactional space: emerging strategies for power and control in the new media. *Telecommunication Policy*, **21**, 365–373.

Winner, L. (1985) Do artefacts have politics? In MacKenzie, D. and Wajcman, J. (eds), *The Social Shaping of Technology*. Open University Press, Milton Keynes.

28

The Wired – and Wireless – Japanese: Webphones, PCs and Social Networks

Kakuko Miyata, Barry Wellman and Jeffrey Boase

28.1 The Mobile-ization of Society

28.1.1 From Wired to Wireless

Once upon a time, not so long ago, people were rooted to their homes and workplaces by computers that were wired in place by electric and Internet cables. In those days, the magic book was called *Wired* magazine. Its vibrant pages told all that was avant-garde about the Internet and other forms of computer-mediated communication. Its name evoked how computer-mediated communication electronically connected people to the world.

Then mobility came. First, portable computers – originally 11 kg back-breakers – shrank into portable notebooks weighing 2–4 kg. Software became standardized and the Internet globally available, so wherever people went they could connect with the network. Personal digital assistants, such as the *Palm*, became capable of accessing the Internet. However, the most widespread change was the birth of mobile phones, which became as common in people's pockets and handbags as their keys. By the turn of the 21st century, mobile phones had become *webphones* (our term): capable of connecting to the internet to use the web and exchange e-mail and short text messages.[1] Meanwhile, *Wired* magazine had become a ghost of itself, declining 29% from 240 pages in September 1996 to 170 pages in June 2004, with the editors noting ruefully that their

---

[1] We include SMS ("short message service", sometimes known as "texting") in addition to regular e-mail in all of our analyses. Mobile phones that can access the Internet have been so rare in the English-speaking world that we had to coin a new word, "webphone", to refer to them.

magazine "used to be as thick as a phone book" (*Wired*, 2004, p. 23). Indeed, the name *Wired* has itself become anachronistic in the increasingly wireless society.

Webphones have become the most individualized and intimate of information and communication technologies (Srivastava and Kodate, 2004; Wellman and Hogan, 2004). It is time to consider a new era: how the peripatetic mobile users of the Internet communicate with the members of their social networks and communities (see also Rheingold, 2002).

28.1.2 Japan as a Leading-Edge Case Study

Japan has been at the forefront of the turn towards mobility, with widespread use of webphones. These are much smarter phones than the norm because of the ease of Internet use. Not surprisingly, Japan boasts the highest percentage of mobile Internet users as a proportion of total mobile users. Even casual visitors note the many Japanese pedestrians walking with phones (*keitai*) to their ears, sending text messages on trains and silently scanning their inboxes during get-togethers and meetings. Although Japan is advanced, it is not unique in East Asia, with heavy use in China (Yan, 2003), South Korea (Chae and Kim, 2003) and elsewhere.

The widespread use in Japan of advanced mobile connectivity to the Internet provides a case study to provide better knowledge of the future mobile-ized society (see also Ito et al.'s book on keitai (Ito et al., 2005). This chapter uses survey data from Yamanashi Prefecture to address ongoing debates about the effects of Internet use on community and social support. We focus on three research questions:

1. *Who uses webphones and PCs to send e-mail?* We examine the social characteristics and the social relationships of the users of Internet-connected webphones and PCs.
2. *How do people use these media?* We compare communication via webphones and PCs.
3. *To what extent are webphones and PCs used in social networks?* We compare strong, supportive and weak ties and local and long-distance relationships.

We conclude this chapter by discussing the implications of webphone and PC use for the nature of Japanese communication and social networks, and the turn towards *networked individualism* that is happening in Japan and in other developed societies (Wellman, 2001, 2002). The ubiquitous Internet – along with personal mobile communication – is fostering a societal turn away from groups and toward people connected to each other as individuals rather than as members of households, communities, kinship groups, workgroups and organizations (Wellman and Hogan, 2004). These technologies enable individuals to have personalized communication with whoever, whenever and – with the advent of the mobile Internet – wherever they want.

28.1.3 Transformations of Community

28.1.3.1 From Door-to-Door to Place-to-Place

Although the impacts of the Internet and mobile phones are new, the trend is not. For more than a century, the developed world has been experiencing a shift away from communities based on villages and neighborhoods towards flexible partial communities based on networked households and individuals. One transition was the 19th/20th century move from "door-to-door" to "place-to-place" community relationships. This transition was driven by revolutionary developments in both transportation and communication. It was a move away from a solidary group in a single locale – where people normally walked through the village to each others' homes – to contact between people in different places and multiple social networks – where people used cars, planes, trains and telephones to connect with each other.

Instead of group membership, people became members of multiple social networks where boundaries are more permeable, interactions are with more diverse others and linkages switch among multiple networks (Wellman, 1999, 2001; see also Castells, 2000; Putnam, 2000). Hence, many people can communicate with others in ways that ramify across group boundaries. Rather than relating to one group, they cycle through interactions with a variety of others: at work or in the community. Their work and community networks are diffuse and sparsely knit, with vague, overlapping, social and spatial boundaries. The structure and composition of these networks affect people's control over their lives, and people's structural positions in these networks affect the kinds of resources to which they have access. In this place-to-place world of the early 21st century, groups have become less important.

28.1.3.2 From Place-to-Place to Person-to-Person

Another shift is under way: from place-to-place networks to person-to-person networks in which the individual – and not the household (or workgroup) – is the primary unit of linkage. Until fairly recently, transportation and communication have fostered place-to-place community, with expressways and airplanes speeding people from one location to another (without much regard to what is in between). Telephone and postal communication have been delivered to specific, fixed locations.

The change from place-based inter-household ties to individualized person-to-person interactions and specialized role-to-role interactions has been facilitated by the Internet and especially by wireless personal communication: mobile phones, PDAs and webphones (Wellman, 2001). At present, communication is taking over many of the functions of transportation for the exchange of messages. Communication itself is becoming more mobile, with mobile phones and wireless computers proliferating. Although the turn towards person-to-person networks happened well before the development of cyberspace (Wellman and Wetherell, 1996; Wellman, 1999a), the rapid emergence of computer-mediated communications means that relations in cyberplaces are joining with relations on the ground (Wellman and Haythornthwaite, 2002).

Changes in the nature of computer-mediated communication both reflect and foster the development of networked individualism in net-worked societies. Complex social networks have always existed, but recent technological developments in communication have afforded their emergence as a dominant form of social organization. The technological development of computer networks and the societal flourishing of social networks are affording the rise of networked individualism in a positive feedback loop. Just as the flexibility of less-bounded, spatially dispersed, social networks creates a demand for collaborative communication and information sharing, the rapid development of computer communications networks nourishes societal transitions from group-based societies to network-based societies (Castells, 1996, 2000; Wellman, 2002).

In the early days of the Internet, there were fears that it would destroy communities by drawing people away from face-to-face and now-traditional telephonic communication to leading lonely inauthentic lives online (the debate is reviewed in Wellman and Gulia, 1999; Wellman and Haythornthwaite, 2002). By now, a good deal of research has shown that the Internet has not destroyed communities. Rather, it adds on to existing relationships with community members: friends, acquaintances, relatives and even neighbors (see the chapters in Wellman and Haythornthwaite, 2002). The more means of communication that people have, the more they communicate. Furthermore, the Internet appears to be fostering a shift in the means of connectivity from transportation to communication: from airport terminals and road networks to computer terminals, mobile phones and cyber-networks.

Although community has not declined with the advent of the Internet and mobile communication, neither has the nature of community remained the same. The spread of computer-mediated communication media is facilitating social changes that have been developing for decades in the ways in which people contact, interact and obtain resources with each other. Internet and mobile phone connectivity is to persons and not to jacked-in telephones that ring in a fixed place for anyone in the room or house to pick up. The developing personalization, wireless portability and ubiquitous connectivity of the Internet all facilitate networked individualism as the basis of community. Because connections are to people and not to places, the technology affords shifting of work and community ties from linking people-in-places to linking people at any place. Computer-supported communication is *everywhere*, but it is situated *nowhere*. It is I-alone that is reachable wherever I am: at a home, hotel, office, highway or shopping center. The person has become the portal.

This shift facilitates "*personal communities*" (Wellman, 1979) that supply the essentials of community separately to each individual: support, sociability, information, social identities and a sense of belonging. The person, rather than the household or group, is the primary unit of connectivity. Just as 24/7/365 Internet computing means the high availability of people in specific places, the spread of mobile phones and wireless computing is increasingly coming to mean an even higher availability of people without regard to place. Supportive convoys travel ethereally with each person (Ling and Ytrri, 2002; Katz and Aakhus, 2002).

With networked individualism, each person is a switchboard between ties and networks. People remain connected, but as individuals, rather than being rooted in the home bases of work unit and household. Each person operates a separate personal community network, and switches rapidly among multiple sub-networks. The inherently personal and individualistic webphone makes this even more convenient. In effect, the Internet and other new communication technology are helping each individuals to personalize their own communities. This is neither a prima facie loss nor gain in community, but rather a complex, fundamental transformation in the nature of community.

28.2 Japanese Webphone Use

28.2.1 The Spread of Webphones in Japan

Japan has been in the midst of these changes towards networked individualism. Although it is unlikely that Japan was ever as stably group-centered as stereotypes have portrayed it, recent research has shown that many Japanese engage in personal communities similar to those in other developed countries. Their networks are sparsely knit, not very local, and consist of both friends and kinfolk (Nozawa, 1996; Otani, 1999). Indeed, the very epidemic of mobile phone use in Japan shows how physically dispersed relationships have become.

Accessing the Internet through the use of mobile phones has already become integrated into daily life for a significant proportion of the Japanese population (Barnes and Huff, 2003). By the end of May 2001, more than 40 million Japanese were able to access the Internet through their mobile phones, with the number rising by 55% to more than 62 million by the end of March 2003 (MPMHAPT, 2004).

Japan was one of the first countries to launch third-generation mobile services, in October 2001, and the first country to launch commercial services based on the W-CDMA standard. The number of mobile phone users exceeded 80 million by March 2004, with most (nearly 70 million) capable of connecting to the Internet from their phones. Seventeen million had advanced third-generation service, whose features include quicker Internet connections. About 90% of Japan's mobile phones could connect to the Internet in September 2003, the highest percentage in the world. By March 2004 more than 60% of all mobile phone subscribers could bring a visual element to their communication via built-in digital cameras (MPMHAPT, 2004).

Since 1999, NTT DoCoMo's Internet mode services (i-mode) have made it possible for subscribers to access web sites specially designed for mobile phones. As a result, mobile phones have expanded from a conventional device for voice transmission into a much broader mobile channel for information and entertainment. Subscribers to i-mode services can not only exchange e-mails, but also read breaking news, reserve tickets and buy newly released pop songs.

The four major Japanese providers of webphone access to the Internet are NTT DoCoMo, KDDI, Vodafone and Tsu-ka (in order of number of subscribers, January 2004). Each uses a variety of Internet protocols: DoCoMo's i-mode is the most popular, followed by WAP (Wireless Application Protocol) and WAP2. Although WAP2 is the least popular, it is rapidly gaining a foothold in the market as it permits advanced "3G" (third-generation) services that provide GPS (global positioning system), video clips, higher speed and other advanced features (Kageyama, 2003).

Japanese webphones have relatively large screens compared with all but the most recent American mobile phones. Sending e-mail through Japanese webphones is similar to sending e-mail through PCs, although users have to cope with less user-friendly telephone keypads. Webphones can send and receive e-mails to and from PCs, and also to other webphones. Users enter the e-mail address of the recipient, a subject line and then the contents of their message. There are helpful typing shortcuts for commonly used words and icons.

28.2.2 Young Adults are Being Served

Young Japanese are heavy users of mobile phones, and regard their gadgets as a personal digital assistant powered by telephone technology (Srivastava and Kodate, 2004). The percentage of young adults in Japan who use webphones to e-mail is much higher than in the USA and many parts of Europe, where webphone e-mail has failed to attract a majority of people from any age group. This difference is partly due to marketing strategies taken by Japanese mobile phone providers that have catered to the desires of youth and young adults. Japanese providers initially sold webphones as entertainment devices for the younger generation, rather than trying to sell them as practical tools for older business people. By gearing webphones to the younger generation as something fun and relatively inexpensive, they were able to capture the group that was already the largest consumers of mobile phones. As Japanese youth were the first adopters of webphones, webphone use has diffused so quickly and become so ubiquitous among them (Habuchi, 2005; Ito, this volume, Chapter 9). After gaining a foothold in the youth market, webphone providers beefed up bandwidth and web interfaces, making their services more attractive to a wider audience.

Cultural differences and marketing tactics may have driven the quick and ubiquitous adoption of this new technology by younger Japanese. They were probably predisposed to send e-mail through webphones by their extensive use of pagers in 1990s to contact friends and organize social activities. (Parents who wanted to contact their mobile children also spurred the use of pagers.) This incorporation of pagers into everyday routines set the stage for the adoption of webphones with their advantage of smoothly integrating voice and message contact.

Only a few ethnographic studies have investigated younger Japanese use of pagers and mobile phones. One study reports that mobile phones afford Tokyo youth important advantages (Ito, this volume, Chapter 9). The ability to send short messages at any time allows users to keep in frequent contact with friends, strengthening their social networks and providing a feeling of "ultraconnectedness". This sort of communication typically occurs frequently but with only a small number of 2–5 friends. At the same time, typing quick messages gives a new kind of freedom, as it often can be done somewhat covertly without alerting parents. Contacting friends can occur late at night while parents are sleeping, something not easily done through wired landline phones that rarely reside in Japanese children's bedrooms. Complementing this have been Japanese concerns since the 1960s to develop more communication between spouses and between parents and children (Matsuda, 2004). The result is a concentrated, active use of mobile phones to expand and enhance contact between close friends and immediate family.

In contrast to young adults and youth, older Japanese adults first encountered the Internet by using personal computers (PCs) to e-mail and use the web (Miyata, 2002). The mobile phones they first used were not able to access the Internet. Hence, some older adults have not developed the habit of using mobile phones to access the Internet even when their new webphones have this capability. However, many are embracing it, especially when their (often small) homes do not have PCs. Thus Japanese housewives are using webphones to alleviate loneliness, manage family relationships and gain empowerment through Internet access to communication and information (Miyata, 2002; Dobashi, 2005).

28.3 The Yamanashi Study

Our study of Internet users is based on a random sample survey of 1,320 adults, conducted in November–December 2002 in Yamanashi prefecture in Japan. Yamanashi is a mixed rural and urban area, located in the center of Japan, more than 100 km west of central Tokyo. It is typical of Japan (outside of the Tokyo and Osaka urban agglomerations) in the characteristics of its population and Internet users, and it is famous because Mount Fuji rises in it.

Forty neighborhoods in the Yamanashi prefecture were randomly selected by postal code, with a further random selection of 33 individuals within each of those neighborhoods. These potential respondents were chosen from a voters' list of people aged to 20–65 years. Surveys were in paper form and delivered in person. They were also collected in person 3 weeks after being dropped off. Three-quarters (76%) of the selected individuals completed the survey, providing a sample size of 1,002 respondents.

We divided respondents into three types because those using only webphones or PCs may have different characteristics and patterns of use than those who use both media:

- Those who use only webphones;
- Those who use only PCs;
- those who use both webphones and PCs.

28.4 Webphone and PC Contact with Social Networks

28.4.1 Frequency of E-mail Contact

About half of the Yamanashi respondents do not use webphones at all to send e-mail, and 2% use it only for work and do not send personal messages (Figure 28.1). About 15% of the respondents who have used webphone e-mail did not send any "yesterday" (the day before they completed the questionnaire). One-quarter of all respondents sent 1–5 e-mails through webphones and 9% sent more than 5.

Slightly more e-mails are sent by webphones than by PCs. Fewer respondents send e-mail via PCs (56%) than by webphones (51%) (Figure 28.1), and fewer sent e-mail by PCs "yesterday": 21% vs 14%. Figure 28.1 shows that at every level of e-mail activity, more e-mails are sent by webphones than by PCs.

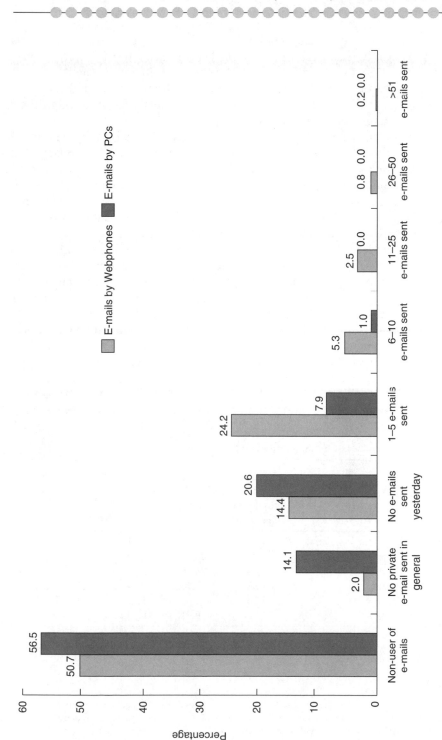

Figure 28.1 Percentage of respondents using e-mails.

Those who only use webphones send an average of about six e-mails per day, whereas those who only use PCs send two-thirds fewer, an average of about two e-mail per day.[2] The greatest number of e-mails are sent by those using both webphones and PCs, about six e-mails per day through their webphone, plus about two e-mails per day through their PC, for a total of eight e-mails per day.

In short, using both media, webphones and PCs, adds to the number of e-mails sent: one medium does not replace the other. This finding, that using both a PC and a webphone is associated with more frequent e-mail use, is congruent with other studies' findings that the more media are used, the greater is the overall amount of communication (Haythornthwaite and Wellman, 1998; Quan-Haase et al., 2002). It further suggests that different sorts of e-mail are being sent by webphones and PCs.

Do people substitute webphones for PCs when sending e-mail, or does e-mail communication by webphones amplify email communication by PCs? For example, people might start a webphone e-mail exchange while away from their PCs, but continue it later by PC when they are at home or work. To address this matter, we focus on the 25% of the Yamanashi respondents who use both webphones and PCs to send e-mail ($N = 251$).

We did not find any significant correlation between the frequencies of webphone and PC e-mail. That is, the amounts of webphone and Internet e-mail use are independent of each other and add on to each other.

When people can use both media to send e-mail, they tend to send more e-mail by webphones than by PCs (Table 28.1). For example, one-

Table 28.1   Frequencies of sending e-mail (%)

Webphones	PCs				
	No private e-mail sent in general	No e-mails sent yesterday	11 to 15 e-mails sent	6 to 10 e-mails sent	Total
No private e-mail sent in general	1.23	1.65	1.23	0.00	4.12
No e-mails sent yesterday	4.94	24.28	2.88	0.82	32.92
1 to 5 e-mails sent	9.88	20.58	13.99	0.00	44.44
6 to 10 e-mails sent	2.88	4.94	2.47	0.82	11.11
11 to 25 e-mails sent	2.06	2.06	1.23	0.41	5.76
26 to 50 e-mails sent	0.00	0.41	0.82	0.41	1.65
Total	14.07	20.56	7.88	1.00	100.00

$N = 243$; $\chi^2 = 40.88$, $p < 0.01$

---

[2] The per day estimates of frequencies of contact were translated from respondent-reported frequency codes. The original values were: 0 = no e-mail sent in general or no e-mail sent or received yesterday; 1 = 1–5 e-mails yesterday; 2 = 6–10 e-mails yesterday; 3 = 11–25 e-mails yesterday; 5 = 26–50 e-mails yesterday; 6 = more than 51 e-mails yesterday.

quarter of those using both webphones and PCs who sent 1–5 e-mails by their webphones did not send any e-mails by PCs, and 14% of dual web phone–PC users sent the same number of e-mails by webphones as by PCs. As we do not have process information, we cannot trace the sequencing of Internet communication between webphones and PCs.

28.4.2 Webphone and PC E-mail Users

There are large age differences in use of e-mail by mobile phone (Table 28.2). The frequency of sending personal e-mail by webphones declines dramatically with age. By contrast, the frequencies of sending e-mails by PC are not different across age.

Table 28.2 Demographic variables and perceived ability to use technology (multiple regression)

	Frequencies of sending e-mail by webphone		Frequencies of sending e-mail by PC	
	B	β	B	β
Gender (0 = female, 1 = male)	−0.241**	−0.159	−0.025	−0.037
Age (years) (reference = 20–29):				
30–39	−0.483**	−0.256	−0.045	−0.054
40-49	−0.579**	−0.317	−0.002	−0.002
50-59	−0.743**	−0.425	0.023	0.030
≥60	−0.688**	−0.293	0.035	0.034
Education (reference = middle school):				
High school	−0.017	−0.011	−0.048	−0.072
Some college	−0.065	−0.036	−0.043	−0.054
Undergraduate degree or more	−0.104	−0.055	0.036	0.043
Employment status (reference = full-time worker):				
Part-time worker	0.011	0.005	−0.010	−0.010
Self-employment	0.024	0.008	0.100**	0.074
Student	0.691**	0.130	−0.174**	−0.074
Home maker	−0.132	−0.060	0.011	0.011
Other type	−0.076	−0.028	0.016	0.014
Unemployed	−0.079	−0.020	−0.010	−0.006
Partner (0 = no, 1 = partner)	−0.234**	−0.138	0.004	0.005
Kids living together (0 = no, 1 = yes)	0.052	0.035	−0.026	−0.039
Perceived ability to use technology	0.033**	0.207	0.026**	0.369
Constant	0.786		−0.231	
Adjusted R^2	0.350		0.122	
N	969		969	

**Significant at 0.01.

The percentage of people using both webphones and PCs declines markedly with age. A large majority (92%) of young adults in their 20s access e-mail through the use of webphones, About half (46%) of the respondents in their 20s use both the webphone and PC, and an equal percentage, 46%, only use webphones. In sum, 92% of all respondents in their 20s use webphones for e-mail, i.e. are 1.9 times more likely to be webphone users than the average respondents. Somewhat older adults in their 30s also are disproportionately high webphone users, comprising 29% of all webphone users even though they are only 20% of the sample, and are 1.4 times more likely to be webphone users than the average respondents. The disproportionate use of webphones by younger adults is so great that adults aged 20–39 years own two-thirds (68%) of all web-phones, even though they comprise only two-fifths (41%) of the sample (for more details on the characteristics of the sample, see Miyata et al., 2005)

In contrast to low webphone e-mail use by adults over 40 years old, the percentage of respondents using only PCs to send e-mail increases until the age of 60 years. Moreover, older adults are more apt to use only PCs for e-mail, whereas younger adults are more apt to use both webphones and PCs. Not only do the ways of accessing the Internet vary by age, but so does the frequency of sending e-mail. Older Yamanashi residents, aged 50 years or more, are much less likely to send e-mail or to use the Internet at all.

Gender is also associated with how people send e-mail. Women tend to send more private e-mails by webphones. Their higher level of webphone use may well counterbalance their historically low PC use (Ono and Zavodny, 2004). By contrast, many middle-aged men make heavy use of their PCs at work and have less need for webphones. Nevertheless, both women and men in their 20s are equally likely to use webphones – and to use them frequently. However, it is a different story for those who are no longer in their 20s. Men over 30 years old are somewhat more likely than similarly aged women to use both webphones and PCs: 38% of men in their 30s use both webphones and PCs to e-mail, compared with 33% of women in their 30s. Whereas 35% of men in their 40s use both media, only 21% of women in their 40s use both (Miyata et al., 2005).

Respondents with higher levels of technological skill are more likely to send greater numbers of e-mails – both by webphones and by PCs.[3] This is in accord with a national Japanese survey that found that people using both webphones and PCs report higher levels of skill in using keyboards

---

[3] Self-perceived technical confidence was measured by a scale ranging from 7 to 21. The scale was compiled from seven questions that asked respondents to rate their ability to do certain technical tasks: sending a fax, recording television programs using a VCR, sending e-mails by computer or webphone, typing with a keyboard, using a search engine from a computer or webphone, downloading a file from a computer or webphone, and installing a computer program.

(Ikeda, 2002). However, the respondents' self-perception of technological skill has a stronger effect on the number of e-mails sent by PCs than on that sent by webphones. The higher the respondents rated their ability to use various kinds of technology, the more they tend to send e-mails by PC: the effect of self-perceived technical confidence on the number of e-mails sent by PCs is stronger than the effect on the number of e-mails sent by webphones.

Students are likely to send more webphone e-mails than full-time workers, but less PC e-mail than full-time workers. Self-employed persons sent more e-mails by their PCs than full-time workers.

Thus webphone e-mails tend to be sent by young people who often do not have much technological skill. By contrast, PC e-mails tend to be sent by older men who are more apt to be self-employed and to have more technical skill.

28.4.2 Strong, Supportive and Weak Ties

Does the Internet help maintain the social networks that are so important in the age of networked individualism? In a society that has moved from groups to networks, we would expect that both webphone and PC e-mails would be used to contact large numbers of somewhat specialized ties: both strong intimate ones and weaker ones of acquaintanceship.

As an approach to this matter, we examine two properties of social networks: the number of addresses kept in webphones or PCs and the number of supportive ties. The respondents were asked to report the number of e-mail addresses kept in their webphones and their PCs.

The respondents reported an average of 30.2 addresses in their webphones and 16.9 addresses in their PCs. Most of the ties in both their webphone and PC address books are weaker ties of acquaintanceship rather than stronger ties with close friends and relatives. This is because respondents report that they exchange supportive e-mail with only one to three network members, presumably those with whom they have strong ties. Although this may be a low estimate, as earlier Japanese data show a mean of five network members to be strong ties (Otani, 1999), as do Canadian data (Wellman, 1979; Wellman and Wortley, 1990), the inference is clear: only a few names in address books are strong, supportive ties. Most ties are weaker, but still significant enough to be entered into address books.

Regression analysis shows that the more e-mails people send by webphones, the more names they keep in their webphone address books (Table 28.3). Moreover, heavy PC e-mail users (those who sent 6–10 messages "yesterday") also keep more names in their webphone address books, indicating that they too have a larger number of weak ties in their networks.

Table 28.3 Number of social ties through webphones or PCs (regression analysis)

Predictors	Number of addresses kept in webphone		Number of addresses kept in PC		Number of supportive ties that have an e-mail connection	
	B	β	B	β	B	β
Gender (0 = female, 1 = male)	−4.967	−0.056	4.622	0.065	−0.150	−0.024
Age (years) (reference = 20–29)						
30–39	−14.017*	−0.145	4.514	0.057	−0.409	−0.060
40–49	−9.289	−0.086	2.341	0.028	−0.799	−0.110
50–59	−20.764*	−0.133	−4.798	−0.046	−1.079*	−0.116
60–65	−35.077	−0.067	−1.534	−0.005	−1.839	−0.039
Education (reference = Middle School):						
High school	−7.208	−0.081	5.035	0.069	0.382	0.061
Some college	0.546	0.006	3.789	0.046	0.845	0.121
Undergraduate degree or more	−3.360	−0.032	10.066	0.135	0.773	0.111
Employment status (reference = full-time worker):						
Part-time worker	9.929	0.078	−4.395	−0.036	−0.068	−0.007
Self-employment	24.986*	0.094	7.976	0.050	0.576	0.038
Student	29.341**	0.127	−10.832	−0.058	0.854	0.050
Home maker	0.054	0.000	−6.347	−0.051	1.686**	0.151
Other type	−1.697	−0.010	−8.512	−0.059	0.327	0.027
Unemployed	−3.790	−0.015	−4.815	−0.023	−0.858	−0.046
Partner (0 = no, 1 = partner)	−3.208	−0.036	2.850	0.039	−0.369	−0.058
Kids living with parents (0 = no, 1 = yes)	−7.602	−0.086	1.889	0.027	−0.920*	−0.149
Associations	33.028**	0.186	8.693	0.060	1.688**	0.139
Number of e-mails sent by webphone yesterday (reference = non-users)						
1–5	6.744	0.076	0.246	0.003	0.749*	0.119
6–10	23.220**	0.166	8.706	0.073	1.662**	0.156
>11	24.503**	0.141	6.947	0.047	2.557**	0.193
Number of e-mails sent by PC yesterday (reference = non-users):						
1–5	1.812	0.013	29.630**	0.332	0.857*	0.091
6–10	37.491*	0.092	45.258**	0.179	0.663	0.022
Constant	−4.003		−13.011		−0.094	
Adjusted R^2	0.189		0.161		0.186	
N	421		347		505	

*, Significant at 0.05; **, significant at 0.01.

Does the immediacy of portable webphones or the range of PCs facilitate the availability of social support? To measure the number of supportive ties that have an e-mail, respondents were asked to report the number of network members that would give them words of encouragement (emotional support), provide them with a small amount of money (financial support) or aid them in tasks such as moving or providing goods and services (instrumental support). Respondents reported that an average of 2.8 network members were in contact by e-mail.

We are particularly interested in the characteristics of e-mail-using people who have a high number of strong, supportive ties in their network. Regression analysis shows that age, the number of e-mails exchanged, employment status and household composition are all significantly associated with the number of supportive strong e-mail ties (Table 28.3).

E-gregariousness and supportiveness are related. The more e-mails sent "yesterday" by either webphones or PCs, the greater is the number of supportive ties that are linked by email (although not necessarily in contact "yesterday"). This suggests that the Internet facilitates larger, more actively supportive social networks. It is also congruent with Canadian data showing that people with larger social networks get more social support from network members, both per capita and in aggregate (Wellman and Gulia, 1999; Wellman and Frank, 2001).

Webphone e-mails have a different characteristic than PC e-mails. Not only are they usually used to contact close friends and relatives (Miyata et al., 2005), but Table 28.3 shows that the frequency of webphone contact is more strongly associated with socially supportive ties than is the frequency of PC contact.

There are also demographic correlates with socially supportive e-mail ties. People in their 50s have more strong, supportive ties than those in their 20s. Perhaps they accumulate over time like barnacles! Those who send a high number of e-mails per day by webphones or PCs also have a relatively higher amount of strong, supportive e-mail ties. People with children living at home have a relatively lower number of strong, supportive ties compared with those without children living at home. Homemakers have more strong supportive ties than full-time workers.

28.4.4 Local and Long-distance Contact

The Internet has frequently been touted as supporting far-flung ties, and even "global villages" [to use Marshall McLuhan's phrase (McLuhan, 1962)]. In practice, physical body needs and the tendency for birds of a feather to flock together (and with similar needs) mean that the Internet has been used extensively for both local and long-distance relationships (Hampton and Wellman, 2003).

Table 28.4   Distance between sender and receiver of e-mails

E-mails by webphone	By both webphone and PC		By webphone only	By PC only
	E-mails by webphone	E-mails by PC		
Living together	19.1	11.5	18.1	0.0
Less than10 minutes away by car	12.7	4.9	13.5	3.8
Less than 1 hour away by car	42.7	39.3	46.8	34.6
Less than 5 hours away by car	19.7	34.4	18.1	38.5
More than 5 hours away by car	5.1	4.9	2.9	15.4
Living abroad	0.6	4.9	0.6	7.7

In Yamanashi, e-mail sent by webphone tends to be more local than e-mail sent by PC. Those who only use webphones send e-mail to people living an average of about 10 minutes away by car (Table 28.4). By contrast, those who only use PCs send e-mail to people living about 1 hour away by car. Those respondents who own both a webphone and a PC use their webphone to e-mail people living less than a 60-minute drive away, and use their PC to e-mail people living between 1 and 5 hours drive away.

28.5 Japan and the Turn Towards Networked Individualism

28.5.1 Mobile-ized Japan

There is significant variation in the amount and kinds of contact with social networks by webphone and PC. More e-mails per day are made through the use of webphones than through PCs. People use webphones more than PCs to send e-mails, even when they have both webphones and PCs available.

Using webphones and PCs to send e-mails is additive: high (or low) use of one medium does not mean high (or low) use of the other. This is similar to North American research showing that e-mail use adds on to face-to-face and telephone contact (Quan-Haase et al., 2002). The result is a greater overall volume of contact. In Japan, where webphone e-mails have been added to the communication mix, they apparently increase the overall amount of contact by adding on to PC e-mail and presumably to face-to-face and telephone contact. This suggests that the Japanese may be among the most communication-connected people in the world.

Yet the two media are used somewhat differently. E-mail that is exchanged via webphones tends to be with people who are nearby, whereas e-mail that is exchanged via PCs tends to be with people who are further away. Webphone address books are larger than PC address books, but webphone e-mailing is done more selectively. Webphone e-mail tends to be with people who are nearby, whereas PC e-mail tends to be with people who are further away as well as nearby. Webphones are most often used to send short, quick messages to close friends and family, allowing them to keep emotionally connected and organize meetings, or to those who are nearby, facilitating arrangement of everyday activities. By contrast, PC-based e-mail messages are more apt to be communications that are not as connected to imminent physical get-togethers.

Those who send many webphone e-mails have more supportive ties. Here, too, webphones appear to be especially used for supporting intensive relationships with loved ones and other strong ties. They are interfaces for intimate contact, and enable intimates to be accessible anywhere and anytime.

The use of webphones varies by age and gender. Those who sent more e-mails by webphones tend to be in their 20s and 30s. More women than men use webphones to send e-mail. In addition to webphones, people in their 20s and 30s are also heavy users of PCs. Many are dual users of both webphones and PCs. Between the ages of 30 and 59 years, there is an increase in the proportion of respondents who exchange e-mail through the use of PC only.

Our statistical findings fit ethnographic research showing that young Japanese use webphones to send quick e-mails to nearby good friends. For example, Ito (this volume, Chapter 9) reports that mobile phones afford Tokyo youth a few important advantages. The ability to send short messages at any time allows them to keep in constant contact with friends, strengthening their social networks, giving the feeling of "ultra-connectedness". This sort of communication typically occurs with only a small number of friends (between two and five), at an extremely high frequency. At the same time, typing quick messages gives a new kind of freedom, as it can be done somewhat covertly, without alerting parents. Contacting friends can occur late at night while parents are sleeping, something not so easily done through a regular landline phone. These kinds of contact may simply be about trivial matters, used to maintain a feeling of connectedness, or about arranging things such as asking a spouse to pick up food on the way home from work. The mobile nature of webphones also makes them perfect for arranging meetings or changing plans at the last second (Smith, 2000; Ling and Yttri, 2002). As typing messages is more difficult through a webphone than a PC, we believe that respondents reserve PC e-mail for richer, in-depth contact with those who are living at a distance.

The advantages that webphones offer Japanese youth are probably similar to the advantages that ordinary mobile phones (those that cannot be used to access the Internet) offer young people in other countries. Youths adopt mobile phones worldwide to increase their autonomy and the quality of their ties with friends. For example, European youths are more likely than their parents to use mobile phones to build their social networks and to tell parents their whereabouts (Ling, 2004). Furthermore, in many developed countries, mobile phones have become so incorporated into youth culture that exchanges of text messages, airtime and even mobile phones themselves have become heavily reciprocated, binding youth together. Text message exchanges are often incorporated into face-to-face contact with peers during "hang-out time". When messages and phones are shared among the group, they add to the interaction of the entire group rather than only of their owners (Taylor and Harper, 2003).

Will young users continue to rely on webphones as they grow older? On the one hand, the desire to be in constant contact with friends may dissipate as young adults enter more instrumental relationships at the workplace and save their recreational time for contact with spouses and family at home. On the other hand, heavy habitual use of this technology between friends and family may continue as people age and continue to integrate webphones into their work and domestic relationships.

It is probable that within a short time, the great majority of Yamanashi residents will use both webphones and PCs to send messages. Hence those respondents who mix media by using both webphones and PCs are a harbinger of this future. Not only do such dual-mode users have more strong and weak ties with whom they can exchange e-mail, but they are also in more frequent contact with them. Moreover, those who exchange higher numbers of e-mails tend to have a greater number of strong ties.

Our Yamanashi research is in accord with research done in North America and Europe showing that the Internet is not a self-contained world (Castells et al., 2003; Chen et al., 2002; Quan-Haase et al., 2002; Boase et al., 2003). The Internet is another means of communication that is being integrated into the regular patterns of social life. Rather than operating at the expense of the "real" face-to-face world, the Internet is an extension, with people using all means of communication to connect with friends and relatives.

The Yamanashi study highlights how different forms of computer-mediated communication are used for different purposes. Webphones are most often used to send short, quick messages to those who are physically close by. Webphones are also used to maintain strong ties with people who are socially close (see also Rivière and Licoppe, 2003). However, webphones are not used much to contact weaker ties or to develop more diverse networks. This may be because webphones are not well suited to provide connections to Internet sites where weak tie relationships may be formed, such as chat rooms and issue-oriented sites. Then again, there may be a population cohort effect because heavy users of e-mails by webphones are younger adults who may not be as interested in discussing issues as are middle-aged and older Japanese adults.

Our results call into question the traditional stereotype of Japan as a closed, bounded society (see also Otani, 1999). We find people on the move, getting information, making arrangements and contacting friends and relatives through wireless webphones and wired PCs. It is a trend towards mobile connectivity – which we call mobile-ization – that is becoming increasingly prevalent throughout Japanese society. As those in their 20s grow into middle-age, we expect their mobile communication to continue, although tempered by a heavier reliance on faster and more informative big-screen PCs at work and at home.

28.5.2 Changing Communication Networks; Changing Social Networks

Not only has the volume of communication increased, we suspect that the velocity of communication has also increased – in Japan and elsewhere in the Internet-using world. Although e-mail is asynchronous and does not necessitate instantaneous response, in practice many people respond quickly just as they would respond to voicemail. This may be especially true when e-mail is delivered by highly personal webphones that are becoming treated as people's "third skins" (Fortunati, this volume, Chapter 13). Moreover, distant network members who did not have much contact when limited to face-to-face, telephone (wired, mobile or web-phone) or postal communication now keep in frequent touch: they rely on the Internet for a higher proportion of their contact than do community members who live nearby (Chen et al., 2002; Quan-Haase et al., 2002).

Hence the impact of computer-mediated communication will be that people have larger scale social networks: more people, more communication and at greater speed. Those Japanese who use both webphones and PCs to communicate have larger social networks, bigger address books and more frequent communication with these networks.

It is not clear whether the high use of computer-mediated communication will foster more densely knit communities – good for conserving resources – or more sparsely knit communities – good for obtaining new information and other resources. On the one hand, some characteristics of the Internet foster denser networks: the ability of Internet users to communicate simultaneously with multiple others, and the ease of copying and forwarding messages to others. In such cases, it is more likely for the friend of my friend to become my friend. On the other hand, as social networks become larger, it is often more difficult for them to maintain their density. As the size of the network increases arithmetically, the number of ties must increase geometrically to maintain the same level of density.

The turn towards networked individualism before and during the age of the Internet suggests more people maneuvering through multiple communities of choice, where kinship and neighboring contacts become more of a choice than a requirement. This phenomenon started in Japan before the advent of the Internet (Nozawa, 1996; Otani, 1999), but webphones and PCs are probably accelerating it. Webphone users have the possibility of contacting whoever they want, whenever they want and wherever they are located. This suggests a fragmentation of community, with people increasingly operating in a number of specialized communities that rarely grab their entire, impassioned or sustained attention. The multiplicity of communities should reduce informal social control and increase autonomy. It is easier for people to leave unpleasantly controlling communities and increase their involvement in other, more accepting ones.

The personal communities of a networked world are both homogeneous and heterogeneous. An individual's partial communities will often be homogeneous, because search engines and discussions tend to find and link others with shared interests. Yet an individual's overall community will be heterogeneous because people have multiple interests, and those who share interests in one area are unlikely to share interests in others. Moreover, the properties of computer-mediated communication easily allow the inclusion of others in conversations through multiple address lines and chatting. This might ostensibly expand the scope of homogeneous discussion. In practice, the larger is the network, the more heterogeneous are the participants (Feld, 1982). Durkheimian social cohesion will be in dynamic tension with Simmelian interconnected marginality.

28.5.3 Dividing Digitally

Webphone and PC use may be fostering a complex digital divide. Social scientists have been discussing the digital divide for at least a decade: the gap between users and non-users of the Internet. More recently, they have highlighted the gap between those who merely have marginal access to the Internet and those who are active, informed users: what Castells (1996) calls "the interactors" and "the interacted" (see also Chen and Wellman, 2005). The Yamanashi study shows us that even the "interactors" may themselves have limited use of the power of the Internet when they only use webphones. Those less technically skilled use webphones rather than PCs. However, the limited screen size and access speed of webphones restrict the use of web sites, keyboard limitations constrain the length and complexity of messages and a more limited range of people are routinely contacted. Moreover, webphone messages are overwhelmingly segregated exchanges between two persons, whereas PC-based e-mail involves bringing multiple others into conversations. The result is a mixture of segregated, bilateral web conversations integrated with group-based chats with physically present friends (Ito and Okabe, 2005),

Networked individualism should have substantial effects on social cohesion. Rather than people being a part of a hierarchy of encompassing polities like nesting Russian dolls, they belong to multiple, partial communities and polities. Some communities may be widely dispersed, such as those found in electronic diasporas linking far-flung members of emigrant ethnic groups (Mitra, 2003). Some may be traditional, local groups of neighbors with connectivity enhanced by listservs and other forms of computer-mediated communication. In a "glocalized" world, local involvements fit together with far-flung communities (Wellman, 2003), because the McLuhanesque "global village" (McLuhan, 1962) complements traditional communities rather than replaces them. This is especially true today when almost all computers are physically wired into the Internet, rooting people in their desk chairs. Even as the world goes wireless, the persistence of tangible interests, such as neighborly get-togethers or local intruders, keep the local important (Hampton and Wellman, 2003). Local and long-distance – webphone and PC – it is all one fluid and complex social network.

28.6 Acknowledgements

Our research has been supported by grants from the Japan Society for the Promotion of Science, KAKENHI15330137, the Matsushita International Foundation and the Social Science and Humanities Research Council of Canada. The Centre for Urban and Community Studies, University of Toronto, hosted Kakuko Miyata during her research leave, 2002–03. We thank Mitsuhiro Ura, Hiroshi Hirano, Ken'ichi Ikeda, Tetsuro Kobayashi and Kaichiro Furutani for their collaboration in the design of the survey, Bonnie Erickson, Bernie Hogan, Ken'ichi Ikeda, Mizuko Ito and Rachel Yould for their advice on our research and Monica Prijatelj, Julie Wang and Natalie Zinko for their editorial assistance.

28.7 References

Boase, J., Chen, W., Wellman, B., and Prijatelj, M. (2003) Is there a place in cyberspace: the uses and users of the Internet in public and private spaces. *Cultures et Géographie* 46 (Eté) 5–20.
Barnes, S.J. and Huff, S.L. (2003) Rising sun: iMode and the wireless internet. *Communications of the ACM,* **46,** 79–84.
Castells, M. (1996) *The Rise of the Network Society.* Blackwell, Oxford.
Castells, M. (2000) *The Rise of the Network Society,* 2nd edn. Blackwell, Oxford.
Castells, M., Tubella, I., Sancho, T., and Wellman, B. (2002) *La Sociedad Red en Catalunya: un Analysis Empirico.* Universitat Oberta Catalunya, Barcelona.
Castells, M., Tubella, I., Sancho, T., Diaz de Isla, I. and Wellman, B. (2003) *The Network Society in Catalonia: an Empirical Analysis.* Universitat Oberta de Catalunya, Barcelona.

Chae, M. and Kim, J. (2003) What's so different about the mobile Internet. *Communications of the ACM*, **46**, 240–247.

Chen, W., Boase, J. and Wellman, B. (2002) The global villagers: comparing Internet users and uses around the world. In Wellman, B. and Haythornthwaite, C. (eds), *The Internet in Everyday Life*, 1st edn. Blackwell, Oxford, pp. 74–113.

Chen, W. and Wellman, B. (2004) Charting digital divides: within and between countries. In Dutton, W., Kahin, B., O'Callaghan, R. and Wyckoff, A. (eds), *Transforming Enterprise*. MIT Press, Cambridge, MA, pp. 467–498.

Dobashi, S. (2005) Gendered use of *keitai* in domestic circumstances. In Ito, M., Matsuda, M. and Okabe, D. (eds), *Portable, Personal, Pedestrian: Mobile Phones in Japanese Life*. MIT Press, Cambridge, MA, in press.

Feld, S. (1982) Social structural determinants of similarity among associates. *American Sociological Review*, **47**, 797–801.

Habuchi, I. (2005) Accelerating reflexivity. In Ito, M., Matsuda, M. and Okabe, D. (eds), *Portable, Personal, Pedestrian: Mobile Phones in Japanese Life*. MIT Press, Cambridge, MA, in press.

Hampton, K. and Wellman, B. (2003) Neighboring in netville: how the Internet supports community and social capital in a wired suburb. *City and Community*, **2**, 277–311.

Haythornthwaite, C. and Wellman, B. (2002) The Internet in everyday life: an introduction. In Wellman, B. and Haythornthwaite, C. (eds), *The Internet in Everyday Life*. Blackwell, Oxford, pp. 3–44.

Ikeda, K. (2002) Patterns of a mobile phone use: a social psychological viewpoint. In Cabinet Office of Japan, *The Information Society and Youth: Report of the Fourth National Survey on "The Information Society and Youth"*, pp. 287–301.

Ito, M., Matsuda, M. and Okabe, D. (eds) (2005) *Portable, Personal, Pedestrian: Mobile Phones in Japanese Life*. MIT Press, Cambridge, MA, in press.

Kageyama, Y. (2003) NTT tests superfast mobile phone. Associated Press, 7 December, http://news.yahoo.com/news?tmpl=story2&cid=528&ud=/ap/20031207.ap_on_hi_te/.

Katz, J. and Aakhus, M. (eds) (2002) *Perpetual Contact*. Cambridge University Press, Cambridge.

Ling, R. (2004) *The Mobile Connection: the Cell Phone's Impact on Society*. Morgan Kaufmann, New York.

Ling, R. and Yttri, B. (2002) Hyper-coordination via mobile phones in Norway. In Katz, J. and Aakhus, M. (eds), *Perpetual Contact*. Cambridge University Press, Cambridge, pp. 139–169.

Matsuda, M. (2004) *Mobile Communication and Selective Sociality*, 1–41.

McLuhan, M. (1962) *The Gutenberg Galaxy: the Making of Typographic Man*. University of Toronto Press, Toronto.

Mitra, A. (2003) Ethnic groups and e-diasporas online. In Christensen, K. and Levinson, D. (eds), *Encyclopedia of Community*. Sage, Thousand Oaks, CA, pp. 119–120.

Miyata, K. (2002) Social support through computer networks for child-raising mothers in Japan. In Wellman, B. and Haythornthwaite, C. (eds), *The Internet in Everyday Life*. Blackwell, Oxford, pp. 520–548.

Miyata, K., Boase, J., Wellman, B. and Ikeda, K. (2005) The mobile-izing Japanese: connecting to the Internet by PC and webphone in Yamanashi. In Ito, M., Matsuda, M. and Okabe, D. (eds), *Portable, Personal, Pedestrian: Mobile Phones in Japanese Life*. MIT Press, Cambridge, MA, pp. 143–69.

MPMHAPT (2004) *White Paper: Information and Communications in Japan 2004*. http://www.johotsusintokei.soumu.go.jp/whitepaper/eng/WP2004/press_information01.pdf

Nozawa, S. (1996) Aspects spatiaux de liens personnels dans le Japon moderne. *Bulletin de la Societe Neuchateloise de Geographie*, **40**, 83–97.

Ono, H. and Zavodny, M. (2004) *Gender Differences in Information Technology Usage: a U.S.-Japan Comparison*. Working Paper 2004-2. Federal Reserve Bank of Atlanta, Atlanta, GA.

Otani, S. (1999) Personal community networks in contemporary Japan. In Wellman, B. (ed.), *Networks in the Global Village*. Westview Press, Boulder, CO, pp. 279–297.

Putnam, R. (2000) *Bowling Alone: the Collapse and Revival of American Community*. Simon and Schuster, New York.

Quan-Haase, A., Wellman, B., Witte, J. and Hampton, K. (2002) Capitalizing on the Internet: network capital, participatory capital, and sense of community. In Wellman, B. and Haythornthwaite, C. (eds), *The Internet in Everyday Life*. Blackwell, Oxford, pp. 291–324.

Rheingold, H. (2002) *Smart Mobs: the Next Social Revolution.* Perseus, Cambridge, MA.

Rivière, C.A. and Licoppe, C. (2003) From voice to text: continuity and change in the use of mobile phones in France and Japan. Presented at the International Sunbelt Social Network Conference, Cancun, Mexico, February.

Smith, M. (2000) Some social implications of ubiquitous wireless networks. *ACM Mobile Computing and Communications Review,* 4(2), 25–36.

Srivastava, L. and Kodate, A. (2004) Shaping the future mobile information society: the case of Japan. Presented at the ITU/MIC workshop, Shaping the Future Mobile Information Society, Seoul, March;

http://www.itu.int/osg/spu/ni/futuremobile/general/casestudies/JapancaseLS.pdf.

Taylor, A.S. and Harper, R. (2003) The gift of the gab? A design oriented sociology of young people's use of mobiles. *Computer Supported Cooperative Work,* 12, 267–296.

Wellman, B. (1979) The community question. *American Journal of Sociology,* 84, 1201–1231.

Wellman, B. (1999) From little boxes to loosely-bounded networks: the privatization and domestication of community. In Abu-Lughod, J. (ed.), *Sociology for the Twenty-first Century: Continuities and Cutting Edges.* University of Chicago Press, Chicago, pp. 94–114.

Wellman, B. (1999a) The network community: an introduction to networks in the global village. In Wellman, B. (ed.), *Networks in the Global Village.* Westview Press, Boulder, CO, pp. 1–47.

Wellman, B. (2001) Physical place and cyber-place: changing portals and the rise of networked individualism. *International Journal for Urban and Regional Research,* 25, 227–252.

Wellman, B. (2002) Little boxes, glocalization, and networked individualism. In Tanabe, M. and van den Besselaar, P. and Ishida, T. (eds), *Digital Cities II: Computational and Sociological Approaches.* Springer, Berlin, pp. 10–25.

Wellman, B. (2003) Glocalization. In Christensen, K. and Levinson, D. (eds), *Encyclopedia of Community.* Sage, Thousand Oaks, CA, pp. 559–62.

Wellman, B. and Frank, K. (2001) Network capital in a multilevel world: getting support from personal communities. In Lin, N., Burt, R.S. and Cook, K. (eds), *Social Capital.* Aldine De Gruyter, Hawthorne, NY, pp. 233–273.

Wellman, B. and Gulia, M. (1999) A network is more than the sum of its ties: the network basis of social support. In Wellman, B. (ed.), *Networks in the Global Village.* Westview, Boulder, CO, pp. 83–118.

Wellman, B. and Haythornthwaite, C. (eds) (2002) *The Internet in Everyday Life.* Blackwell, Oxford.

Wellman, B. and Hogan, B. (2004) The Internet in everyday life. In Bainbridge, W. (ed.), *The Encyclopedia of Human Computer Interaction.* Berkshire Publishing, Great Barrington, MA.

Wellman, B. and Wetherell, C. (1996) Social network analysis of historical communities: some questions from the present and the past. *The History of the Family,* 1(1), 97–121.

Wellman, B. and Wortley, S. (1990) Different strokes from different folks: community ties and social support. *American Journal of Sociology,* 96, 558–588.

Wired (2004) Hypelist. *Wired,* February, p. 23.

Yan, X. (2003) Mobile data communications in China. *Communication of the ACM,* 46(12), 81–85.

Subject Index